Photonic Sensors: Advances and Applications

Photonic Sensors: Advances and Applications

Edited by Juan Landers

CLANRYE
INTERNATIONAL
www.clanryeinternational.com

Clanrye International,
750 Third Avenue, 9th Floor,
New York, NY 10017, USA

ISBN: 978-1-63240-854-9

Cataloging-in-Publication Data

Photonic sensors : advances and applications / edited by Juan Landers.
    p. cm.
Includes bibliographical references and index.
ISBN 978-1-63240-854-9
1. Photonics. 2. Optical detectors. I. Landers, Juan.
TA1520 .P46 2019
621.36--dc23

For information on all Clanrye International publications
visit our website at www.clanryeinternational.com

CLANRYE
INTERNATIONAL

Contents

Preface

I am honored to present to you this unique book which encompasses the most up-to-date data in the field. I was extremely pleased to get this opportunity of editing the work of experts from across the globe. I have also written papers in this field and researched the various aspects revolving around the progress of the discipline. I have tried to unify my knowledge along with that of stalwarts from every corner of the world, to produce a text which not only benefits the readers but also facilitates the growth of the field.

A photonic sensor is a device that is made of a p-n junction, which converts photons into current. Some examples of photosensors are photodiodes and phototransistors. Polarization, photoelectric effect, photochemical changes and weak interaction effects are the mechanisms fundamental to the working of photonic sensors. A number of performance characteristics are important for the characterization of photonic sensors. These include response time, spectral response, detectivity, responsivity and quantum efficiency, among others. Depending on the mechanisms or the technology involved in the detection of light, photosensors can be classified into semiconductor, photoelectric, photovoltaic, photochemical and thermal photosensors. Some photoelectric photosensors are phototubes and photomultiplier tubes. Active-pixel sensors, charge-coupled devices, etc. are some of the semiconductor photosensors. This book includes some of the vital pieces of work being conducted across the world, on various topics related to photonic sensors. The various advancements in this field are glanced at and their applications as well as ramifications are looked at in detail in the book. It will prove to be immensely beneficial to students and researchers in this field.

Finally, I would like to thank all the contributing authors for their valuable time and contributions. This book would not have been possible without their efforts. I would also like to thank my friends and family for their constant support.

Editor

Highest Achievable Detection Range for SPR Based Sensors Using Gallium Phosphide (GaP) as a Substrate: A Theoretical Study

Rajneesh K. VERMA[1] and Akhilesh K. MISHRA[2]

[1]*Department of Physics, Central University of Rajasthan (India), BandarSindri, Ajmer, 305817, India*

[2]*Department of Electrical Engineering, Technion - Israel Institute of Technology, Haifa 32000, Israel*

[*]Corresponding author: Rajneesh K. VERMA   E-mail: rkverma@curaj.ac.in

Abstract: In the present study, we have theoretically modelled a surface plasmon resonance (SPR) based sensing chip utilizing a prism made up of gallium phosphidee. It has been found in the study that a large range of refractive index starting from the gaseous medium to highly concentrated liquids can be sensed by using a single chip in the visible region of the spectrum. The variation of the sensitivity as well as detection accuracy with sensing region refractive index has been analyzed in detail. The large value of the sensitivity along with the large dynamic range is the advantageous feature of the present sensing probe.

Keywords: Optical fiber sensors; evanescent waves; gallium phosphide; surface plasmon

1. Introduction

The surface plasmon resonance (SPR) technique is the most promising tool for the detection of various chemical and biological species [1–12]. At the present time, the SPR technique is being used not only in the biochemical species detection and gas sensing but also in imaging, terahertz plasmonics, artificially structured materials lithography, and many other areas [2–4]. In SPR based sensors, the famous Kretschmann-Reather configuration is utilized [1]. In this configuration, a thin layer of metal such as silver or gold is deposited onto the base of a prism as shown in Fig. 1. A p-polarized light is allowed to fall from the substrate side. The medium to be sensed is kept in contact with the metal layer.

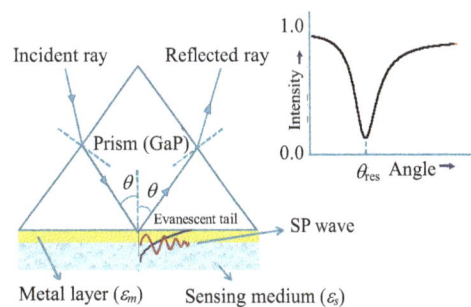

Fig. 1 Schematic diagram of the sensing probe.

Under the resonance condition, the reflected light beam causes a sharp dip into the intensity when the wavevector of the incident wave matches with the wave vector of the surface plasmon wave (Fig. 1 insight). If n_s and ε_m represent the refractive index and the dielectric function of the sensing medium and metal layer, respectively, the resonance condition can be written as

$$\frac{2\pi}{\lambda_o} n_{\text{substrate}} \sin \theta_{\text{SPR}} = \text{Re}\left[\frac{2\pi}{\lambda_o}\left(\frac{\varepsilon_m n_s^2}{\varepsilon_m + n_s^2}\right)^{1/2}\right]. \quad (1)$$

The expression on the left hand side represents the propagation constant (k) of the evanescent wave, and the right hand side represents the propagation constant of the surface plasmon wave existing at the gold sensing layer interface. θ_{SPR} is the resonance angle, i.e. the angle at which two wave vectors match. Here, λ_o is the wavelength of light being used.

In SPR based sensors, two interrogation schemes are generally used. One is called the angular interrogation, and the other is called the spectral interrogation. In the angular interrogation technique, we fix the wavelength of the incident light, i.e. we use a monochromatic light source and vary the angle of incidence from the critical angle to 90°. This change in the angle changes the propagation constant of the incident wave. At a particular value of the angle of incidence, this propagation constant becomes equal to the SPR propagation constant leading to the condition of the resonance. This resonance angle is quite sensitive to the changes in the sensing region environment. A slight change in the surrounding refractive index causes a corresponding change in the resonance angle. By measuring the change in the resonance angle, the change in the surrounding region refractive index can be measured.

In another scheme called the wavelength interrogation scheme, we use a polychromatic source of light and fix the angle of incidence to any particular value between θ_c and 90°. This again changes the value of the propagation constant of the incident beam. At a particular value of the incident wavelength, the propagation constant of the incident wave may become equal to the surface plasmon wave vector. This is called the resonance condition, and the wavelength at which this occurs is called the resonance wavelength. At this wavelength, the intensity of the reflected light shows the minimum.

For a change in the refractive index of the sensing medium, the resonance wavelength changes. By measuring the change in the resonance wavelength, the corresponding change in the refractive index of the medium can be measured. There are other techniques also for measuring changes in the refractive index such as the Mach-Zhander interferometer and Febry-Perot interferometer, however, SPR sensors are also getting progressed with the equal pace such as in the sensing of hydrogen gas using palladium [13−15].

In the present work, an angular interrogation technique is used to exploit the famous Krestchmann-Reather configuration. People have devised various sensing probes which are applicable either for the gas sensing or for the liquid medium. We here are proffering a sensing chip which can detect various gases as well as highly concentrated liquid in the visible region of the spectrum thereby producing a large dynamic range of the sensing medium refractive index. This could be possibly realized by utilizing the sensor chip which can be fabricated by depositing a thin layer of gold layer (50 nm) onto the base of a prism made up of a semiconducting material called gallium phosphide. Gallium phosphide is a transparent semiconducting material having refractive index around 3.3 in the visible range of the spectrum. Also it is a wide band gap material. The detailed theoretical analysis is carried out in terms of the sensitivity and detection accuracy. The most advantageous feature of having large detection range has been addressed.

In order to realize the practical design of the sensor, consider a GaP prism coated with a gold layer of about 50 nm as shown in Fig. 1. For coating the gold layer, any vacuum deposition technique such as thermal evaporation, sputtering or electron beam deposition can be used. The chemical or the gaseous medium which is to be sensed is kept in contact with the gold layer. A p-polarized light with the intensity I and wavelength 632 nm from He-Ne laser is allowed to fall on the substrate. The intensity of the reflected light is measured with respect to the angle of incidence. A sharp dip is observed at a particular value of the angle of the incidence. This will be the resonance angle. A slight change in the sensing region will reflect in terms of the corresponding change in the resonance angle.

In mathematical modelling, we require the value of the dielectric constant of the metal layer for the He-Ne laser wavelength which can be calculated from the Drude formula [6]:

$$\varepsilon_m(\lambda) = \varepsilon_{mr} + i\varepsilon_{mi} = 1 - \frac{\lambda^2 \lambda_c}{\lambda_p^2(\lambda_c + i\lambda)} \qquad (2)$$

where ε_{mr} and ε_{mi} are the real and imaginary parts of the dielectric constants of the gold layer. Also λ_p and λ_c are the plasma and collision frequencies of the gold layer. Their values are 1.6826×10^{-7} m and 8.9342×10^{-6} m, respectively.

To get a feel of the refractive index variation, one should note that most of the gases possess the refractive index close to unity whereas highly concentrated liquid chemicals such as solutions of $C_7H_6Cl_2$ have their refractive index values around 1.6 to 1.7 [11]. For the calculation of the reflected light beam intensity, one should be very precise because it is the resonance angle which decides the sensitivity of the sensor. In the present case, we have used a N-layer matrix method for the calculation of the reflection coefficient. The first medium is the GaP prism, the second medium is the gold layer, and the third medium is the sensing region itself which is to be sensed. These layers are assumed to be stacked along the z axis as shown in Fig. 2.

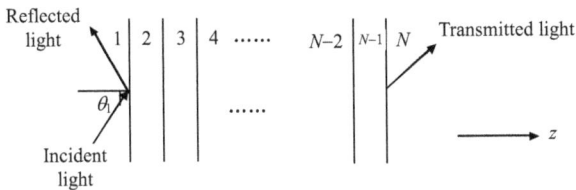

Fig. 2 Stack of N-layers arranged along the z axis.

The arbitrary medium layer has the thickness d_i, the dialectic constant ε_i, and the permeability μ_i. The relationship between the tangential electric field and

magnetic field at $z=z_1=0$ and at $z= z_{N-1}$ are related by

$$\begin{bmatrix} E_1 \\ H_1 \end{bmatrix} = \mathbf{M} \begin{bmatrix} E_{N-1} \\ H_{N-1} \end{bmatrix} \qquad (3)$$

where E_1 and H_1 are the tangential electric and magnetic fields at the first boundary. E_{N-1} and H_{N-1} are the similar fields at the Nth layer. M is called the characteristic matrix of the combined structure and is given by

$$\mathbf{M} = \prod_{i=2}^{N-1} M_i = \begin{bmatrix} M_{11} & M_{12} \\ M_{21} & M_{22} \end{bmatrix} \qquad (4)$$

where

$$q_i = \left(\frac{\mu_i}{\varepsilon_i}\right)^{1/2} \cos\theta_i = \frac{\left(\varepsilon_i - n_1^2 \sin^2\theta_1\right)^{1/2}}{\varepsilon_i} \qquad (5)$$

and

$$\beta_i = \frac{2\pi}{\lambda} n_i \cos\theta_i (z_i - z_{i-1}) = \frac{2\pi d_i}{\lambda}\left(\varepsilon_i - n_1^2 \sin^2\theta_1\right)^{1/2}. \qquad (6)$$

In the present case, there are three media and two layers N=2. Therefore, n_1 is the refractive index of the substrate, $n_2 = \sqrt{\varepsilon_m}$ is the refractive index of the metal layer (gold), n_3 is the refractive index of the sensing medium, and d_2 will be the thickness of the gold layer which is 50 nm in the present case. If θ_1 is the angle of incidence and λ_o is the wavelength of light in the free space which is 632 nm for He-Ne laser, the amplitude reflection coefficient r_p for the p-polarized incident wave is given as

$$r_p = \frac{(M_{11} + M_{12}q_N)q_1 - (M_{21} + M_{22}q_N)}{(M_{11} + M_{12}q_N)q_1 + (M_{21} + M_{22}q_N)}. \qquad (7)$$

All the M_{11}, M_{12}, M_{21}, and M_{24} are the four matrix coefficients of the matrix in (4). This will be a complex number. The reflection coefficient is the absolute square of r_p as

$$R_p = |r_p|^2. \qquad (8)$$

We shall evaluate the performance of the sensor in terms of the two parameters:

Sensitivity: if the refractive index of the sensing medium is altered by δn_s, the resonance angle will also change, and if the change in the corresponding angle is $\delta\theta_{res}$, we define the sensitivity as

$$S_n = \frac{\delta\theta_{\text{SPR}}}{\delta n_s}. \qquad (8)$$

Detection accuracy: for each SPR curve, we have to determine the exact location of the SPR angle which will be more correct if the SPR curve is sharp. So we define a parameter called the detection accuracy which is just the reciprocal of the full width at half maximum (FWHM) of an SPR curve:

$$DA = \frac{1}{\Delta\theta_{\text{FWHM}}}. \qquad (10)$$

In this particular section of the article, we shall present various theoretical results obtained from the theory discussed above.

Now since we are using here the technique of angular interrogation, so we shall check first the possibility of the SP wave excitation by incident light. For this, we have plotted in Fig. 3 the propagation constant of the surface plasmon wave with the angle of incidence along with the propagation constant of the incident wave from glass as well as from the GaP substrate.

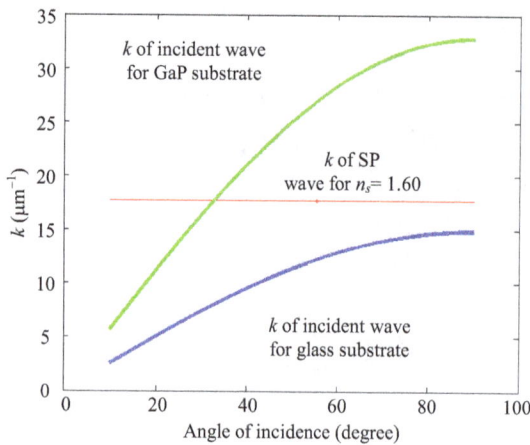

Fig. 3 Variation of propagation constants (k) with angles.

As it is quite clear from (1) that the surface plasmon wave vector is independent of the angle of incidence, hence the curve comes out to a straight line parallel to the angle axis as shown in Fig. 3. The surface plasmon wave vector is plotted for n_s=1.62. In the same graph, we have also plotted the propagation constant of the direct light, i.e. light in the glass prism, i.e. the evanescent wave and also the propagation constant of the wave in the GaP

prism. It is quite obvious from the figure that the propagation constant of light through the glass prism does not intersect the SP wave propagation constant which indicates that the SP wave cannot be excited for these waves. However, the SP dispersion curve intersects the light wave for GaP showing the possibility that the SP can be excited by these waves at a particular angle of incidence called the resonance angle. Thus, it is clear that the SP wave at the gold and high refractive index medium interface can be excited by using a GaP substrate. Now we shall calculate the reflection coefficient.

In Fig. 4, we have plotted the variation of reflected light intensity with the angle of incidence. As it is obvious, the intensity shows a sharp dip at a particular value of the angle of incidence. For n_s=1.30, the resonance angle comes out to be 24.9788°. As we increase the value of n_s from 1.30 to 1.50, i.e. δn_s=0.2, the SPR dip shifts to the higher value of angle of incidence θ_{SPR}= 29.9641° giving rise to a shift of $\delta\theta$= 4.9853°. The larger the value of shift is, the greater the sensitivity is. This cannot be compared for the Si based prism as the SPR can not be excited for such a high value of sensing region refractive index. In the same figure, we have also plotted the SPR curve for n_s= 1.7 and 1.9, and the four curves are well depict the quite high value of the sensitivity separately.

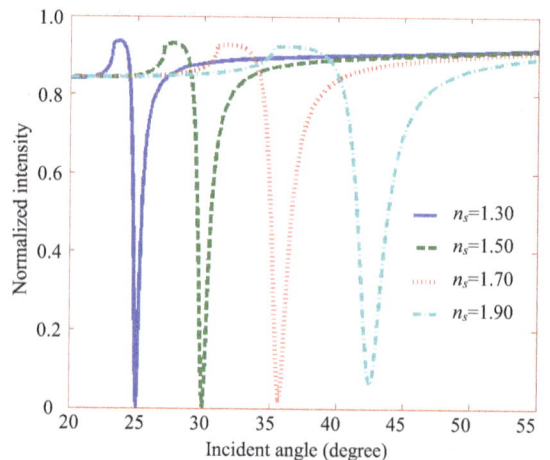

Fig. 4 Variation of normalized intensity with the angle of incidence (SPR curves).

To check the variation of sensitivity with the sensing region refractive index, we have calculated the resonance angle for a large range of refractive index starting from the gaseous medium to highly concentrated liquids.

The curve between the sensitivity and sensing region refractive index is shown in Fig. 5. The sensitivity increases with an increase in the sensing region refractive index as for higher concentrations the resonance condition will be satisfied at higher values of SPR angles. One more parameter is the detection accuracy. For the calculation of the detection accuracy, first we have calculated the full width at half maximum FWHM of each SPR curve, and then the DA is evaluated as per the definition given in (10).

or uses a buffer layer to bring the SPR dip in the visible region, but in the present case, we have modelled a single sensing chip which can be used to sense the gases as well as the liquid medium in the visible region of the spectrum. The large range of n_s is the most advantageous feature of the present study. The theoretical study given in the current study has already been verified experimentally by Motogaito et al. [11]. However, a detailed theoretical analysis in terms of detection accuracy and sensitivity is missing. The current study provides a material to fill that gap.

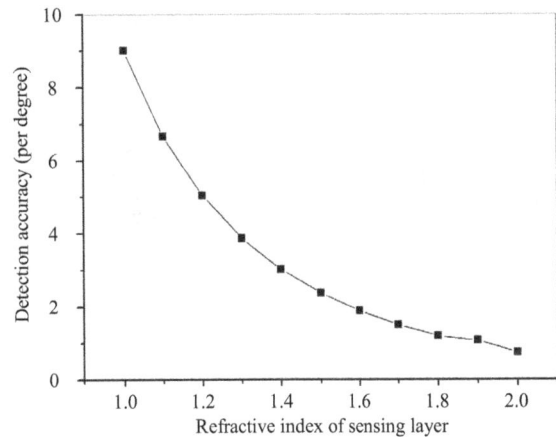

Fig. 6 Variation of detection accuracy with sensing region refractive index.

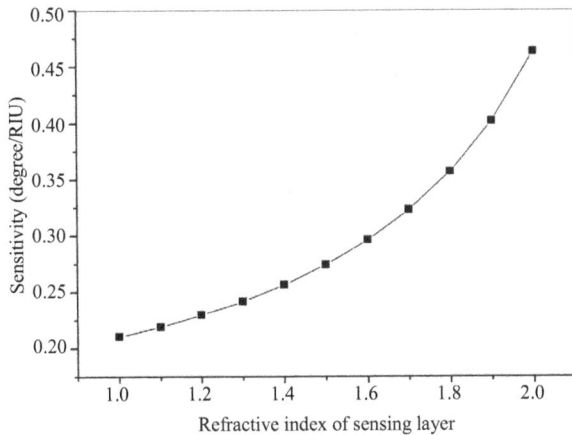

Fig. 5 Sensitivity variation with sensing region refractive index.

It is visible from the SPR curves that with an increase in the sensing region the refractive index, the SPR curves get broadened giving rise to poor detection accuracy. However, the broadening in SPR curves as given in Fig. 4 is not too much so that it may affect θ_{SPR}. The variation of detection accuracy with refractive index of the sensing layer is plotted in Fig. 6.

More importantly, we are using the visible light source, i.e. a He-Ne laser with the wavelength 632 nm, and we are able to sense a large spectrum of the sensing region refractive index. The sensor reported till now either uses the infra red (IR) source

References

[1] R. C. Jorgenson and S. S. Yee, "A fiber-optic chemical sensor based on surface plasmon resonance," *Sensors and Actuators B: Chemical*, 1993, 12(3): 213−220.

[2] O. Esteban, R. Alonso, C. Navarrete, and C. A. G. Cano, "Surface plasmon excitation in fiber-optic sensors: a novel theoretical approach," *Journal of Lightwave Technology*, 2002, 20(3): 448−453.

[3] L. K. Chau, Y. F. Lin, S. F. Cheng, and T. J. Lin, "Fiber-optic chemical and biochemical probes based on localized surface plasmon resonance," *Sensors and Actuators B: Chemical*, 2006, 113(1): 100−105.

[4] X. Chen, R. R.Wei, M. Shen, Z. F. Zhang, and C. F. Li, "Bistable and negative lateral shifts of the reflected light beam from Kretschmann configuration with nonlinear left handed metamaterials," *Applied Physics B*, 2010, 101(1): 283−289.

[5] S. A. Zynio, A. V. Samoylov, E. R. Surovtseva, V. M. Mirsky, and Y. M. Shirsov, "Bimetallic layers increase sensitivity of affinity sensors based on surface plasmon resonance," *Sensors*, 2002, 2(2): 62–70.

[6] R. K. Verma and B. D. Gupta, "Surface plasmon resonance based fiber optic senor for IR region using conductin metal oxide," *Journal of the Optical Society of America A*, 2010, 27(4): 846–851.

[7] G. G. Nenninger, P. Tobiska, J. Homola, and S. S. Yee, "Long range surface plasmons for high resolution surface plasmon resonance sensors," *Sensors and Actuators B: Chemical*, 2001, 74(1–3): 145–151.

[8] Rajan, A. K. Sharma, and B. D. Gupta, "Fiber optic sensor based on long range surface plasmon resonance," *Journal of Optics A: Pure and Applied Optics*, 2007, 9(7): 682.

[9] A. K. Sharma, Rajan, and B. D. Gupta, "Influence of dopants on the performance of a fiber optic surface plasmon resonance sensor," *Optics Communications*, 2007, 274(2): 320–326.

[10] S. K. Mishra and B. D. Gupta, "Surface plasmon resonance based fiber optic pH sensor utilizing Ag/ITO/Al/hydrogel layers," *Analyst*, 2013, 138(9): 2640–2646.

[11] A. Motogaito, S. Nakamura, J. Miyazaki, H. Miyake, and K. Hiramatsu,"Using surface plasmon polariton at GaP-Au interface in order to detect chemical species in high refractive index media," *Optics Communications*, 2015, 341: 64–68.

[12] S. Singh, S. K. Mishra, and B. D. Gupta, "SPR based fibre optic biosensor for phenolic compounds using immobilization of tyrosinase in polyacrylamide gel," *Sensors and Actuators B: Chemical*, 2013, 186: 388–395.

[13] J. Zhou, Y. Wang, C. Liao, B. Sun, J. He, G. Yin, *et al.*, "Intensity modulated refractive index sensor based on optical fiber Michelson interferometer," *Sensors and Actuators B: Chemical*, 2015, 208: 315–319.

[14] M. Deng, C. P. Tang, T. Zhu, Y. J. Rao, L. C. Xu, and M. Han, "Refractive index measurement using photonic crystal fiber-based Fabry-Perot interferometer," *Applied Optics*, 2013, 49(9): 1593–1598.

[15] P. Tobiska, O. Hugon, A. Trouillet, and H. Gagnaire, "An integrated optic hydrogen sensor based on SPR on palladium," *Sensors and Actuators B: Chemical*, 2001, 74(1–3): 168–172.

A High Sensitivity Pressure Sensor Based on Two-Dimensional Photonic Crystal

Shangbin TAO, Deyuan CHEN[*], Juebin WANG, Jing QIAO, and Yali DUAN

Nangjing University of Posts and Telecommunications, Nanjing, 210009, China

[*]Corresponding author: Deyuan CHEN   E-mail: chendy@njupt.edu.cn

Abstract: In this paper, we propose and simulate a pressure sensor based on two-dimensional photonic crystal with the high quality factor and sensitivity. The sensor is formed by the coupling of two photonic crystal based waveguides and one nanocavity. The photonic crystal with the triangular lattice is composed of GaAs rods. The detailed structures of the waveguides and nanocavity are optimized to achieve better quality factor and sensitivity of the sensor. For the optimized structures, the resonant wavelength of the sensor has a linear redshift as increasing the applied pressure in the range of 0 – 2 GPa, and the quality factor keeps unchanged nearly. The optimized quality factor is around 1500, and the sensitivity is up to 13.9 nm/GPa.

Keywords: Photonic crystal; waveguide; nanocavity; pressure sensor

1. Introduction

Eli Yablonovitch and Sajeev John proposed the conception of photonic crystal (PhC) in 1980s [1], which can mold the flow of light. Photonic crystal is a structure in which a periodic variation in refractive index occurs at the scale of the wavelength of light in one, two or three directions. PhCs have many attractive characteristics such as photonic bandgap (PBG), localization effect, and refraction effect. By introducing point defect and/or line defect, PhC-based devices can be made [2], such as PhC-based waveguide [3–5], nanocavity [6–8], low threshold lasers [9–12], PhC-based filter [13–15], PhC fibers [16–19], optical sensors, and so on.

In last few years, a large number of research and investigations have shown that PhC can also be used to make optical sensors. Yang *et al.* presented an electro-optical sensor based on a slotted 2D-PhC waveguide which exhibited over 30-times sensitivity enhancement and about 6.6 times improvement in quality factor compared with the W1-PhC waveguide based electro optical sensor [20]. Xu *et al.* designed a micro displacement sensor based on the 2D-PhC line-defect resonant cavity structure which could be integrated in a micro-electro-mechanical system [21]. Besides, PhC can be used to design a pressure sensor. Shanthi *et al.* demonstrated a 2D-PhC based pressure sensor whose sensitivity could reach 2 nm/GPa and dynamic range could reach 7 GPa [22]. Olyaee *et al.* designed a 2D-PhC pressure sensor which had the sensitivity of 8 nm/GPa and wide linearity range between 0 and 10 GPa.

In this paper, we design and simulate a 2D-PhC based pressure sensor using the coupling of two

PhC-based waveguides with a PhC-based nanocavity. The sensor is made of GaAs with the stronger photo-elastic effect. The refractive index of GaAs varies under pressure, and the resonant wavelength of the microcavtiy made of GaAs rods shifts. The sensor properties such as the quality factor, resonant wavelength, transmittance, sensitivity, and dynamic range are investigated. The electromagnetic analysis of the sensor has been conducted by applying a commercial finite element method code, i.e. COMSOL multiphysics.

2. Pressure effect analysis

The sensor principle is based on the change in the refractive index of the material induced by the photoelastic effect, piezoelectric effect, and eletrooptic effect. The change in the refractive index will modify the resonant wavelength of the structure. And it is found that the variation of the refractive index is mainly due to the photo-elastic effect [23].

The relationship between the refractive index and the stress can be expressed as follows [22–24]:

$$
\begin{pmatrix} n_{xx} \\ n_{yy} \\ n_{zz} \\ n_{yz} \\ n_{xz} \\ n_{xy} \end{pmatrix} = \begin{pmatrix} n_0 \\ n_0 \\ n_0 \\ 0 \\ 0 \\ 0 \end{pmatrix} - \begin{pmatrix} C_1 & C_2 & C_2 & 0 & 0 & 0 \\ C_2 & C_1 & C_2 & 0 & 0 & 0 \\ C_2 & C_2 & C_1 & 0 & 0 & 0 \\ 0 & 0 & 0 & C_3 & 0 & 0 \\ 0 & 0 & 0 & 0 & C_3 & 0 \\ 0 & 0 & 0 & 0 & 0 & C_3 \end{pmatrix} \begin{pmatrix} \sigma_{xx} \\ \sigma_{yy} \\ \sigma_{zz} \\ \sigma_{yz} \\ \sigma_{xz} \\ \sigma_{xy} \end{pmatrix}
$$

$$(1)$$

where C_1, C_2, and C_3 are defined as

$$C_1 = n_0^3(p_{11} - 2\upsilon p_{12})/2E \tag{2}$$

$$C_2 = n_0^3\left[p_{12} - \upsilon(p_{11} + p_{12})\right]/2E \tag{3}$$

$$C_3 = n_0^3 p_{44}/2G \tag{4}$$

where p_{11}, p_{12}, and p_{44} are strain-optic constants，E, G, and υ represent the Youngs modulus, shear modulus, and Poissons ratio. σ_{ij} stands for the pressure along ij direction, and n_{ij} denotes the refractive index along ij direction.

We assume that the pressure is orthogonal to the photonic crystal plane and applied only in one direction, which means that the pressure is

distributed uniformly and the whole structure is under the hydrostatic pressure state, therefore:

$$\sigma_{xx} = \sigma_{yy} = \sigma_{zz} = \sigma \tag{5}$$

$$\sigma_{xy} = \sigma_{xz} = \sigma_{yz} = 0 \tag{6}$$

So (1) can be simplified to

$$n = n_o - (C_1 + 2C_2)\sigma \tag{7}$$

where n_0 is the refractive index without the applied pressure.

For GaAs, we have [23]

$$n = 3.43 \tag{8}$$

$$p_{11} = -0.165 \tag{9}$$

$$p_{12} = -0.140. \tag{10}$$

3. Design and analysis of 2D-PhC pressure sensor structure

3.1 Design of 2D-PhC

The structure of the 2D-PhC is shown in Fig. 1(a). It is of 15×20 triangular lattice arrays of GaAs rods in air. The photonic crystal lattice constant and rods radius are 540 nm and 130 nm, respectively.

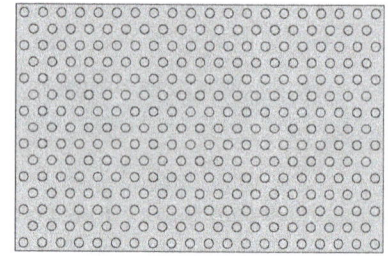

(a)

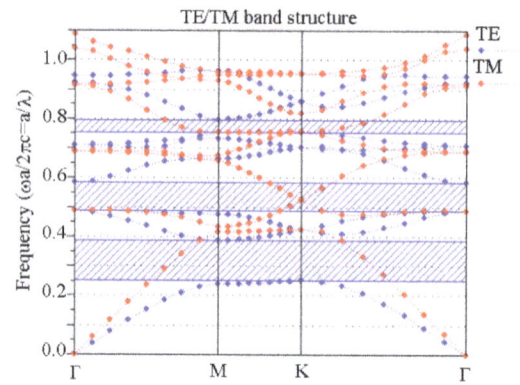

(b)

Fig. 1 Two-dimensional photonic crystal: (a) schematic diagram and (b) TE/TM mode band structure of the 2D-PhC.

The electromagnetic analysis of the 2D-PhC

structure has shown a large bandgap for TE polarization between $0.251(a/\lambda)$ and $0.386(a/\lambda)$ as shown in Fig. 1(b).

Two line defects are introduced by removing GaAs rods in order to form the input and output waveguides as shown in Fig. 2. These two waveguides are coupled through a nanocavity which is formed by removing one GaAs rod between the two waveguides. The port marked "input" with an arrow in Fig. 2 is the input port. The light source is put at the input port in order to excite the structure. The port marked "output" in Fig. 2 is the output port. The photonic detector is placed at the output port to detect the output light so that the resonant wavelength of the structure can be analyzed. The part of the sensor, i.e., the waveguides ends and the nanocavity marked in Fig. 2 by the dotted lines are studied and optimized in order to achieve better performance.

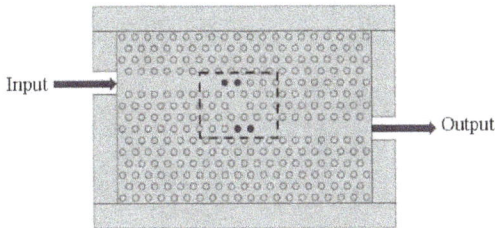

Fig. 2 Schematic diagram of the 2D-PhC sensor.

3.2 Design of waveguides and nanocavity

The detailed structure which is marked by dotted lines in Fig. 2 can influence the performance of the sensor. So the waveguides are optimized at their ends as shown in Fig. 3, the enlarged view of the coupling part. Three cases are considered. In Case (a), the rods numbered 1 and 1', and 2 and 2' are kept. In Case (b), rods numbered 1 and 1' are removed, and 2 and 2' are kept. In Case (c), rods numbered 1 and 1', and 2 and 2' are all removed.

The transmission spectra of the sensors with different waveguides ends are displayed in Fig. 4. It is obvious that Case (c) cannot be used for its worst transmittance, multi-modes, and lowest quality

factor. So no more GaAs rods should be removed. For Case (a), the quality factor is better up to 480, while the transmittance is low; for Case (b), the quality factor is low to 130, while transmittance is better.

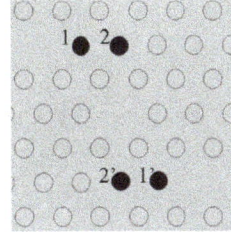

Fig. 3 Enlarged views of the sensor which is marked by dotted lines in Fig. 2.

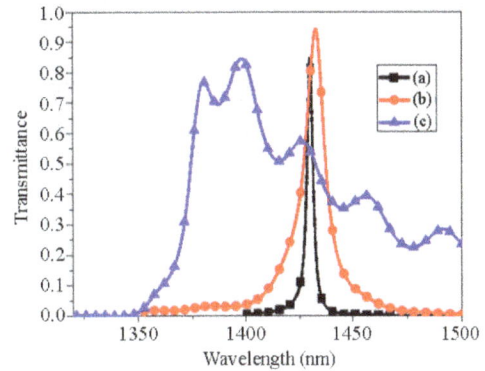

Fig. 4 Transmission spectrum of the 2D-PhC sensor with different coupling parts of Cases (a), (b), and (c).

We also analyze the case that less GaAs rods are removed to form the waveguide. The corresponding transmittance is even lower because most of the light is reflected.

Besides, the structure of the nanocavity determines the quality factor when the lattice constant and radius of rods are decided. By properly designing the nanocavity, the high quality factor and sensitivity may be achieved.

Removing one GaAs rod as is shown in Fig. 2 can be the simplest nanocavity. But the quality factor is too low, so it is necessary to optimize the structure of the nanocavity. In order to increase the quality factor and the sensitivity, the structure of the nanocavity is changed. And two small GaAs rods with the radius of 65 nm are placed inside the nanocavity as is shown in Fig. 5.

(a)   (b)

Fig. 5 Enlarged coupling part of Cases (a) and (b).

3.3 Performance of the sensor

For both the structures of Cases (a) and (b) shown in Fig. 5, the pressure is applied with a step of 0.5 GPa from 0 to 2 GPa to study the performance of the sensors.

Figure 6 shows the transmittance of the sensor for Case (a) under different pressures from 0 to 2 GPa. It is clear that with increasing the pressure, the coupling resonant mode red-shifts, the transmittance decreases from 96% to 72%, and the quality factor decreases from 1140 to 838. The resonant mode wavelength with pressure has been dotted in Fig. 7. An obvious linear relationship can be observed, allowing linear measurements of the pressure. Therefore, it is relatively proper to estimate the sensor sensitivity S as the ratio of the resonant wavelength variation to the applied pressure variation, i.e., the slop of the curve:

$$S = \frac{\Delta\lambda}{\Delta P}. \tag{11}$$

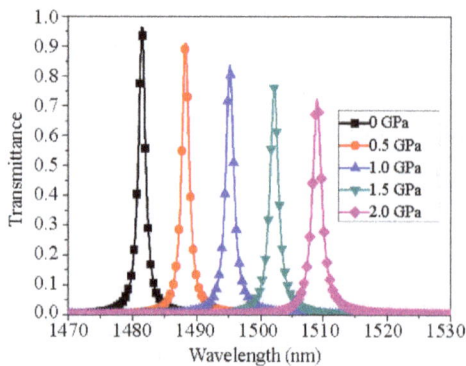

Fig. 6 Transmission spectrum for different applied pressures.

In the case, S equals 13.7 nm/GPa. It seems to have the better performance parameter, but the transmittance drops when increasing pressure. When

applying 2 GPa pressure, the transmittance drops to 72%, which is unacceptable.

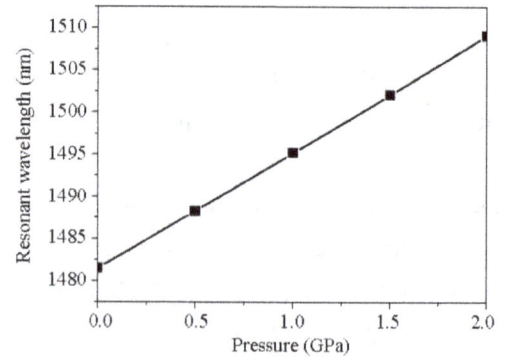

Fig. 7 Linear relationship between the resonant wavelength and pressure in the range of 0 to 2 Gpa.

For Case (b), the transmittance spectra of the sensor under different pressures are shown in Fig. 8. With increasing pressure, the resonant wavelength has a linear redshift, and the transmittance and the quality factor change slightly.

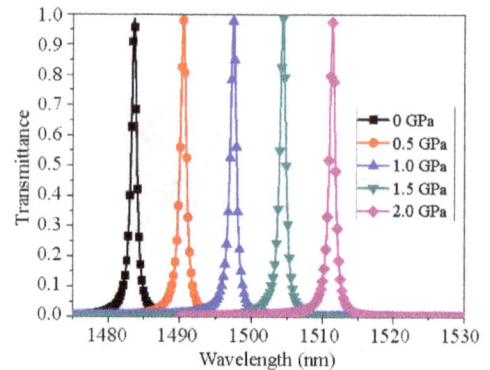

Fig. 8 Transmission spectrum for different applied pressures.

Also, the relationship between the resonant mode wavelength and pressure in the range of 0 to 2 GPa has been plotted in Fig. 9. It is clear that the resonant wavelength shifts linearly with the applied pressure, allowing linear measurements of the pressure. So we can calculate the quality factor to be around 1500, and the sensitivity is up to 13.9 nm/GPa.

According to the quality factor and the sensitivity of the above two cases, Case (b) is the better one for the pressure sensor. The parameters are summarized in Table 1.

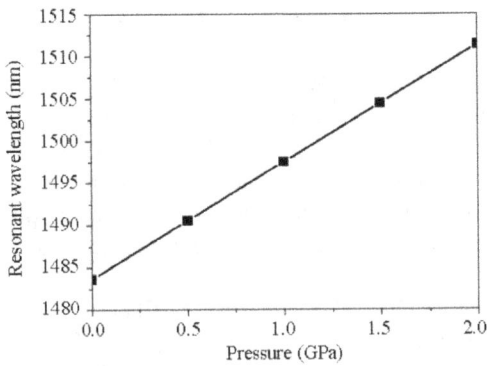

Fig. 9 Linear relationship between resonant wavelength and pressure in the range of 0 to 2 Gpa.

Table 1 Parameters of the sensor.

Parameters	Values
Lattice constant a	540 nm
Radius of the rod r	130 nm
Radius of the rod in the cavity	65 nm
Size	20×15
Refractive index of GaAs rod n_o	3.43
Refractive index of background	1
Polarization	TE
PBG	0.251–0.386(a/λ)

The comparison between the proposed pressure sensor and similar sensors is presented in Table 2.

Table 2 Comparison between the proposed sensor and other similar sensors.

Parameters	Quality factor	Sensitivity (nm/GPa)
Ref.[22]	75.5	2
Ref.[24]	1470	8
Proposed sensor	1500	13.9

4. Conclusions

We have proposed the design of a pressure sensor based on the two-dimensional photonic crystal with the triangular lattice of GaAs rods surrounded by air. This sensor is constructed with PhC-based waveguides coupling with a nanocavity. The quality factor increases to 1500 by modifying the lattice constant, radius of GaAs rods, and the structure of waveguides and nanocavity. The refractive index of GaAs changes along with the applied pressure due to the photo-elastic effect. Through simulation, we have found that the resonant wavelength shifts and follows a linear law. The quality factor is around 1500, and the sensitivity is about 13.9 nm/GPa under the dynamic range between 0 and 2 GPa. This sensor can be used in detecting occations such as long-distance pipeline strain monitoring or other sensing applications because of its high sensitivity and fast responding.

Acknowledgment

This work was supported by the NSFC (National Natural Science Foundation of China) (Nos. 61302101 and 61306003), the NSF (Natural Science Fundation) of Jiangsu Province of China (No. BK20130874), the NSF of the Jiangsu Higher Education Institutions (No. 13KJB510025), GSRIP (Graduate Student Research and Innovation Project) of Jiangsu Province (Nos. KYLX_0801 and SJLX15_0376), NUPTSP (Nanjing University of Posts and Telecommunications Science Project) (No. NY214051).

References

[1] E. Yablonovitch, "Inhibited spontaneous emission in solid-state physics and electronics," *Physics Review Letters*, 1987, 58(20): 2059–2062.

[2] J. D. Joannopoulos, R. D. Meade, and J. N. Winn, *Photonic crystal: modling of flow of light.* Princeton, NJ: Princeton University Press, 1995.

[3] A. Mekis, J. C. Chen, I. Kurland, S. Fan, P. R. Villeneuve, and J. D. Joannopoulos, "High transmission through sharp bends in photonic crystal waveguides," *Physical Review Letters*, 1996, 77(18): 3787–3790.

[4] M. Tokushima, H. Kosaka, A. Tomita, and H. Yamada, "Lightwave propagation through a 120° sharply bent single-line-defect photonic crystal waveguide," *Applied Physics Letters*, 2000, 76(8): 952–954.

[5] A. Lavrinenko, P. Borel, L. Frandsen, M. Thorhauge, A. Harpth, M. Kristensen, *et al.*, "Comprehensive FDTD modelling of photonic crystal waveguide components," *Optics Express*, 2004, 12(2): 234–248.

[6] N. Susumu, C. Alongkarn, and I. Masahiro, "Trapping and emission of photons by a single defect in a photonic bandgap structure," *Nature*, 2000, 407(6804): 608–610.

[7] Y. Akahane, T. Asano, B. Song, and N. Susumu, "High-Q photonic nanocavity in a two-dimensional photonic crysal," *Nature*, 2003, 425(6961): 944–947.

[8] K. Srinivasan, P. Barclay, O. Painter, J. Chen, A. Y. Cho, and C. Gmach, "Experimental demonstration of a high quality factor photonic crystal microcavity," *Applied Physics Letters*, 2003, 83(10): 1915–1917.

[9] M. Loncar, T. Yoshie, A. Scherer, P. Gogna, and Y. Qiu, "Low-threshold photonic crystal laser," *Applied Physics Letters*, 2002, 81(15): 2680–2682.

[10] H. G. Park, S. H. Kim, S. H. Kwon, Y. Ju, J. Yang, J. Baek, *et al.*, "Electrically driven single-cell photonic crystal laser," *Science*, 2004, 305(5689): 1444–1447.

[11] O. Painter, A. Husain, A. Scherer, P. T. Lee, I. Kim, J. D. O'Brien, *et al.*, "Lithographic tuning of a two-dimensional photonic crystal laser array," *IEEE Photonics Technology Letters*, 2000, 12(9): 1126–1128.

[12] K. Inoue, M. Sasada, J. Kawamata, K. Sakoda, and J. W. Haus, "A two-dimensional photonic crystal laser," *Applied Physics Letters*, 1999, 38(2B): 157–159.

[13] N. Mec, P. Kuzel, L. Duvillaret, A. Pashkin, M. Dressel, and M. T. Sebastian, "Highly tunable photonic crystal filter for the terahertz range," *Optics Letters*, 2005, 30(5): 549–551.

[14] W. Li, Y. Fu, Q. Zhang, and D. F. Shi, "Filtering performance comparision of two types of photonic crystal filter," *Laser & Infrared*, 2010, 40(7): 762–765.

[15] Y. Kanamori, N. Matsuyama, and K. Hane, "Resonant wavelength tuning of a pitch-variable 1-D photonic crystal filter at telecom frequencies," *IEEE Photonics Technology Letters*, 2008, 20(13): 1136–1138.

[16] J. C. Knight, T. A. Birks, P. S. Russell, and D. M. Atkin, "All-silica single-mode optical fiber with photonic crystal cladding," *Optics Letters*, 1996, 21(19): 1547–1549.

[17] R. F. Cregan, B. J. Mangan, and J. C. Knight, "Single-mode photonic band gap guidance of light in air," *Science*, 1999, 285(5433): 1537–1539.

[18] S. Kunimasa, S. Yuichiro, and K. Masanori, "Coupling characteristics of dual-core photonic crystal fiber couplers," *Optics Express*, 2003, 11(24): 3188–3195.

[19] P. Russell, "Photonic crystal fibers," *Journal of Lightwave Technology*, 2007, 24(12): 4729–4749.

[20] D. Yang, H. Tian, and Y. Ji, "The study of electro-optical sensor based on slotted photonic crystal waveguide," *Optics Communications*, 2011, 284(20): 4986–4990.

[21] Z. Xu, L. Cao, C. Gu, Q. He, and G. Jin, "Micro displacement sensor based on line-defect resonant cavity in photonic crystal," *Optics Express*, 2006, 14(1): 298–305.

[22] K. V. Shanthi and S. Robinson, "Two-dimensional photonic crystal based sensor for pressure sensing," *Photonic Sensors*, 2014, 3(3): 248–253.

[23] M. Huang, "Stress effects on the performance of optical waveguides," *Solids & Structures*, 2003, 40(7): 1615–1632.

[24] S. Olyaee and A. A. Dehghani, "High resolution and wide dynamic range pressure sensor based on two-dimensional photonic crystal," *Photonic Sensors*, 2012, 2(1): 92–96.

[25] optical waveguides," *Solids & Structures*, 2003, 40(7): 1615–1632.

[26] S. Olyaee and A. A. Dehghani, "High resolution and wide dynamic range pressure sensor based on two-dimensional photonic crystal," *Photonic Sensors*, 2012, 2(1): 92–96.

3

Study on 3D CFBG Vibration Sensor and its Application

Qiuming NAN[1,2*] and Sheng LI[1,2]

[1]*National Engineering Laboratory for Fiber Optic Sensing Technology, Wuhan University of Technology, Wuhan, 430070, China*

[2]*Key Laboratory of Fiber Optic Sensing Technology and Information Processing, Ministry of Education, Wuhan, University of Technology, Wuhan, 430070, China*

[*]Corresponding author: Qiuming NAN   E-mail: 197114012@qq.com

Abstract: A novel variety of three dimensional (3D) vibration sensor based on chirped fiber Bragg grating (CFBG) is developed to measure 3D vibration in the mechanical equipment field. The sensor is composed of three independent vibration sensing units. Each unit uses double matched chirped gratings as sensing elements, and the sensing signal is processed by the edge filtering demodulation method. The structure and principle of the sensor are theoretically analyzed, and its performances are obtained from some experiments and the results are as follows: operating frequency range of the sensor is $10\,\mathrm{Hz} - 500\,\mathrm{Hz}$; acceleration measurement range is $2\,\mathrm{m\cdot s^{-2}} - 30\,\mathrm{m\cdot s^{-2}}$; sensitivity is about $70\,\mathrm{mV/m\cdot s^{-2}}$; crosstalk coefficient is greater than $22\,\mathrm{dB}$; self-compensation for temperature is available. Eventually the sensor is applied to monitor the vibration state of radiation pump. Seen from its experiments and applications, the sensor has good sensing performances, which can meet a certain requirement for some engineering measurement.

Keywords: Three-dimensional (3D); matched; chirped fiber Bragg grating (CFBG); edge filtering demodulation; crosstalk

1. Introduction

Vibration measurement is one of the most effective methods for mechanical equipment condition monitoring and fault diagnosis. However, vast majority of vibration measurement is based on electromagnetic sensors, although this technology has been used for some of its advantages, its application is greatly limited in some special occasion [1–4]. For example, in the flammable, explosive, and electromagnetic-interference situation, a large network is required, and the signal is required to transmit over a long distance. The fiber Bragg grating sensing technology is a newly minted one, which has such characteristics, such as nonelectric detection, anti-interference of electromagnetic, ease to use in large-scale network, large amount of information, and long-distance transmission. It provides a new technical means for the vibration monitoring of mechanical equipment [5, 6]. According to statistics, there are being done some researches on a one-way fiber grating vibration sensor [7], but in practical application, it is frequently demanded to detect the spatial vibration of a certain part of the machine at the same time. Due to the limitation of the installation space or other reasons, three one-way vibration sensors can be installed in the identical position so difficultly that the authentic state of equipment can not be monitored completely and correctly [8–10]. Consequently, it is very essential to conduct the research on a three-dimensional chirped fiber Bragg grating (3D CFBG) vibration sensor.

2. Structure and principle of sensor

2.1 Structure design

According to the application requirements, the overall design goals of the sensor are as follows:

(1) Integration and seal design
(2) Working frequency $\geq 500\,\text{Hz}$
(3) Acceleration range $\geq 20\,\text{m}\cdot\text{s}^{-2}$
(4) Ambient temperature range: $-20\,℃ - 80\,℃$
(5) Single dimensional size $\leq 100\,\text{mm}$

According to this goal, we design a structure of the three-dimensional fiber grating vibration sensor as shown in Fig. 1. The sensor is composed of three unidirectional sensing units that are mutually vertical. Each sensing unit is composed of a base, an elastomer, a mass block, and two CFBGs, and its structure and principle are shown in Fig. 2. Two CFBGs written in the same optic fiber are formed into precise matching by accurate fabrication process, and they are symmetrically fixed between the base and the mass block. In order to avoid dead zone, we give them 2 nm/s pretension. When the sensor receives the vibration signal from the outside world, the mass supported by the elastic membrane will vibrate up and down to drive the CFBGs stretch along the axial direction, and the reflection peaks of CFBG1 and CFBG2 will change, as shown in Fig. 3. The light emitting diode (LED) bandwidth is approximately 40 nm. The detected signal by the photodetector is the change in the envelope spectrum of the two CFBGs, which is negatively correlated with that of the overlapped spectrum. To take the upward acceleration of the mass block as an example, CFBG2 elongates, CFBG1 shrinks, the overlapped area becomes bigger but the detected signal is negative, and vice versa.

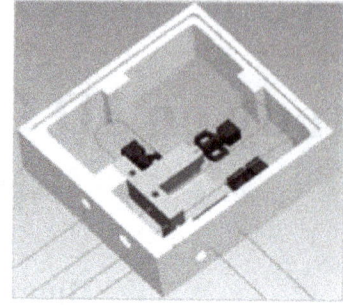

Fig. 1 Structure of 3D CFBG accelerometer.

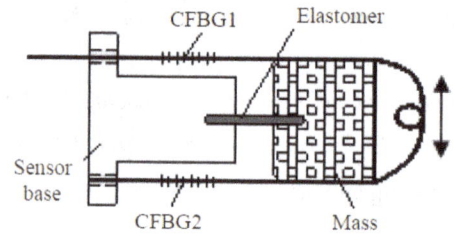

Fig. 2 Structure and principle of sensing unit.

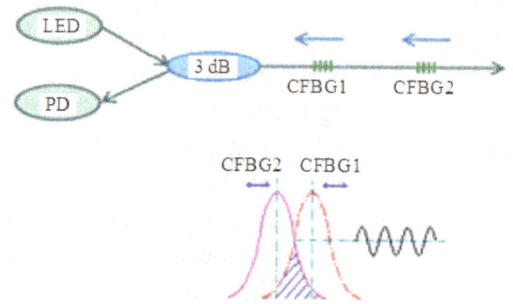

Fig. 3 Principle of matching demodulation.

From the structure and principle of the sensor, the sensor has the following two notable advantages. Firstly, when the external temperature changes, the reflection spectra of the two gratings will move to the same direction, and the envelope area remains constant, so the self-compensation for temperature change can be realized. Secondly, when the sensor is forced to vibrate, the moving direction of two reflection spectrums are opposite, and the envelope area is increased significantly. Accordingly, the sensitivity of the sensor is effectively improved.

2.2 Theoretical analysis

In order to further illustrate the working principle of the sensor, the authors have conducted theoretical calculation. As shown in Fig. 4, the upward exciting force exerted to the mass block is $F_a = Ma$, and it makes the elastomer deform, while CFBG2 is elongated to generate a left tension F_f.

For the elastomer, the stress state can be transformed into equivalent to the force F_a and a clockwise torque $M = F_f \times d$, where d is the distance between mass block and elastic node, namely the distance between A and B in Fig. 4.

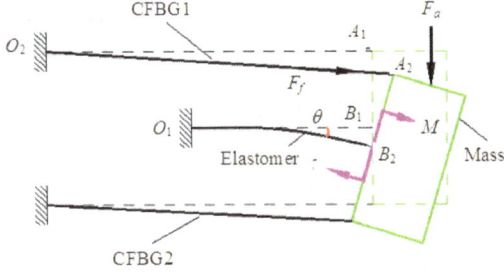

Fig. 4 Schematic diagram of the force exerted to the sensor.

According to the calculation of mechanics of materials for the deformation of bending beam, the deflection ω of elastic body can be expressed as

$$\omega = \frac{F_a l^3}{3EI} - \frac{Ml^2}{2EI} \quad (1)$$

where l is the length of the elastomer, E is the elastic modulus, and I is the moment of inertia.

$$\Delta L = O_2 A_2 - O_2 A_1$$
$$= \sqrt{\left(O_2 A_1 + A_2 B_2 \sin\theta\right)^2 + \left(A_1 B_1 + B_1 B_2 - A_2 B_2 \cos\theta\right)^2}$$
$$- O_2 A_1. \quad (2)$$

The tension ΔL of the optical fiber subjected to external vibration can be expressed as (2). As shown in Fig. 4, θ is the angle of the elastomer caused by the stress. As the rotation angle is very small, (2) can be expressed as

$$\Delta L = \sqrt{L^2 + \omega^2} - L = \omega \quad (3)$$

where L is the original length of the fiber.

The axial stress ε of the fiber and the change $\Delta\lambda$ in wavelength caused by the stress can be expressed respectively as

$$\varepsilon = \frac{\Delta L}{L} = F_f / E_f A \quad (4)$$

where E_f, A, and K_f are the elastic modulus, cross-sectional area, and stiffness of optical fiber, respectively.

$$\Delta\lambda = (1 - P_e)\lambda\varepsilon \quad (5)$$

where P_e and λ are the effective elastic optical coefficient and initial wavelength of fiber grating,

respectively.

From (1) to (5), the sensitivity S of the sensor can be expressed as

$$S = \frac{\Delta\lambda}{a} = (1 - P_e)\lambda\frac{1}{L}\frac{2ml^2}{6EI + 3K_f dl^2} \quad (6)$$

where m is the quality of the mass. From (6), it can be observed that there is a linear relationship between the wavelength change $\Delta\lambda$ of FBG and the outside acceleration a.

The elastomer can be regarded as a uniform cantilever beam, so its stiffness K_e is expressed as

$$K_e = \frac{3EI}{l^3}. \quad (7)$$

According to the kinematics equation of the object structure, the first-order resonant frequency f of the sensor is calculated as

$$f = \frac{1}{2\pi}\sqrt{\frac{K}{m}} = \frac{1}{2\pi}\sqrt{\frac{K_e + (d/L)^2 K_f}{m}}. \quad (8)$$

From (6) and (8), if m is larger and l is longer, then S is bigger, and f is smaller; vice versa. Therefore, the structure parameters of the sensor should be reasonably designed according to the requirement of the measurement.

3. Performance tests of sensor

In accordance with the above designs, we have made some samples of 3D CFBG vibration sensor. Next, we test the sensing performances of the sensor by means of the experimental apparatus, as shown in Fig. 5. These sensing performances mainly include amplitude-frequency characteristics, acceleration characteristics, anti-crosstalk, and anti-temperature influence performance.

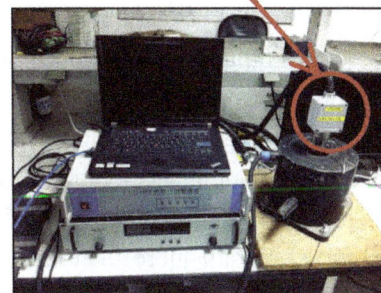

Fig. 5 Vibration testing system.

3.1 Amplitude-frequency characteristics

In the amplitude-frequency characteristic test, the output acceleration of the exciter is kept constant, $3\,m/s^{-2}$ as the input vibration signal. The frequency of the input signal starts from $10\,Hz$, incrementing $50\,Hz$ as a step, and when the frequency value reaches $800\,Hz$, the step size of the frequency is adjusted to $20\,Hz$. The experimental data of each step should be recorded. During the test, the amplitude-frequency characteristics of three sensing units (X, Y, Z) are measured, respectively, as shown in Fig. 6. The results show that their first-order resonant frequencies are $910\,Hz$, $890\,Hz$, and $890\,Hz$, respectively, the flat segments of the curves are all the range from $10\,Hz$ to $500\,Hz$, namely, and the operating frequency range of the sensor is $10\,Hz$ – $500\,Hz$.

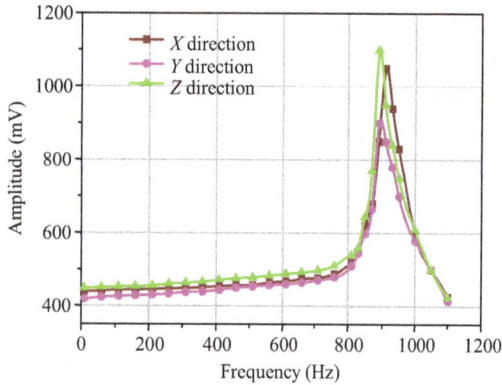

Fig. 6. Amplitude-frequency characteristics of 3 sensing units.

3.2 Linear calibration

Linear calibration is a good method to test the sensor's performances like sensitivity, linearity, and linear range. The frequency of the input signal is $200\,Hz$, and measurement acceleration is within $2\,m\cdot s^{-2}$ – $30\,m\cdot s^{-2}$, which are read out by the piezoelectric standard acceleration sensor, type 4371. Write down the output voltage values of the 3D CFBG vibration sensor at each setting value. The calibration curves are shown in Fig. 7. From Fig. 7, we can see when the input acceleration is within the range of $2\,m\cdot s^{-2}$ – $30\,m\cdot s^{-2}$, the linear degrees of calibration curves are all more than 0.999, and the three sensitivities are $73\,mV/m\cdot s^{-2}$, $69\,mV/m\cdot s^{-2}$, and $75\,mV/m\cdot s^{-2}$, respectively. With an increase in the acceleration, the output voltage increases slowly, and the sensitivity decreases. When the acceleration is more than $30\,m\cdot s^{-2}$, the output voltage becomes saturated. Therefore, the acceleration measurement range of the sensor can be considered as $2\,m\cdot s^{-2}$ – $30\,m\cdot s^{-2}$.

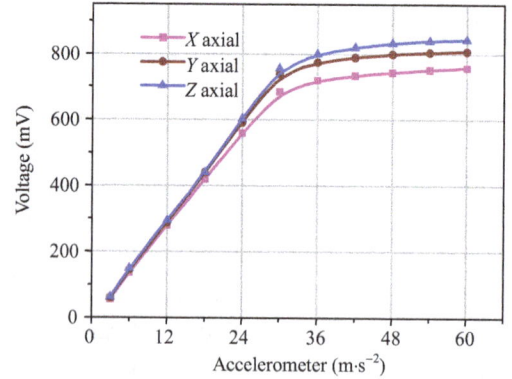

Fig. 7 Calibration curve of 3D FBG accelerometer.

3.3 Anti-crosstalk performance

First of all, let the exciter vibrate in the X axis direction, the input signal is a sine wave signal, of which the vibration frequency is $60\,Hz$, and the acceleration is $6\,m\cdot s^{-2}$. Three-direction response data are collected at the same time, and the effective values of output voltage are $450\,mV$, $32\,mV$, and $34\,mV$, respectively. Similarly, under the condition of the invariable vibration signal, let Y and Z axes be the main vibration directions, respectively, repeat the above experiments, and the results are shown in and Figs. 8, 9, and 10. From Table 1, the maximum crosstalk coefficient can be calculated about $22\,dB$, which indicates that crosstalk has little effect on the measurement.

Table 1 Test results of crosstalk coefficient (mV).

Main vibration direction		X	Y	Z
Output response	X	450	36	38
	Y	32	462	39
	Z	34	35	470

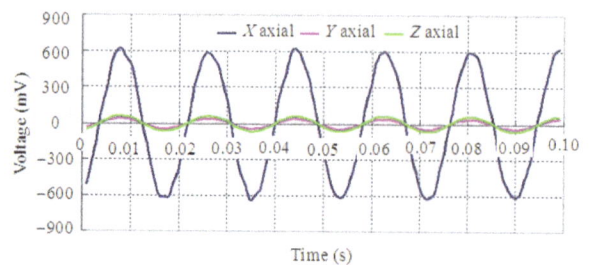

Fig. 8 Response curves of X shaft as main vibrating direction.

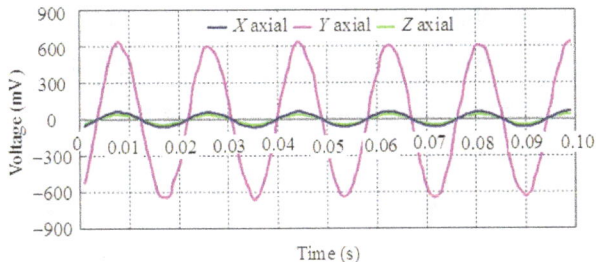

Fig. 9 Response curves of Y shaft as main vibrating direction.

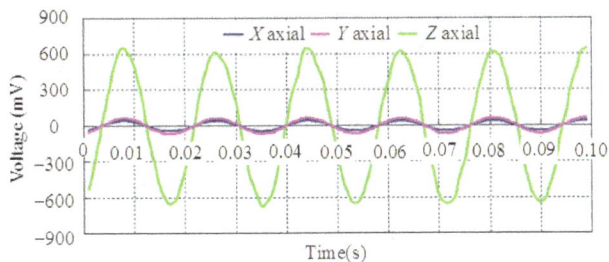

Fig. 10 Response curves of Z shaft as main vibrating direction.

3.4 Anti-temperature performance

The sensing performance of the sensor based on edge filtering demodulation is largely determined by the matching state of the two gratings. When the ambient temperature changes, the reflection peaks of the two gratings will shift; if the two reflection peaks move synchronously, the matching state keeps constant, but if they are not synchronized, the matching state will change. The results of the anti-temperature influence experiment are shown in Fig. 11.

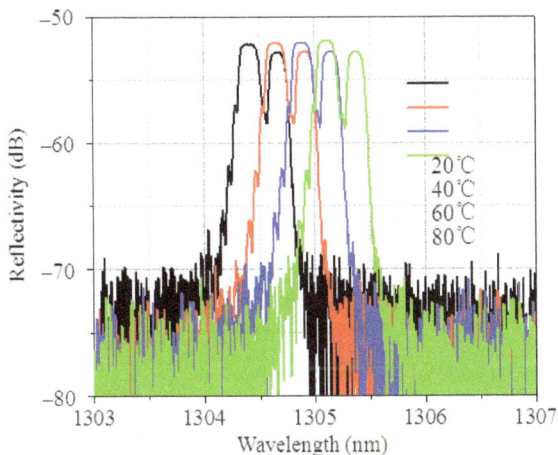

Fig. 11 Matching state of double gratings.

From Fig. 11, it can be seen that the matching state of the double gratings has no change when the temperature varies from $20\,°C$ to $80\,°C$, and the output is stable, which shows that the sensor is insensitive to the change of the temperature and has a strong ability to resist temperature interference.

4. Application

4.1 Application description

The 3D CFBG vibration sensor developed in the paper has been used in the vibration monitoring of the radiation pump in petrochemical industry. In the industrial test, 3 pumps were monitored and 2 sensors were installed on each pump. The monitoring parameters included the vibration of the bearing seat in the three directions, the horizontal (V), vertical (H), and axial (A). In the installation, the sensor was installed on the bearing seat by bolts. Site installation conditions are shown in Fig. 12.

Fig. 12 Sensor site installation photos.

4.2 Application results

On Jan 20th, 2014, the online monitoring system issued a warning signal. It was found through the query that the vibration speed of 3H of the No.3 pump was about 5.2 mm/s, as shown in Fig. 13, and this value was increasing. Further analysis of the vibration data in other directions, which were larger than the normal values, is shown in Fig. 14, which showed that the working condition of the pump was deteriorating. Moreover, further studies found that there were the fault characteristic frequencies of the inner ring and cage of the bearing in the velocity spectrum. Through the comprehensive analysis of the above test results, the preliminary diagnosis was that there were peeling defect on the inner ring and severe wear on the cage.

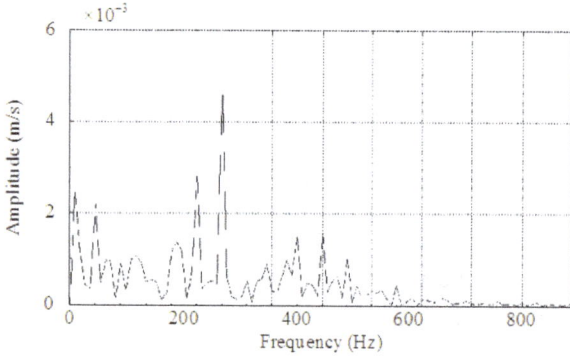

Fig. 13 Frequency spectrum of 3H measuring point.

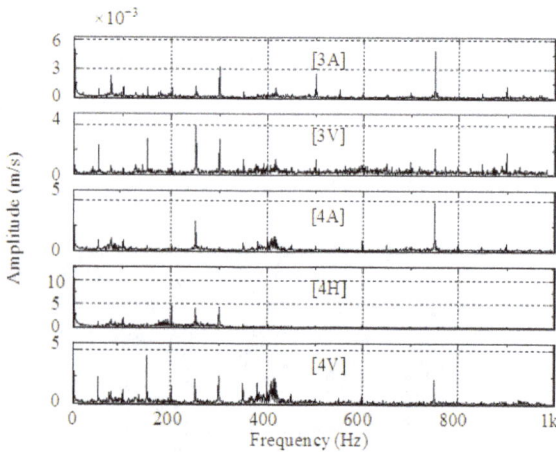

Fig. 14 Frequency spectrum of multimeasuring points.

Fig. 15 Fault bearing physical picture.

On Jan 23th, the radiation pump was removed for maintenance and it was found that there were two larger and a dozen smaller exfoliations on the inner ring of the bearing, and some obvious damages on the surface of the shaft, as shown in Fig. 15. According to the above situation, the bearing was replaced and the system showed the pump had been into working in the normal state after maintenance, which adequately demonstrated the accuracy and reliability of the above fault diagnosis.

5. Conclusions

A novel variety of 3D vibration sensor based on CFBG is developed to measure 3D vibration in the mechanical equipment field. Theoretical analysis and experimental tests have been performed for the sensor, and it has been applied to engineering practice. The main conclusions are as follows.

(1) In order to have a good performance, the following innovative design is implemented. Firstly, the sensing unit is designed as an improved cantilever structure to increase the operating frequency of the sensor. Secondly, the two matched chirped gratings are used as a sensing element to increase the measurement range of acceleration. Finally, the demodulation method based on the edge filtering is used to solve the problem of high frequency signal acquisition.

(2) The results of performance test are as follows. The operating frequency range of the sensor is 10 Hz – 500 Hz, the acceleration measurement range is 2 m·s^{-2} – 30 m·s^{-2}, the sensitivity is about 70 mV/m·s^{-2}, the crosstalk coefficient is greater than 22 dB, and it has a strong ability to resist the temperature interference. These experimental results are in good agreement with the theoretical analysis.

(3) The 3D CFBG vibration sensor has been applied to monitor the working state of the radiation pump and the process of the bearing fault has been successfully monitored, which may be an important basis for fault analysis.

Acknowledgment

This research was funded by the Key Project of National Science Foundation of China, Award Number: 61290311.

References

[1] Y. Zheng, "Vibration monitoring and analysis for rotating machinery," *Gas Tribine Technology*, 2010, 23(1): 39-48.

[2] D. Fu and Y. Wang, "The research of on-line vibration monitoring system of rotating machinery," *Machine Tool & Hydraulics*, 2005, (1): 152-153.

[3] Q. M. Nan, "Study on dynamic sensing and monitoring methods for petrochemical facilities based on optical fiber grating," Ph.D. dissertation, Dept. Wuhan University, China, 2014.

[4] Z. Zhou, Q. Liu, Q. Ai, and C. Xu, "Intelligentmonitoring and diagnosis for modern mechanical equipment based on the integration of embedded technology and FBGS technology," *Measurement*, 2011, 44(9): 1499-1511.

[5] Q. M. Nan, "Study and application of CFBG vibration sensor with symmetrical push-pull configuration," in *the 22nd International Conference on Optical Fiber Sensor*, Beijing, China, 15-19, 2012.

[6] G. Wang and B. Xie, "Improving the performance of chirped fiber grating with cladding being etched as sinusoidal function," *Optik International Journal for Light and Electron Optics*, 2010, 122(6): 557-559.

[7] Y. Zhao and Z. Li, "Study on tri-axial accelerometer based on FBG," *Chinese Journal of Scientific Instrument*, 2006, 27(6): 299-301.

[8] J. Wang, C. Chen, D. Tang, C. Zhang, and Y. Cui, "Three-component photo-elastic fiber optic accelerometer based on the photo-elastic effect," *Chinese Journal of Sensors and Acuators*, 2006, 19(3): 804-806.

[9] L. Ding, J. Zhao, J. Wang, and Q. Sui, "Design of FBG vibration sensor based on matching filter demodulation," *Electro-Optic Technology Application*, 2009, 24(3): 36-39.

[10] Q. Nan, "Research on 3D FBG Accelerometer and demodulation method," *Chinese Optics Letters*, 2014, (A01): 149-153.

4

Alumina Ceramic Based High-Temperature Performance of Wireless Passive Pressure Sensor

Bo WANG[1], Guozhu WU[2, 3*], Tao GUO[2, 3], and Qiulin TAN[2, 3]

[1]*School of Computer Science and Control Engineering, North University of China, Taiyuan, 030051, China*

[2]*Key Laboratory of Instrumentation Science & Dynamic Measurement, Ministry of Education, North University of China, Taiyuan, 030051, China*

[3]*Science and Technology on Electronic Test and Measurement Laboratory, North University of China, Taiyuan, 030051, China*

[*]Corresponding author: Guozhu WU   E-mail: pillar921@163.com

Abstract: A wireless passive pressure sensor equivalent to inductive-capacitive (LC) resonance circuit and based on alumina ceramic is fabricated by using high temperature sintering ceramic and post-fire metallization processes. Cylindrical copper spiral reader antenna and insulation layer are designed to realize the wireless measurement for the sensor in high temperature environment. The high temperature performance of the sensor is analyzed and discussed by studying the phase-frequency and amplitude-frequency characteristics of reader antenna. The average frequency change of sensor is 0.68 kHz/℃ when the temperature changes from 27℃ to 700℃ and the relative change of twice measurements is 2.12%, with high characteristic of repeatability. The study of temperature-drift characteristic of pressure sensor in high temperature environment lays a good basis for the temperature compensation methods and insures the pressure signal readout accurately.

Keywords: Pressure sensor; wireless passive; high temperature; zero drift; resonant frequency

1. Introduction

Pressure signal acquisition has been greatly demanded in harsh environment, featuring turbine engine design, and petroleum exploration, whereas all of them require the sensor capable of operating at more than 200 ℃ [1–3]. Existing sensors are able to meet the requirements in the measuring range, frequency response, and accuracy, while they can only play roles in some of "moderate" environments [4, 5]. For now, the pressure sensor application is still deficient in at least the following two aspects, or just meets the minimum demands: one is ultra-high temperature environment, the other is metallic environment, which causes serious electromagnetic interference. Currently, pressure sensors applied for high temperature have been carried out in-depth research, including silicon on insulator (SOI) high temperature pressure sensor [6, 7], SiC high temperature pressure sensor [8], silicon-sapphire pressure sensor [9], fiber optic pressure sensor [10], [11], surface acoustic wave (SAW) pressure sensor [12, 13], and the pressure sensor equivalent to inductive-capacitive (LC) resonance circuit based pressure sensor [14–16].

A pressure sensor based on LC resonance circuit

uses alumina ceramic as the sensor substrate, which can survive in high temperature above 1000℃. The sensor is under test at different temperatures to analyze and discuss the impact on the pressure signal readout, which provides valuable reference to the research of temperature compensation.

2. Working principle

The pressure sensor can be considered as an LC resonance circuit, which consists of inductance connected in series with capacitance and resistance. L_s is inductance coil, R_s is equivalent resistance, and C_s is variable capacitance that responses to outer pressure. The resonance frequency of sensor f_0 and quality factor Q, neglecting the parasitic capacitance, can be approximately expressed as

$$f_0 = \frac{1}{2\pi\sqrt{L_sC_s}} \tag{1}$$

$$Q = \frac{1}{R}\sqrt{\frac{L_s}{C_s}}. \tag{2}$$

The wireless sensing scheme is presented here by utilizing near-field magnetic coupling between the reader antenna and sensor. By extracting a certain impedance characteristic parameter (amplitude and phase), the variation of sensor frequency can be detected with high resolution. The equivalent wireless coupling circuit is shown in Fig. 1, where L_a is the inductance of antenna coil, R_a is the equivalent resistance of antenna, and M is the coefficient of mutual inductance between reader antenna and sensor.

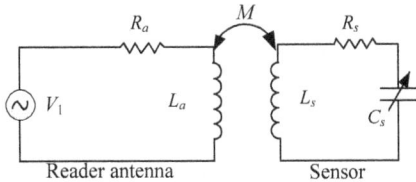

Fig. 1 Equivalent circuit of wireless coupling scheme.

3. Design and fabrication

The pressure sensor was fabricated separately by using high temperature ceramic sintering and post-fire metallization process. ESL 44007 green tape was chosen as the sensor substrate, and ESL 9598-G cermet silver/platinum conductor was selected as the screen printing paste. The properties of green tape composed of 99 percent alumina are listed in Table 1. The main fabrication process of the pressure sensor can be roughly divided into punching, via filling and carbon membrane filling, lamination, high-temperature sintering, screen printing, and low temperature sintering as shown in Fig. 2. A punching machine was used to drill a cavity in Layer 2 and via holes in all three layers firstly. Next, via holes were filled with metal paste to achieve electrical connection throughout the multilayer tapes. After via holes filling, Layers 2 and 3 were precisely stacked by the collating machine. Then, a carbon membrane that had the same dimensions as the cavity was placed into the cavity of the stacked two-layer substrate to support the pressure-sensitive membranes avoiding from collapsing and cracking during the lamination process. After carbon membrane filling, all three layers were stacked together to form a complete cavity. Then, the ceramic tape was placed into the lamination machine under a pressure of 21 MPa and a temperature of 70℃ for 20 minutes to make the multilayered tapes sufficiently bonded. Finally, the laminated substrate was sintered in a furnace at temperature of 1500℃ for 2 hours. It should be noticed that in the high temperature sintering process, with an adequate temperature-time curve, the carbon membrane would volatilize through the apertures of the ceramic tapes and the cavity would remain intact. After sintering the alumina substrate embedded with air cavity, the inductor and capacitor patterns were printed on the surface of substrate by the screen-printing technique. Afterwards, the semi-finished sensor was placed into a furnace at 125℃ for 15 minutes to dry the metal paste and sintered at peak temperature of 850 ℃ for 10 minutes. The profile of sensor structure and fabricated pressure sensor with thickness of 0.38 mm is shown in Fig. 3.

Table 1 Characteristics of the ESL 44007 green tape.

Quantity	Value
Unfired thickness	~200 μm
Tape uniformity	150±10%
X, Y shrinkage	15.8±1%
Relative permittivity (25℃, 1 MHz)	9
Dielectric loss (25℃, 1 MHz)	0.0001

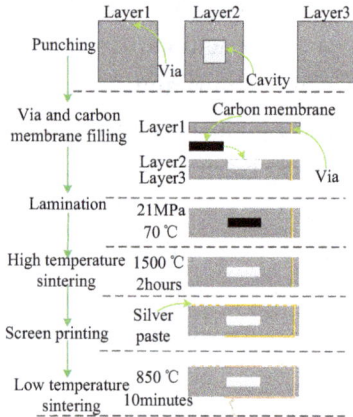

Fig. 2 Fabrication process of pressure sensor.

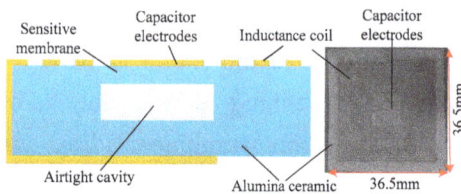

Fig. 3 Profile of pressure sensor structure and fabricated pressure sensor.

To wireless coupling with the sensor, a proper reader antenna was designed to enhance the sensing distance and signal strength of sensor. Copper was used to enwind as a cylindrical spiral antenna owing to the characteristics of low cost and excellent mechanical and electrical performance.

High temperature would change some characteristic parameters of reader antenna (inductance, resistance, etc.), thus affecting the signal readout of the sensor (signal strength and frequency deviation, etc.). To improve the accuracy of sensor signal measurement, an insulation layer was designed to insure that the reader antenna was working in a normal temperature range, and therefore, minimized the influence of reader antenna. The sensor was put in the high temperature zone while spiral antenna was placed in the room temperature zone, and two of them were separated by the insulation layer. The schematic diagram is shown in Fig. 4.

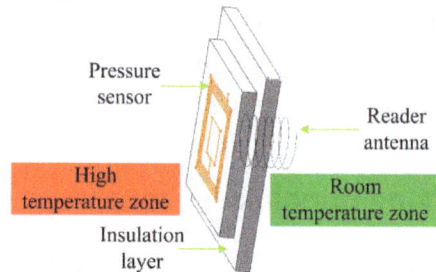

Fig. 4 Schematic of wireless coupling measurement.

4. Measurement and discussion

The high temperature performance of sensor measurement system is shown in Fig. 5. Muffle furnace worked as a heat source in atmospheric environment. Impedance analyzer was used to detect the phase-frequency and amplitude-frequency characteristics of reader antenna.

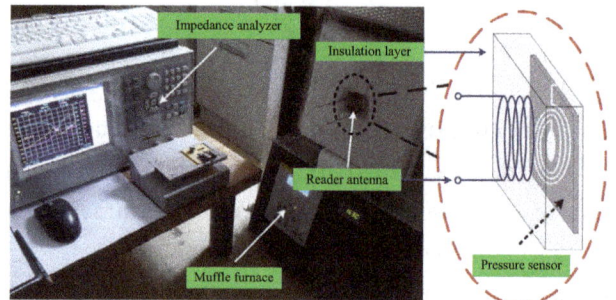

Fig. 5 High temperature measurement system of pressure sensor.

The sensor was measured at a distance of 10 mm (maximum measurement distance is 50 mm) away from reader antenna to get a clear frequency peak signal. Figure 6 presents the temperature of the insulation layer at different sides. The temperature rises uniformly inside the muffle furnace, while the outside of furnace goes up slowly, which ensures the accuracy readout of sensor signal.

The amplitude and phase of reader antenna corresponding to frequency are presented in Figs. 7 and 8 respectively when temperature rose from 27 ℃ to 700 ℃. The resonant frequency and coupling strength of pressure sensor decreased when the temperature increased. The average change in the resonance frequency versus temperature was 0.68 kHz/℃, namely zero drift 0.68 kHz/℃.

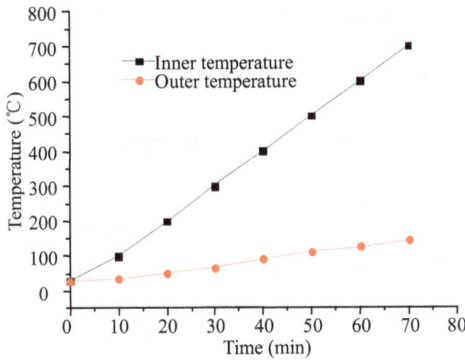

Fig. 6 Inner and outer temperature of insulation layer.

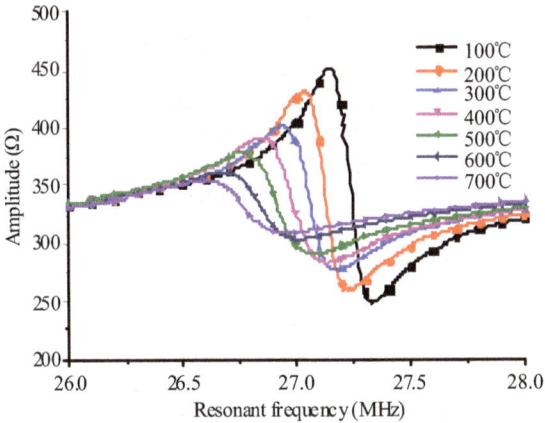

Fig. 7. Amplitude-frequency characteristic of reader antenna at different temperatures.

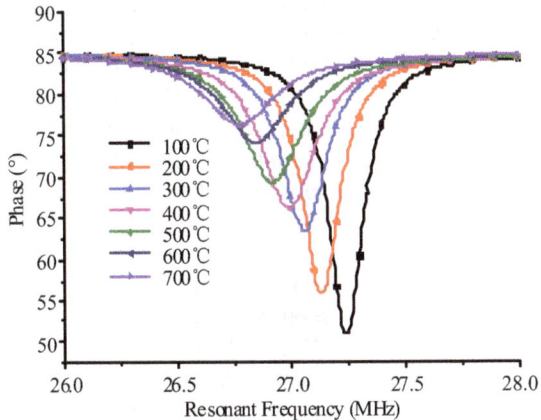

Fig. 8 Phase-frequency characteristic of reader antenna at different temperatures.

With the evaluated temperature, the alumina ceramic relative permittivity increased, thus increasing the C_s, while inductance L_s merely changed due to the low coefficient of thermal expansion of alumina ceramic. Therefore, the resonant frequency of the sensor changed accordingly with the temperature according to (1). The inductive-capacitive resonance circuit was

formed by the silver conductor. The resistivity of silver changes with temperature and can be expressed as

$$R_1 = R_2 * [a * (T_1 - T_2)] \qquad (3)$$

where R_1 is the resistivity value adjusted to T_1, R_2 is the resistivity value known or measured at temperature T_2, a is the temperature coefficient, for Ag being 0.0038, T_1 is the temperature at which resistivity value needs to be known, and T_2 is the temperature at which known or measured value was obtained. The resistance of the sensor at 27 ℃ measured by a multimeter is 5.7 Ω; by using (3), the resistance of the sensor at 700℃ is 14.58 Ω. The increased resistivity causes a decrease in Q value, which will make the coupling strength gradually weakened and the frequency curve unrecognizable.

When the muffle furnace cools down while other conditions remain unchanged, rise temperature to 700 ℃ again to proceed repetitive experiment. Figure 9 represents the extracted value of resonant frequency of sensor versus temperature. From the second measurement result, zero drifts was 0.7 kHz/℃, with a good repetition compared with the first time.

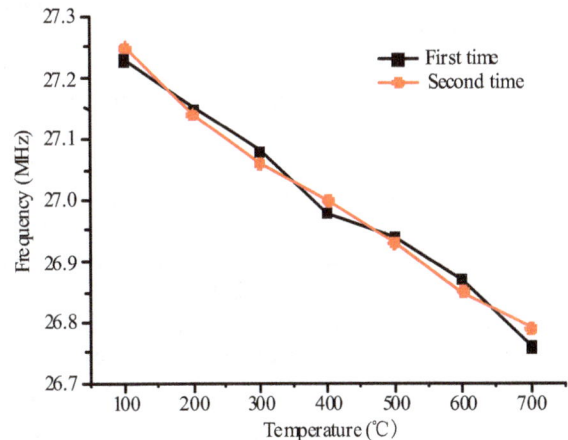

Fig. 9 extracted resonant frequency of senor versus temperature.

5. Conclusions

The wireless passive pressure sensor based on alumina ceramic is presented here to analyze its high temperature performance. The insulation layer was designed to ensure the stability of reader antenna in high temperature and improve the reliability of the performance of sensor in high temperature. When the temperature rose, the resonant frequency of sensor decreased with a trend of quadratic curve, and the coupling strength decreased gradually. Through experimental analysis, zero drift has great impact on the accuracy of sensor signal readout, and it is mainly affected by the relative permittivity of alumina ceramic and the resistivity of silver conductor. Zero-drift could be eliminated by improving the sensitivity of pressure sensor or putting forward a kind of temperature compensation algorithm to reduce the impact on the pressure signal readout.

Acknowledgment

This work was supported by the National Natural Science Foundation of China (Grant No. 61471324) and the Outstanding Young Talents Support Plan of Shanxi province.

References

[1] W. J. Pulliam, P. M. Russler, and R. S. Fielder, "High-temperature high-bandwidth fiber optic MEMS pressure-sensor technology for turbine engine component testing," *SPIE*, 2002, 4578: 229–238.

[2] R. W. Johnson, J. L. Evans, P. Jacobsen, J. R. Thompson, and M. Christopher, "The changing automotive environment: High-temperature electronics," *IEEE Transactions on Electronics Packaging Manufacturing*, 2004, 27(3): 164–176.

[3] C. V. Netten, "Design of a small personal air monitor and its application in aircraft," *Science of the Total Environment*, 2009, 407(3): 1206–1210.

[4] S. E. Zhu, M. K. Ghatkesar, C. Zhang, and G. C. A. M. Janssen, "Graphene based piezoresistive pressure sensor," *Applied Physics Letters*, 2013, 102(16): 161904-1–161904-3.

[5] G. Chitnis, T. Maleki, B. Samuels, L. B. Cantor, and B. Ziaie, "A minimally invasive implantable wireless pressure sensor for continuous IOP monitoring," *IEEE Transactions on Biomedical Engineering*, 2013, 60(1): 250–256.

[6] M. Narayanaswamy, R. J. Daniel, K. Sumangala, and C. A. Jeyasehar, "Computer aided modeling and diaphragm design approach for high sensitivity silicon-on-insulator pressure sensors," *Measurement*, 2011, 44(10): 1924–1936.

[7] S. Li, T. Liang, W. Wang, Y. Hong, T. Zheng, and J. Xiong, "A novel SOI pressure sensor for high temperature application," *Journal of Semiconductors*, 2015, 36(1): 014014-1–14014-5.

[8] T. H. Lee, S. Bhunia, and M. Mehregany, "Electromechanical computing at 500℃ with silicon carbide," *Science*, 2010, 329(5997): 1316–1318.

[9] D. A. Mills, D. Alexander, G. Subhash, and M. Sheplak, "Development of a sapphire optical pressure sensor for high-temperature applications," *SPIE*, 2014, 9113: 91130H-1–91130H-15.

[10] J. Xu, G. Pickrell, X. Wang, W. Peng, K. Cooper, and A. Wang, "A novel temperature-insensitive optical fiber pressure sensor for harsh environments," *IEEE Photonics Technology Letters*, 2005, 17(4): 870–872.

[11] C. M. Jewart, Q. Wang, J. Canning, D. Grobnic, S. J. Mihailov, and K. P. Chen, "Ultrafast femtosecond-laser-induced fiber Bragg gratings in air-hole microstructured fibers for high-temperature pressure sensing," *Optics Letters*, 2010, 35(9): 1443–1445.

[12] A. Binder, G. Bruckner, N. Schobernig, and D. Schmitt, "Wireless surface acoustic wave pressure and temperature sensor with unique identification based on LiNbO3," *IEEE Sensors Journal*, 2013, 13(5): 1801–1805.

[13] J. G. Rodriguez-Madrid, G. F. Iriarte, O. A. Williams, and F. Calle, "High precision pressure sensors based on SAW devices in the GHz range," *Sensors and Actuators A: Physical*, 2013, 189(2): 364–369.

[14] K. G. Ong, C. A. Grimes, C. L. Robbins, and R. S. Singh, "Design and application of a wireless, passive, resonant-circuit environmental monitoring sensor," *Sensors and Actuators A: Physical*, 2001, 93(1): 33–43.

[15] Q. Tan, C. Li, J. Xiong, P. Jia, W. Zhang, J. Liu, *et al.*, "A high temperature capacitive pressure sensor based on alumina ceramic for in situ measurement at 600℃," *Sensors*, 2014, 14(2): 2417–2430.

[16] J. Xiong, C. Li, P. Jia, X. Chen, W. Zhang, J. Liu, *et al.*, "An insertable passive LC pressure sensor based on an alumina ceramic for in situ pressure sensing in high-temperature environments," *Sensors*, 2015, 15(9): 21844–21856.

5

Spectra Power and Bandwidth of Fiber Bragg Grating Under Influence of Gradient Strain

Qinpeng LIU[*], Xueguang QIAO, Zhen'an JIA, and Haiwei FU

Key Laboratory on Photoelectric Oil-Gas Logging and Detecting (Ministry of Education), Xi'an Shiyou University, Xi'an, 710065, China

[*]Corresponding author: Qinpeng LIU   E-mail: lqp1977@163.com

Abstract: The reflective spectrum power and the bandwidth of the fiber Bragg grating (FBG) under gradient strain are researched and experimentally demonstrated. The gradient strain is applied on the FBG, which can induce FBG bandwidth broadening, resulting in the variation of reflective power. Based on the coupled-mode theory and transfer matrix method, the segmental linear relationship between the gradient strain, the reflective power, and the bandwidth is simulated and analyzed, and the influence of the FBG length on the reflective spectrum is analyzed. In the experiment, the strict gradient stain device is designed; the experimental results indicate that the reflective optic power and the bandwidth of the FBG under gradient stain are concerned with the length of the FBG. Experimental results are well consistent with the theoretical analysis, which have important guiding significance in the FBG dynamic sensing.

Keywords: Fiber Bragg grating; gradient strain; reflective spectrum power; reflective spectrum bandwidth

1. Introduction

As an important wavelength sensitive device, a fiber Bragg grating (FBG) has been widely studied and applied [1–5]. Most of the researches focus on the wavelength sensing technology and theory, but when the inhomogeneous field applying on the FBG, such as the gradient stain and gradient temperature, acts on the grating range of the FBG, which can lead to the FBG chirping, and the reflective spectrum shows new features, such as reflective power and reflective bandwidth [6–10]. It has many methods to form inhomogeneous refractive-index modulation, such as designing the grating structure [11, 12], introducing a strain gradient by designing an

especial fabrication structure [13, 14], and etching the FBG [15]. The essence of the inhomogeneous field modulation induces a refractive-index gradient. The spatially varying structure of the fiber grating can be divided into several sections called the effective-interaction sub-regions, each effective-interaction sub-region possesses an independent sensing ability, whose relationship to the gradient distribution of environmental parameters has been researched [16], and the reflective spectrum of the linear chirp FBG is analyzed by using a four-order Runge-Kutta method and transfer matrix method [17]. The results show that the linearly chirped fiber Bragg gratings can broaden the bandwidth. But all of the researches at present do not provide more complete discussion of reflective spectrum power and the bandwidth for the whole range, which is not the accurate response relationship between the bandwidth broadening and the reflective optic spectra power (ROSP), and the relationship between them is necessary to further research, which is a precondition to the FBG wavelength and intensity sensing.

In this paper, the reflective optic spectra power (ROSP) and the bandwidth of the FBG are elaborately researched based on the coupled-mode theory and transfer matrix method (TMM). The bandwidth and the spectra power of the FBG under the gradient strain are theoretically analyzed, and the relationship between the bandwidth and coefficient of chirping is demonstrated. The significant relationship between the ROSP and chirping coefficient is also theoretically simulated, and the relationship between them is fully proposed. In numerous methods, the external modulating method by designing an especial transducer is simple and easy to realize. At the most basic physical quantities, the gradient stain has typically representative meaning. So the gradient-strain device is accordingly designed; the experimental results show a good agreement with the theoretical analysis, which is an important basis for designing an intensity modulation sensor for the dynamic and static measurement.

2. Theory and analysis

According to the coupled-mode theory, the forward (reference) modes and backward propagation modes (signals) satisfy the following equations:

$$\begin{cases} \dfrac{dR}{dz} = i\left(\delta + \sigma - \dfrac{1}{2}\dfrac{d\phi(z)}{dz}\right)R(z) + ikS(z) \\ \dfrac{dS}{dz} = -i\left(\delta + \sigma - \dfrac{1}{2}\dfrac{d\phi(z)}{dz}\right)S(z) - ik^{*}R(z) \end{cases} \quad (1)$$

where $\delta = 2\pi n_{\text{eff}}(1/\lambda - 1/\lambda_{B})$, $\sigma = 2\pi\Delta n/\lambda$, and $d\varphi(z)/dz = 2\pi n_{\text{eff}}zC/\lambda_{B}^{2}$ are the dc-components of

self-coupling coefficient, respectively. k is an alternating current coupling coefficient. λ_{B} is the Bragg wavelength, and C is the chirping coefficient of the fiber grating, which is defined as $d\lambda_{B}/dz$. According to (1), the reflective optic spectra of the fiber grating is given as

$$R(\lambda) = \frac{\sinh^{2}(\sqrt{k^{2}-\sigma^{2}}L)}{\cosh^{2}(\sqrt{k^{2}-\sigma^{2}}L) - \sigma^{2}/k^{2}}. \quad (2)$$

Assuming the gradient strain is applied to the FBG, the period $\Lambda(z)$ of the fiber Grating is defined as

$$\Lambda(z) = \Lambda_{0}[1 - gz] \quad -\frac{L}{2} \leq z \leq \frac{L}{2} \quad (3)$$

where Λ_{0} is the initial period, L is the length of the fiber grating, and g is the strain gradient, which is defined as $g = d\varepsilon(z)/dz$. Based on the above theoretical model, the FBG can be divided into several sections called the effective-interaction sub-regions, and each sub-region possesses an independent sensing ability, so TMM is the best choose to analyze the reflective optic spectra. Combined with the boundary conditions and transfer matrix, we can further research the relationship between the reflective optic spectra power and chirping coefficient. Because the reflective optic spectrum of the FBG is narrow, the spectrum power density of the sensing source should be used as constant, and the ROSP change ΔP of the chirping FBG can be expressed as

$$\Delta P = \int_{\lambda_{1}}^{\lambda_{2}} \eta P_{\text{BBs}} R(\lambda) d\lambda \quad (4)$$

where λ_{1} and λ_{2} is the minimum resonance wavelength and maximum resonance wavelength, respectively. η is the insertion of loss, and P_{BBS} is the spectrum power density of the sensing source. Then the bandwidth and ROSP of the linear chirping FBG can be simulated by using the TMM. Base on the previous analysis, assume the strain gradient is same, that is to say, the coefficient of chirping is same, and the reflective power variation of the FBG should depend on the length of the FBG, because the ROSP is the result within the scope of the integral.

Therefore, the influence of the length should be considered. For comparison, the ROSP is given in the normalized power. Figure 1 presents the changing rule of the 10-dB bandwidth with 10 mm, 20 mm, and 40 mm long FBGs. The simulated results demonstrate that the bandwidth increases with the chirping coefficient for different length FBGs. The longer FBG means the higher sensitivity and the better linearity. Meanwhile, the relationship between the bandwidth and the chirping coefficient is not strictly linear in the whole range, while it is piecewise linear function, and the shorter FBG means more seriously nonlinear. In addition, the focus of this paper is the relationship between the ROSP and the chirped coefficient, because compared with bandwidth sensing, it is more suitable for the dynamic test. To facilitate the analysis, the reflective optic spectra normalized power (ROSPN) is utilized, and based on the previous theory, the ROSPN can be obtained for different FBGs, as shown in Fig. 2, in which the insert shows the reflective optic spectra of 10 mm, 20 mm, and 40 mm long FBGs, respectively. It can be seen that the ROSPN increases as the chirping coefficient increases. The longer the FBG is, the

more sensitivity to the ROSPN it is. The ROSPN is inclined to be saturated with an increase in the chirping coefficient. This phenomenon is due to that the partial sub-region loses the independent sensing ability when the large strain gradient is applied on the FBG. When the chirping coefficient is small, the FBG has been divided into several effective-interaction sub-regions; it can be generated with the linear bandwidth and power in the chirping coefficient. The relationship is extremely important for intensity sensing, which also determines the linear range of the sensor.

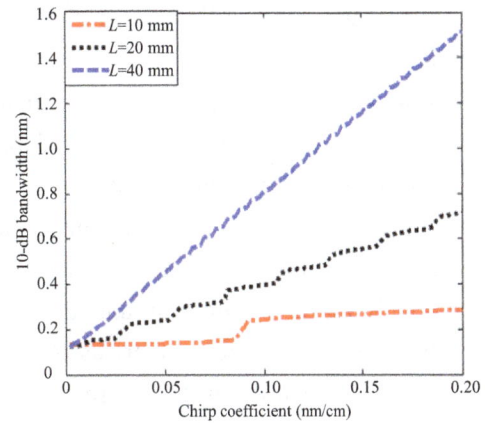

Fig. 1 Relationship between the 10-dB bandwidth and chirping coefficient.

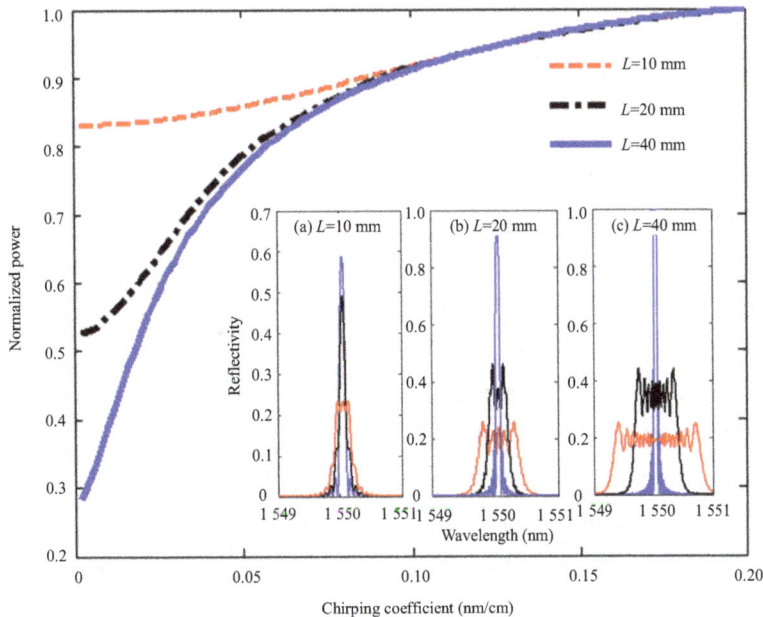

Fig. 2 Relationship between the normalized power and chirping coefficient.

3. Experiments

To demonstrate the relationship between the ROSPN and chirping coefficient, the FBGs used in the experiment are 10 mm, 20 mm, and 45 mm long with the 3-dB bandwidth of 0.28 nm, 0.214 nm, and 0.195 nm. Light from the broad-band source (BBS) is coupled into the FBG via a 3-dB coupler, and the power and the bandwidth of the reflective spectrum are monitored by the optical spectrum analyzer (MS9740A) with the resolution of 0.02 nm. The linear strain is elaborately provided by an especial cantilever. The setup is shown in Fig. 3. The structure of the especial cantilever is designed, and their strain distribution under the influence of pressure is shown in Fig. 4. It can be seen that the strain distributions are gradient in the FBG range, the strain gradient is determined, and the chirping coefficient of the fiber grating is also is accordingly determined. A linear strain distribution is a precondition for the broadening around the central wavelength, FBGs have been fabricated on the upper surface of the cantilever, the geometric center of FBGs is located in the middle position of the cantilever, and the chirping coefficient is modulated by applying side pressure.

Fig. 3 Schematic diagram of the experimental setup.

The response of the ROSP is measured under the side-applied pressure, as shown in Fig. 5. It can be seen that the ROSP increases with an increase in the stain gradient. As the ROSP increases, the ROSP becomes saturated, that is to say, with an increase in the chirping coefficient (strain gradient), the rate of change reduces, and in the meantime, if the grating is longer, its power change become more sensitive to the strain gradient. The change in the ROSP is not strictly linearly varied with the side-applied pressure in the whole range, but it is piecewise linear

functions. The 10-dB bandwidth of the grating is measured under the side-applied pressure, as depicted in Fig. 6. The broadening of the spectra bandwidth varies with applying side pressure. The change in the bandwidth is also not strictly linearly varied with pressure, and FBGs with different lengths have different linear responses. It is obvious that the FBG with the length of 45 mm has the best linear response in the experiment, which has higher bandwidth sensitivity. The longer FBG is, the more sensitive and linear its bandwidth broadening becomes. On the contrary, the shorter the length is, the worse the linearity is, and the lower the sensitivity is. The relationship is extremely important for the intensity sensing, and the linear range dominates the sensing range. Meanwhile, based on the above analysis, the shorter the grating is, the less chirping it will be. It also has great significances to FBG wavelength sensing.

Fig. 4 Surface-strain distribution of the especial cantilever.

Fig. 5 Relationship between the reflective power and pressure.

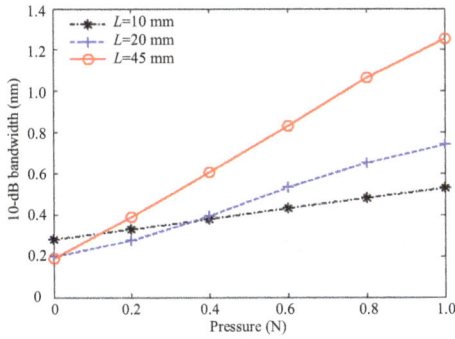

Fig. 6 Relationship between the 10-dB bandwidth and pressure.

4. Conclusions

In this paper, the ROSPN and bandwidth of the FBG under the influence of the linear strain have been theoretically simulated and experimentally demonstrated. The theoretical analysis shows the relationship among the ROSPN, the chirping coefficient, and bandwidth. The main conclusions are as follows:

(1) The relationship among the ROSPN, the chirping coefficient, and bandwidth are not strictly linear in the whole range, and it is piecewise linear functions.

(2) The shorter the grating is, the more insensitivity to the linear strain its bandwidth broadening is. The longer the grating is, the more sensitive its bandwidth broadening becomes under the same condition, and the better the linearity is.

(3) The shorter the grating is, the more insensitivity to linear strain its ROSPN is. The longer the grating is, the more sensitive it becomes under the same condition, and the worse the linearity is. The collusions are very important for FBG intensity sensing, which have guiding significance for the structure design and dynamic range of the FBG intensity sensing. In addition, the analysis indicates that it will have the shorter grating and less chirping. It has also great significances to FBG wavelength sensing in the miniaturization of device.

Acknowledgment

This work was supported in part by the National Natural Science Foundation of China (No. 61077006, 60727004, and 61077060), in part by the Ministry of Education Project of Science and Technology Innovation (No. Z08119), and in part by the Shannxi Province National Science Foundation under Grant 2016JM6055.

References

[1] T. Li, Y. Tan, Z. Zhou, L. Cai, S. Liu, Z. He, *et al.*, "Study on the non-contact FBG vibration sensor and its application," *Photonic Sensors*, 2015, 5(2): 128–136.

[2] L. Li, D. Zhang, H. Liu, Y. Guo, and F. Zhu, "Design of an enhanced sensitivity FBG strain sensor and application in highway bridge engineering," *Photonic Sensors*, 2014, 4(2): 162–167.

[3] J. Wang, T. Liu, G. Song, H. Xie, L. Li, X. Deng, *et al.*, "Fiber Bragg grating (FBG) sensors used in coal mines," *Photonic Sensors*, 2014, 4(2): 120–124.

[4] S. Li and M. Zhou, "Long-term mechanical properties of smart cable based on FBG desensitized encapsulation sensors," *Photonic Sensors*, 2014, 4(3): 236–241.

[5] X. Zhu, "Aluminum alloy material structure impact localization by using FBG sensors," *Photonic Sensors*, 2014, 4(4): 344–348.

[6] Y. G. Han, X. Y. Dong, J. H. Lee, and S. B. Lee, "Simultaneous measurement of bending and temperature based on a single sampled chirped fiber Bragg grating embedded on a flexible cantilever beam," *Optics Letter*, 2006, 31(19): 2839–2841.

[7] T. Guo, B. Liu, and X. Y. Dong, "Linear and Gaussian chirped fiber Bragg grating and its applications in fiber-optic filtering and sensing systems," *IEEE Photonics Technology Letters*, 2007, 19(14): 663–665.

[8] B. Q. Jiang, J. L. Zhao, C. Qin, and F. Fan, "A bandwidth-tuning device based on polymer-packaged fiber Bragg grating," *IEEE Photonics Technology Letters*, 2011, 23(17): 1225–1227.

[9] X. G. Qiao, Y. P. Wang, H. Z. Yang, G. Tuan, Q. Z. Rong, L. Ling, *et al.*, "Ultrahigh-temperature chirped fiber Bragg grating through thermal activation," *IEEE Photonics Technology Letters*, 2015, 27(12): 1305–1308.

[10] M. Abtahi, A. D. Simard, S. Doucet, L. Sophie, and L. A. Rusch, "Characterization of a linearly chirped FBG under local temperature variations for spectral shaping applications," *Journal of Lightwave Technology*, 2011, 23(5): 750–755.

[11] J. L. Cruz, L. Dong, S. Barcelos, and L. Reekie, "Fiber Bragg gratings with various chirp profiles made in etched tapers," *Applied Optics*, 1996, 35(34): 6781–6787.

[12] L. Dong, J. L. Cruz, L. Reekie, and J. A. Tocknott, "Chirped fiber Bragg gratings fabricated using etched tapers," *Optical Fiber Technology*, 1995, 1(4): 363–368.

[13] Y. N. Zhu, P. Shu. C. Lu, M. B. Lacquet, P. L. Swart, A. A. Chtcherbakov, *et al.*, "Temperature insensitive measurements of static displacements using a fiber Bragg grating," *Optics Express*, 2003, 11(16): 1918–1924.

[14] B. Yin, Y. L. Bai, and Y. H. Qi, "Study on tapered chirped fiber grating filter," *Acta Physica Sinica*, 2013, 62(21): 214213–214213.

[15] H. Z. Yang, K. S. Lim, X. G. Qiao, W. Y. Chong, Y. K. Cheong, W. H. Lim, *et al.*, "Reflection spectra of etched FBGs under the influence of axial contraction and stress-induced index change," *Optics Express*, 2013, 21(12): 14808–14815.

[16] G. Tuan , B, Liu, W. G. Zhang, G. Y. Kai, Q. D. Zhao, and X. Y. Dong, "Research on optical fiber grating chirp-sensing technology," *Acta Physica Sinica*, 2008, 28(5): 828–834.

[17] J. J. Wei, Y. P. Liang, and T. L. Dai. "Numerical analysis of reflection spectrum of linearly chirped fiber Bragg gratings," *Laser Technology*, 2012, 36(5): 607–611.

6

Characterization of Miniature Fiber-Optic Fabry-Perot Interferometric Sensors Based on Hollow Silica Tube

Pinggang JIA[1, 2], Guocheng FANG[2], and Daihua WANG[3*]

[1]*Postdoctoral Research Station of Optical Engineering, Chongqing University, Chongqing, 400030, China*

[2]*Key Laboratory of Instrumentation Science & Dynamic Measurement of the Ministry of Education of China, North University of China, Taiyuan, 030051, China*

[3]*Key Laboratory of Optoelectronic Technology and Systems of the Ministry of Education of China, Chongqing University, Chongqing, 400030, China*

*Corresponding author: Daihua WANG   E-mail: dhwang@cqu.edu.cn

Abstract: A miniature fiber-optic Fabry-Perot interferometer (MOFPI) fabricated by splicing a hollow silica tube (HST) with inner diameter of $4\,\mu m$ to the end of a single-mode fiber is investigated and experimentally demonstrated. The theoretical relationship between the free spectrum range and the length of HST is verified by fabricating several MOFPIs with different lengths. We characterize the MOFPIs for temperature, liquid refractive index, and strain. Experimental results show that the sensitivities of the temperature, liquid refractive index, and strain are $16.42\,pm/°C$, $-118.56\,dB/RIU$, and $1.21\,pm/\mu\varepsilon$, respectively.

Keywords: Fiber-optic; Fabry-Perot interferometer; hollow silica tube; characterization

1. Introduction

Miniature fiber-optic interferometers have attracted broad attention for their advantages of simple structure, compact size, immunity to electromagnetic interference, high sensitivity and accuracy, and so on. The commonly used configurations for fiber-optic interferometers include Mach-Zehnder interferometers [1–2], Michelson interferometers [3–4], and Fabry-Perot interferometers [5–7]. Compared with the fiber-optic Fabry-Perot interferometers, miniature fiber-optic Fabry-Perot interferometers (MOFPIs) have also been widely applied in the measurement of temperature, strain, pressure, acceleration, displacement, and ultrasound.

The processing method of MOFPIs includes chemical etching [8, 9], electrical arc discharge [10, 11], laser technology [12], and application of special optical fibers, such as the photonic crystal fiber [13], the tapered optical fiber [14], and the hollow core fiber or hollow silica tube (HST) [15]. Each kind of methods has its special features. For example, chemical etching has low cost but complicated processing. Electric arc discharge is difficult to standardize and repeat. The photonic crystal fiber is high-cost. Compared with these methods, the method using HST has low cost, simple structure, and process without large instruments. Traditional MOFPIs based on the multi-beam inference are fabricated by inserting two well-cut fibers oppositely into an HST and fusing those together [16], or fusing a section of HST between two single mode fibers (SMF) [17].

In recent years, several miniature fiber-optic interferometers based on wavefront splitting are developed. Sun *et al.* [18] presented an optical fiber strain and temperature sensor based on an in-line Mach-Zehnder interferometer using thin-core fiber. Huang *et al.* [19] firstly presented an MOFPI based on HST with small inner diameter. They spliced an HST with 5-μm inner diameter to the tip of an SMF to fabricate a novel MOFPI. Lin *et al.* [20] demonstrated a broadband MOFPI using an SMF end-spliced with a sphered-end hollow core fiber. Frazao *et al.* [21] presented an MOFPI by splicing an HST with a 20-μm inner diameter to the tip of an SMF. Wang *et al.* [22] also fabricated a similar structure by applying the chemical etching.

In this paper, an MOFPI fabricated by splicing a hollow silica tube with inner diameter of 4 μm to the end of an SMF is investigated and experimentally demonstrated. The theoretical relationship between the free spectrum range and the length of HST is verified by fabricating several MOFPIs with different lengths. The MOFPIs for temperature, liquid refractive index, and strain are characterized.

2. Configuration and operating principle

The configuration and operating principle of the proposed MOFPI are illustrated in Fig. 1, and the longitudinal and cross-sectional microscopy images of fabricated MOFPI are shown in Fig. 2. The fabrication process of the MOFPIs is as follows. First, a well-cut SMF (Corning SMF-28, 9/125) and an HST are placed in the left and right holder of fusion splicer (FITE, S183 Version2, Japan). Second, set the intensity and time of the discharge of the fusion splicer as 100 unit and 285 ms, and operate arc discharge once. Finally, cut off the HST at an appropriate distance from the fusion joint under the microscope. According to Fig.2, the inner and outer diameters of HST are approximately 4 μm and

125 μm. When lights propagate along SMF and get to its end, a part of lights in optical fiber core are reflected. Because the inner diameter of HST is smaller than the core diameter of SMF, the remaining lights propagate into HST and are reflected at the end of HST. Two parts of the lights interfere and form Fabry-Perot interference.

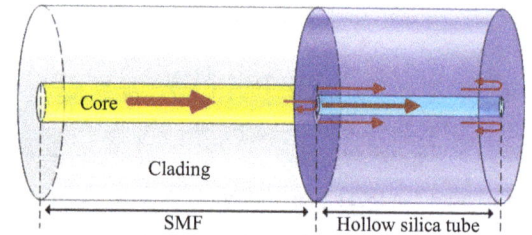

Fig. 1 Configuration of MOFPI.

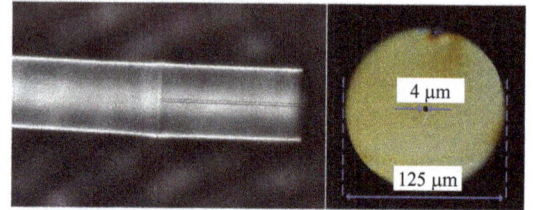

Fig. 2 Longitudinal and cross-sectional microscopy images.

The light can be treated as an electromagnetic wave with time-varying electric fields. The electric field of the reflective lights from the ends of optical fiber and HST can be given by

$$E_1 = r\eta E_0 \cos\left(\omega t - kx_1 + \phi_0\right) \quad (1)$$
$$E_2 = r(1-\eta) E_0 \cos\left(\omega t - kx_2 + \phi_0\right) \quad (2)$$

where E_0 is the electric field of the incident lights, r is the reflective coefficient of the interface of air-silica, η is the energy ratio of lights at the facets of fiber-air and fiber-HST, ω is the angular frequency, t is the time, x_1 and x_2 are the positions of the propagation, k is the propagation constant, and ϕ_0 is the phase constant.

The intensity of reflective lights can be given by

$$I = I_1 + I_2 + 2\sqrt{I_1 I_2}\cos\phi$$
$$= r^2 E_0^2 \left[\eta^2 + (1-\eta)^2 + 2\eta(1-\eta)\cdot\cos\phi\right] \quad (3)$$

where I_1 and I_2 are intensities of reflective lights from the ends of optical fiber and HST, respectively. $\phi = 4\pi nL/\lambda$, where n is the refractivity of silica, L is the length of HST, and λ is the wavelength of the light.

The simulation result is shown in Figs. 3(a) and (b) according to (3) using the software of MATLAB, when $\eta = 0.45$, $r^2 = 0.037$ and $L = 168$ μm and 527 μm.

The interference spectrums of the MOFPIs are obtained by the optical spectrum analyzer (Micron Optics Inc, SM125, USA). The interference spectra are illustrated in Fig. 3 when the lengths of the HSTs are 168 μm, 527 μm, and 2669 μm. According to Fig. 3, the interference spectra agree well with the simulation result. The insert loss is about 40 dB, which may be caused by the loss of light at the connections.

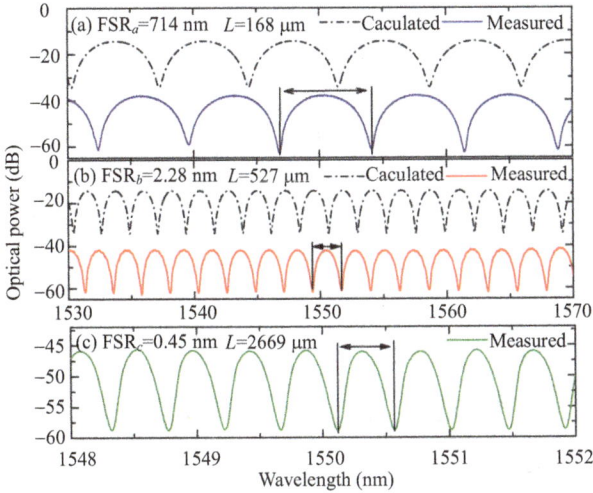

Fig. 3 Interference spectrums of the MOFPIs with HSTs' lengths of (a) 168 μm, (b) 527 μm, and (c) 2669 μm.

The free spectrum range (FSR) can be given by

$$FSR = \lambda_m^2 \big/ 2nL \qquad (4)$$

where λ_m is the wavelength of light.

The FSRs of the MOFPIs with different lengths of HSTs are fabricated and measured around the wavelength of 1550 nm. The theoretical and measured relationship between the length of HST and FSR are shown in Fig. 4. It can be seen that the measured results agree well with the theoretical results based on (4). The FSRs around the wavelength of 1550 nm are 7.14 μm and 2.28 μm when the lengths of the HSTs are 168 μm and 527 μm.

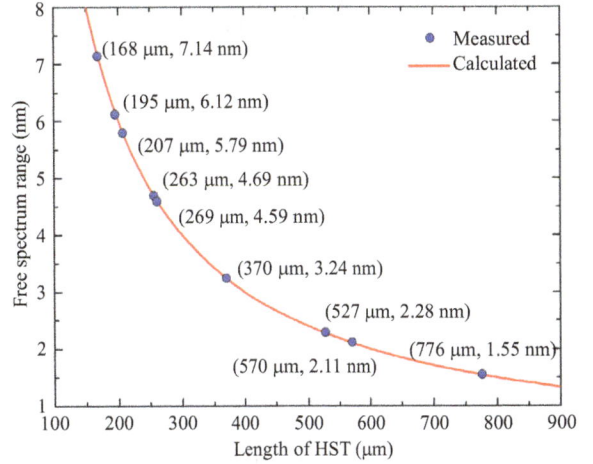

Fig. 4 Relationship between the FSR and the length of HST.

3. Experiment

We characterized the MOFPIs for temperature, liquid refractive index (RI), and strain. Regarding the temperature measurements, a sample MOFPI with the length of 263 μm was placed inside a controllable muffle furnace (Nabertherm, sn209012, Germany). The sample was hold and protected by stainless steel fixture. We increased the temperature from 25 ℃ (room temperature) to 1000 ℃ with 50 ℃ step. The interference spectrum was recorded when temperature was stabilized of 10 minutes at each step. In the experiment, we used the peak tracing method to demodulate the signal. A wavelength shift of the interference spectrum could be observed with an increasing temperature. The wavelength shift is illustrated in Fig. 5 under the temperature from 25℃ to 400℃. The relationship between the wavelength shift of a selected peak and temperature is presented in Fig. 6. The experimental results are well adjusted by the second-order polynomial, given by

$$y = 1552.87 + 7.72 \times 10^{-3} T + 6.61 \times 10^{-6} T^2. \qquad (5)$$

The R^2 (fitting variance) of the fitting curve is 99.88%. However, it is reasonable to divide the temperature range into two different regions, for high (400 ℃ – 1000 ℃) and low temperatures (25 ℃ – 400 ℃) (insets in Fig. 6), where a much well linear approximation can be done in a high temperature. The sensitivities obtained are

16.42 pm/℃. The temperature response is caused by the thermal expansion of the material and thermo-optic effect, and the nonlinearity may be caused by the nonlinear change in the thermal expansion coefficient [23].

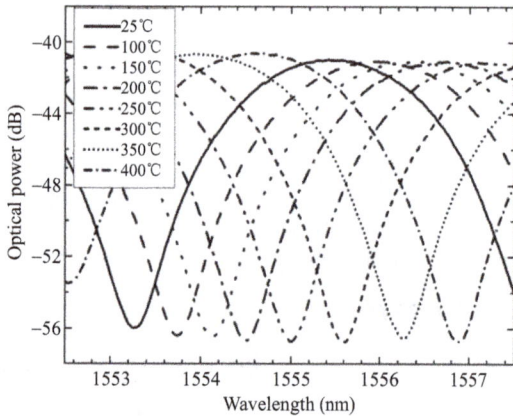

Fig. 5 Wavelength shift of the interference spectrum under the temperature from 25℃ to 400℃.

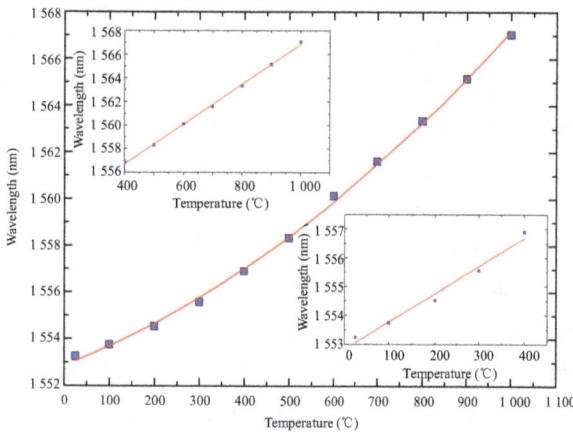

Fig. 6 Temperature response of the sample MOFPI [inset 1 (top left) is the high-temperature response, and inset 2 (bottom right) is the low-temperature response].

In order to test the liquid RI response, a sample MOFPI with the length L of 269 μm was vertically inserted in NaCl solution with different concentrations while the temperature was kept at the room temperature. The RI is related to the concentration of the NaCl solution and varies from 1.3333 to 1.4069. The experiment shows that the optical power gradually decreases with an increase in RI, as shown in Fig. 7. The relationship between the optical power of the sample MOFPI and RI is illustrated in Fig. 8. The obtained sensitivity of RI is −118.56 dB/RIU.

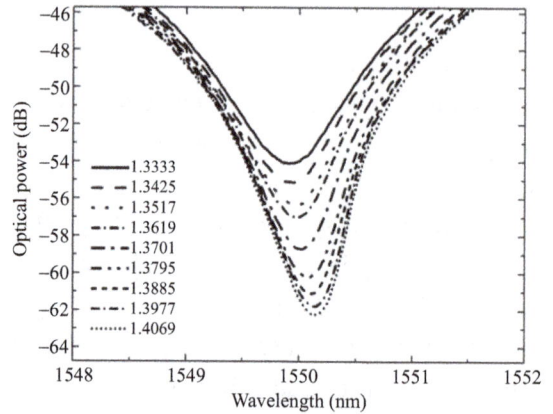

Fig. 7 Interference spectrum when the sample MOFPI was immersed in NaCl solution with different RIs.

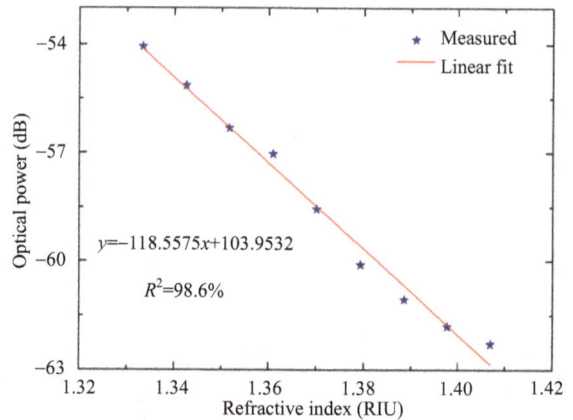

Fig. 8 RI response of the sample MOFPI.

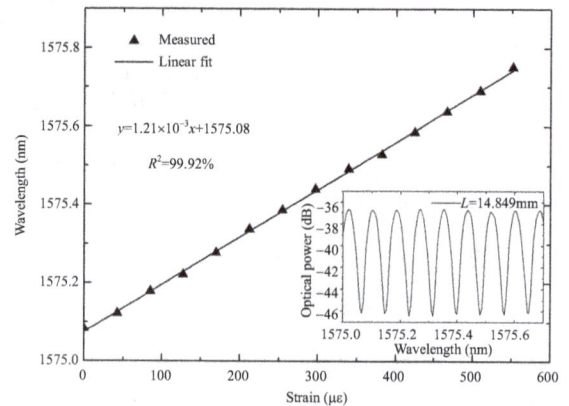

Fig. 9 Strain response of the sample MOFPI [inset (bottom right) is the interference spectrum at around the wavelength of 1575 nm].

In order to test its response to strain, we fabricated an MOFPI with the length of 14.849 mm, whose interference spectrum is shown in Fig. 9 inset. The FSR of the sample is about 83.5 pm at around the wavelength of 1575 nm. We fixed the sample with strain adhesive (KYOWA, #2129, Japan) on a constant-strength brass beam to test strain. A wavelength shift of the interference spectrum was observed when the strain increased, which was caused by the length change of the HST. The result is shown in Fig. 9 and R^2 of the fitting curve is 99.55%. The sample shows a sensitivity of 1.21 pm/$\mu\varepsilon$ under the range of 0 $\mu\varepsilon$ – 550 $\mu\varepsilon$. So the MOFPI can be applied in the measurement of strain.

4. Conclusions

In conclusion, an MOFPI fabricated by splicing a hollow silica tube (HST) with an inner diameter of 4 μm and outer diameter of 125 μm to the end of SMF was investigated and experimentally demonstrated. The theoretical relationship between the free spectrum range and the length of HST was verified by fabricating several MOFPIs with different lengths. We characterized the MOFPIs for temperature, liquid refractive index, and strain. Experimental results showed that the sensitivities of the temperature, liquid refractive index, and strain are 16.42 pm/℃, –118.56 dB/RIU, and 1.21 pm/$\mu\varepsilon$, respectively. Due to its simple fabrication process and low cost, the MOFPI was suitable for mass production. Experimental results showed that the device had a good response to temperature, liquid refractive index, and strain. The MOFPI had advantages of compact size, simple structure and process, and sensitivity to multi-parameters.

Acknowledgment

This work is supported by the National Natural Science Foundation of China under Grant 81127901 and 51405454, and Natural Science Foundation of Shanxi Province under Grant 2015021087.

References

[1] W. Talataisong, D. N. Wang, R. Chitaree, C. R. Liao, and C. Wan, "Fiber in-line Mach-Zehnder interferometer based on an inner air-cavity for high-pressure sensing," *Optics Letters*, 2015, 40(7): 1220–1222.

[2] B. Sun, Y. J. Huang, S. Liu, C. Wang, J. He, C. R. Liao, et al., "Asymmetrical in-fiber Mach-Zehnder interferometer for curvature measurement," *Optics Express*, 2015, 23(11): 14596–14602.

[3] L. B. Yuan, J. Yang, Z. H. Liu, and J. X. Sun, "In-fiber integrated Michelson interferometer," *Optics Letters*, 2016, 31(18): 2692–2694.

[4] A. Zhou, G. P. Li, Y. H. Zhang, Y. Z. Wang, C. Y. Guan, J. Yang, et al., "Asymmetrical Twin-core fiber based Michelson interferometer for refractive index sensing," *Journal of Lightwave Technology*, 2011, 29(19): 2985–2991.

[5] S. Pevec and D. Donlagic, "All-fiber, long-active-length Fabry-Perot strain sensor," *Optics Express*, 2011, 19(16): 15641–15651.

[6] F. C. Favero, G. Bouwmans, V. Finazzi, J. Villatoro, and V. Pruneri, "Fabry-Perot interferometers built by photonic crystal fiber pressurization during fusion splicing," *Optics Letters*, 2011, 36(21): 4191–4193.

[7] M. S. Ferreira, J. Bierlich, J. Kobelke, K. Schuster, J. L. Santos, and O. Frazão, "Towards the control of highly sensitive Fabry-Pérot strain sensor based on hollow-core ring photonic crystal fiber," *Optics Express*, 2012, 20(20): 21946–21952.

[8] X. T. Zou, A. Chao, Y. Tian, N. Wu, H. T. Zhang, T. Y. Yu, et al., "An experimental study on the concrete hydration process using Fabry-Perot fiber optic temperature sensors," *Measurement*, 2012, 45(5): 1077–1082.

[9] P. A. R. Tafulo, P. A. S. Jorge, J. L. Santos, O. Frazão, "Fabry-Pérot cavities based on chemical etching for high temperature and strain measurement," *Optics Communications*, 2012, 285(6): 1159–1162.

[10] S. Liu, Y. P. Wang, C. G. Liao, G. J. Wang, Z. Y. Li, Q. Wang, et al., "High-sensitivity strain sensor based on in-fiber improved Fabry-Perot interferometer," *Optics Letters*, 2014, 39(7): 2121–2124.

[11] D. W. Duan, Y. J. Rao, Y. S. Hou, and T. Zhu, "Microbubble based fiber-optic Fabry-Perot interferometer formed by fusion splicing single-mode fibers for strain measurement," *Applied Optics*, 2012, 51(8): 1033–1036.

[12] Y. Liu and S. L. Qu, "Optical fiber Fabry-Perot interferometer cavity fabricated by femtosecond laser-induced water breakdown for refractive index sensing," *Applied Optics*, 2014, 53(3): 469–474.

[13] T. T. Wang and M. Wang, "Fabry-Pérot fiber sensor for simultaneous measurement of refractive index and temperature based on an in-fiber ellipsoidal cavity," *IEEE Photonics Technology Letters*, 2012, 24(19): 1733–1736.

[14] S. C. Gao, W. G. Zhang, Z. Y. Bai, H. Zhang, W. Lin, L. Wang, et al., "Microfiber-enabled in-line

Fabry-Perot interferometer for high-sensitive force and refractive index sensing," *Journal of Lightwave Technology*, 2014, 32(9): 1682–1688.

[15] G. C. Fang, P. G. Jia, T. Liang, Q. L. Tan, Y. P. Hong, W. Y. Liu, *et al.*, "Diaphragm-free fiber-optic Fabry-Perot interferometer based on tapered hollow silica tube," *Optics Communications*, 2016, 371: 201–205.

[16] K. Mitchell, W. J. Ebel, and S. E. Watkins, "Low-power hardware implementation of artificial neural network strain detection for extrinsic Fabry-Pérot interferometric sensors under sinusoidal excitation," *Optical Engineering*, 2011, 6(4): 495–501.

[17] D. H. Wang, S. J. Wang, and P. G. Jia, "In-line silica capillary tube all-silica fiber-optic Fabry-Perot interferometric sensor for detecting high intensity focused ultrasound fields," *Optics Letters*, 2012, 37(11): 2046–2048.

[18] M. Sun, B. Xu, X. Y. Dong, and Y. Li, "Optical fiber strain and temperature sensor based on an in-line Mach–Zehnder interferometer using thin-core fiber," *Optics Communications*, 2012, 285(18): 3721–3725.

[19] Z. Huang, X. Chen, Y. Zhu, and A. Wang, "Wavefront splitting intrinsic Fabry-Perot fiber optic sensor," *Optical Engineering*, 2005, 44(7): 1–3,.

[20] N. Chen, K. Lu, J. Shy, and C. Lin, "Broadband Micro-Michelson interferometer with multi-optical-path beating using a sphered-end hollow fiber," *Optics Letters*, 2011, 36(11): 2074–2076.

[21] M. S. Ferreira, L. Coelho, K. Schuster, J. Kobelke, J. L. Santos, and O. Frazao, "Fabry-Perot cavity based on a diaphragm-free hollow-core silica tube," *Optics Letters*, 2011, 36(20): 4029–4031.

[22] R. H. Wang and X. G. Qiao, "Intrinsic Fabry-Pérot interferometer based on concave well on fiber end," *IEEE Photonics Technology Letters*, 2014, 26(14): 1430–1433.

[23] J. Mathew, O. Schneller, D. Polyzos, D. Havermann, R. Carter, W. MacPherson, *et al.*, "In-fiber Fabry–Perot cavity sensor for high-temperature applications," *Journal of Lightwave Technology*, 2015, 33(12): 2419–2425.

Research on the Surface Subsidence Monitoring Technology Based on Fiber Bragg Grating Sensing

7

Jinyu WANG[1*], Long JIANG[2], Zengrong SUN[3], Binxin HU[1], Faxiang ZHANG[1], Guangdong SONG[1], Tongyu LIU[1,2], Junfeng QI[4], and Longping ZHANG[4]

[1]*Key Laboratory of Optical Fiber Sensing Technology of Shandong Province, Laser Institute of Shandong Academy of Science, Jinan, 250014, China*
[2]*Shandong Micro-Sensor Photonics Ltd, Jinan, 250014, China*
[3]*Shandong Shenglong Safe Technology Co. Ltd, Jinan, 250032, China*
[4]*Laiwu Mining Co. Ltd of Laiwu Steel Group, Laiwu, 271100, China*
*Corresponding author: Jinyu WANG   E-mail: wangjinyu105@163.com

Abstract: In order to monitor the process of surface subsidence caused by mining in real time, we reported two types of fiber Bragg grating (FBG) based sensors. The principles of the FBG-based displacement sensor and the FBG-based micro-seismic sensor were described. The surface subsidence monitoring system based on the FBG sensing technology was designed. Some factual application of using these FBG-based sensors for subsidence monitoring in iron mines was presented.

Keywords: Fiber Bragg Grating; rock mass displacement; micro-seismic; optical fiber sensing; surface subsidence

1. Introduction

Along with the continuous improvement of the mining technology and the application of large machinery, mining intensity and depth are increasing continuously; rock mass movement caused by mining has produced more and more serious impacts on the production and life, including endangering the life safety of miners and the safety of buildings on the ground, causing contradiction between workers and peasants, and even affecting the enterprise production. So the measurements of the rock mass displacement and the implementation of monitoring and early warning have become the necessary prerequisites for safe production and life.

The stability judgment of traditional field rock mass engineering is based on the observation of basic point of the ground surface displacement [1, 2]. The rock mass subsidence can be obtained by observing the basic point displacement. The basic point displacement is a result of rock movement, but not the movement process of rock mass. It plays an extremely important role in timely warning by monitoring deep rock mass deformation condition to keep safety. The optical fiber sensing technology has advantages such as high sensitivity, large dynamic range, and easy networking [3–5]. So it is possible to realize the high precision dynamic monitoring and early warning of rock mass shift by carrying out the rock displacement monitoring technology research based on the fiber optic displacement and seismic technology.

We designed two types of FBG (fiber Bragg grating)-based sensors. The principles of the FBG-based displacement sensor and the FBG-based micro-seismic sensor were described. We also proposed a surface subsidence monitoring system based on the FBG displacement and micro-seismic sensing technology. And some field test results in iron mine would also be presented.

2. Principles

2.1 FBG-based displacement sensor

According to the principle of fiber Bragg grating sensing [6], we designed a kind of FBG-based displacement sensor. The proposed FBG-based displacement sensor is shown in Fig. 1. The internal elastic element of the sensor is the strain beam with FBG1 pasted on the top and FBG2 pasted on the bottom. The wire-rope which is preloaded by the tension springs and connected to the telescopic slide will produce tensile changes caused by the subsidence, and then the strain beam which is close to the telescopic slides will move up and down, so the FBGs will subject to tensile stress and compressive stress.

The wavelength change of the FBG based on the co-production of temperature and strain can be expressed as

$$\Delta\lambda_B/\lambda_B = (\alpha + \zeta)\Delta T + (1-P_e)\varepsilon \quad (1)$$

where α is the thermal expansion coefficient of the sensor, ζ is the thermo-optical coefficient of the optical fiber, P_e is the stress-optic coefficient, ΔT is the temperature difference, and ε is the axial strain of the fiber.

As two fibers are located in different surfaces of the cantilever beam, the strains of the two fibers are different. The wavelength changes of the two FBGs can be expressed as follows:

$$\Delta\lambda_{B1}/\lambda_{B1} = (\alpha + \zeta)\Delta T + (1-P_e)\varepsilon_1 \quad (2)$$
$$\Delta\lambda_{B2}/\lambda_{B2} = (\alpha + \zeta)\Delta T + (1-P_e)\varepsilon_2 \quad (3)$$

where λ_{B1} and λ_{B2} are the wavelengths of the FBGs. Subtracting (3) by (2), the strain difference can be given by

$$\varepsilon_1 - \varepsilon_2 = (\Delta\lambda_{B1}/\lambda_{B1} - \Delta\lambda_{B2}/\lambda_{B2})/(1-P_e). \quad (4)$$

According to (4), the displacement and strain can be obtained without the influence of temperature.

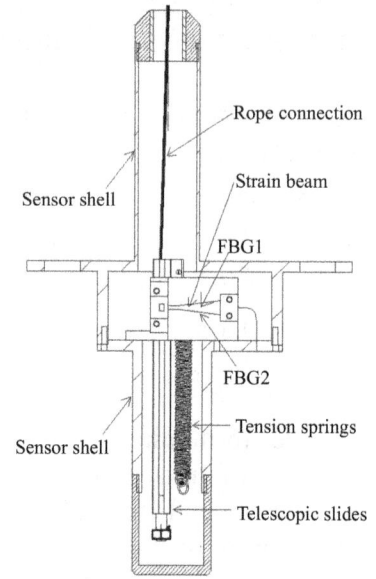

Fig. 1 Schematic of FBG-based displacement sensor.

2.2 FBG-based micro-seismic sensor

The schematic of the proposed FBG-based micro-seismic sensor as illustrated in Fig. 2 is composed of a steel leaf spring with the k_2 elastic coefficient, an L-shaped cantilever with an inert mass m connected to its end, and a fiber Bragg grating with the k_1 elastic coefficient with one end attached on the cantilever and another end attached on the shell. When the micro seismic waves induce vibration of the shell, the FBG will have a central wavelength shift due to the axial strain caused by the inertial force of the mass.

The dimensions a, b, and l are marked in Fig. 2.

The natural angular frequency can be defined as

$$\omega_0 = \sqrt{k/m} = \sqrt{[k_2 + k_1(a/b)^2]/m}. \quad (5)$$

Assuming an external acceleration $a_g = A_g e^{i\omega t}$ and a relative damping coefficient δ, the strain experienced by the FBG can be expressed as

$$\varepsilon = a \cdot a_g \Big/ b \cdot l \cdot \sqrt{(\omega_0^2 - \omega^2)^2 + 4\delta^2 \omega^2} \,. \quad (6)$$

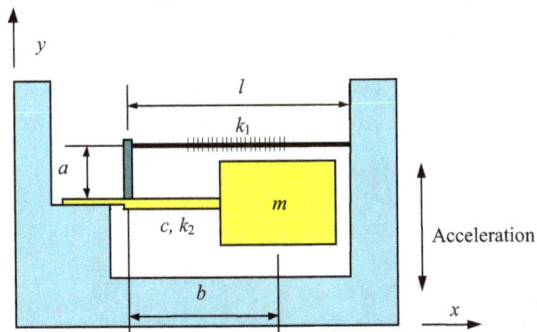

Fig. 2 Schematic of FBG-based micro-seismic sensor.

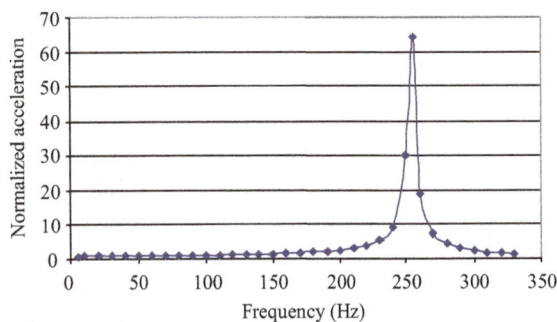

Fig. 3 Photograph of FBG-based micro-seismic sensor.

When the damping ratio $\zeta = \delta/\omega_0 = 0.7$, the strain acceleration sensitivity K can be noted as

$$K = a \big/ bl\omega_0^2 \,. \quad (7)$$

The FBG wavelength shift is proportional to the strain experienced by the FBG. Considering the strain sensitivity for FBGs with peak wavelengths in the C band regime is about $1.2 \, \text{pm}/\mu\varepsilon$ in general, the accelerometer sensitivity (wavelength shift of FBG per unit acceleration) is given by

$$S = \Delta\lambda / a_g = 1.2 \cdot a \big/ bl\omega_0^2 \,. \quad (8)$$

It can be seen from (5) and (8) that the natural angular frequency and the accelerometer sensitivity are determined by five parameters a, b, k_1, k_2, and m. By optimizing the parameters a, b, and m and choosing suitable material k_2, the micro-seismic sensor is designed to fit for micro-seismic measurement in mines. The frequency bandwidth is designed to be 1 Hz to 220 Hz, and the natural frequency is about 260 Hz according to the optimized parameters. The accelerometer sensitivity is 220 pm/g. Figure 3 shows the FBG-based micro-seismic sensor actually used in mine. Figure 4 shows the frequency response of the FBG-based micro-seismic sensor.

The interrogation principle of the FBG-based acceleration sensor depends on the intensity modulation of narrow line width DFB laser [7, 8], when a reflection or transmission spectrum curve of an FBG wavelength shifts due to the strain caused by vibration or acceleration.

Fig. 4 Frequency response of the micro-seismic sensor.

3. Field test and results in iron mine

As the sinkhole of Zhao Zhuang iron mine once had a collapse, this monitoring system was established in iron mine in Zhaozhuang, Laiwu City, Shandong Province to protect nearby buildings, workers, and facilities. The surface subsidence monitoring system contains an FBG demodulator, an FBG micro-seismic demodulator, six FBG-based displacement sensors, four FBG-based acceleration sensors, and optical cables. The signals gathered by these sensors were transmitted to the host computer by cable in real time and analyzed by the software.

3.1 Displacement monitoring

On the ground with the electronic total station monitoring, there are 4 basis points. The positions of the basic points are shown in Fig. 5. Table 1 describes the sensors installation information. The shallow base of the FBG-based displacement sensors was all bellow 35 m.

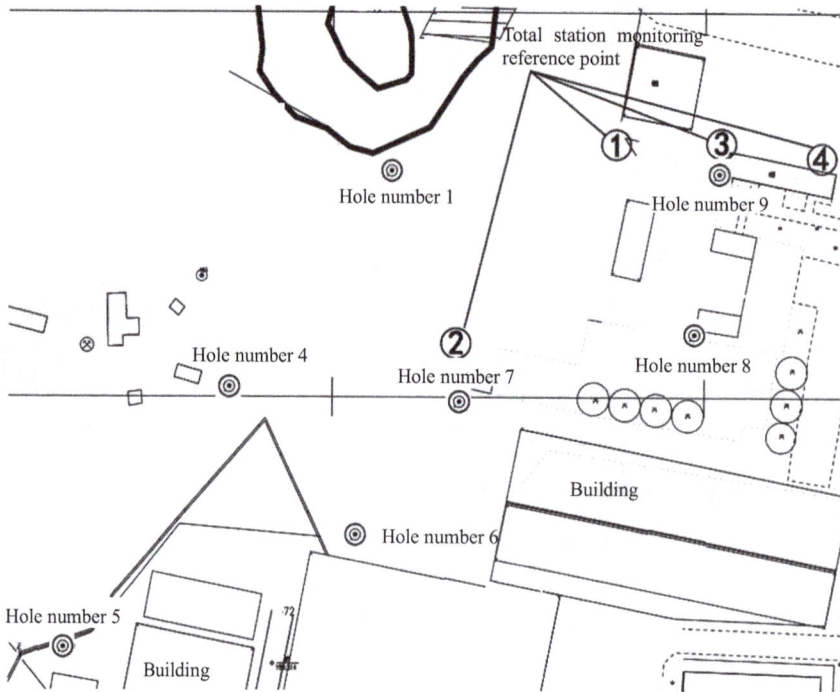

Fig. 5 Sensor installation plan.

Table 1 Placement information of displacement sensors.

Hole number	Sensor type	Installation depth (m)	Sensor number
6#	Shallow base	35	13K04A1001
	Deep base	56	13K04A2001
7#	Shallow base	40	13K04B1002
	Deep base	81	13K04B2002
8#	Shallow base	60	13K04A1002
	Deep base	69	13K04A2002

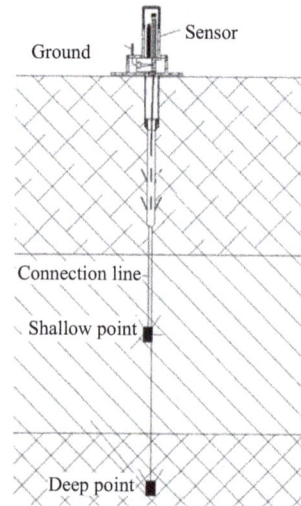

Fig. 6 FBG-based displacement monitoring installation design.

When the FBG-based displacement sensor is installed, it needs to be punched and drilled, and the sensor needs to be installed from the surface of the ground. Figure 6 shows the installation schematic diagram. The deep base, shallow point, and FBG-based displacement sensor are the fixed ends and connected through a connecting line. When there is a relative displacement among them, the FBG-based displacement sensor can detect these minor changes. If the deep base point and shallow point are placed in different rock strata, the changes between the strata of different depths can be monitored. Figure 7 shows the field installation of the FBG-based displacement sensor.

Fig. 7 Field installation of displacement sensor.

Figure 8 shows the displacement data from September 2014 to October 2015. Seen from Fig. 8, the cumulative deformation detected by the displacement sensor is more than 5 cm during this period. The results show that the surface subsidence tended to be stable between December 2014 and May 2015 except the data from Sensor 6. The cumulative data monitored by the deep base of Sensor 6 was reduced after April 2015. As the anchorage end of the sensor was affected by the local rock strata, if the rock at the anchoring part was damaged, the sensor here might fail to monitor. So a better and more reliable fixed way needs to be considered.

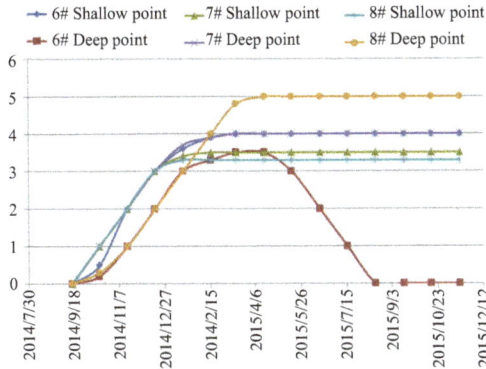

Fig. 8 Monitoring data of FBG-based displacement sensor.

Table 2 shows the displacement read from the electronic total station between June 2013 and December 2014. The average displacement is between 0.67 cm and 0.92 cm. Seen from the analysis result, the average subsidence rate got from the FBG-based displacement sensors is consistent with that obtained from the electronic total station.

Table 2 Statistics of monitoring data.

Hole number	Displacement (cm)	Average settlement (cm/month)
①	16.6	0.92
②	13.3	0.73
③	14.7	0.82
④	12	0.67

3.2 Micro-seismic monitoring

Table 3 describes the installation information of the FBG-based micro-seismic sensors. The FBG-based micro-seismic sensors were all installed bellow 55 m.

Figure 9 shows the field installation of micro-seismic sensors. The sensor was firmly installed on the bottom of the hole by special mounting rod. The grouting pipe, which was a kind of plastic hose and had 25-mm diameter, was inserted into the bottom of the hole. Expansion cements were grouted with manual injection pump, so the micro-seismic sensor could reliably coupled together with the strata. The principal component of expansion cement without shrinkage was Portland cement with the strength of 32.5, whose swelling agent rate was not less than 0.02%.

Table 3 Placement information of micro seismic sensors.

Hole number	Installation depth (m)	Sensor number
1#	60	13K05V1002
4#	84	13K05V1007
5#	62	13K05V1003
9#	55.5	13K05V1010

Fig. 9 Field installation of micro-seismic sensor.

Figure 10 shows the typical micro-seismic events occurred at 16:47:38 on December 27, 2014. The arrival time, energy, and location of the micro-seismic events could be obtained by analyzing the effective micro-seismic signals. Figure 11 shows the processing results of the micro-seismic events between September 2014 and December 2014. The dots and blocks represent different levels of micro-seismic events. represent different levels of micro-seismic events. The block indicates the micro-seismic events that are larger than 10^5 J. And after January 2015, micro-seismic events happened rarely. This measurement result is consistent with the monitoring result of the displacement sensor. Combining these two kinds of detecting results, we can see that there is a strong rock activity.

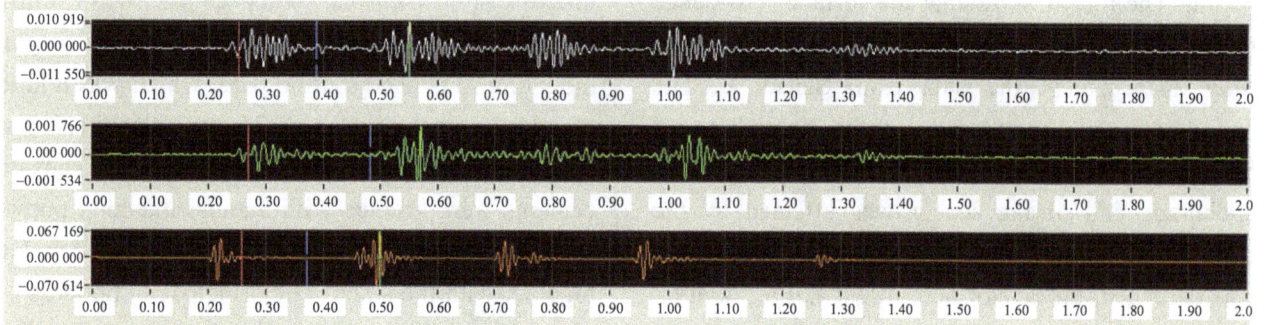

Fig. 10 Typical micro-seismic event.

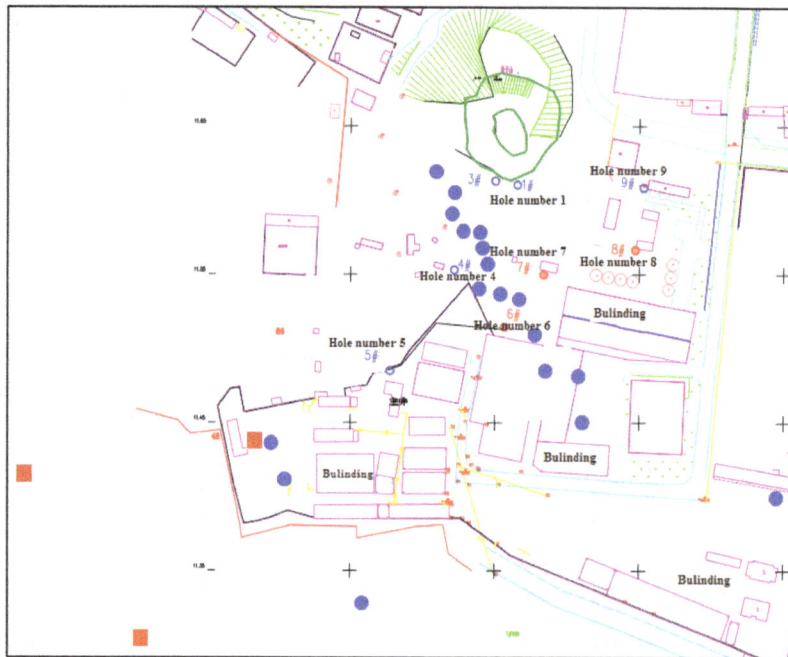

Fig. 11 Plane distribution of micro-seismic events.

4. Conclusions

In summary, we reported two types of FBG-based sensors to monitor the surface subsidence. The results show that the FBG-based displacement sensor and micro-seismic sensor can be combined to apply to the surface subsidence monitoring. The precision of the displacement result depends on the accuracy and reliability of the FBG-based displacement sensors, and the reliability of the analysis result of the micro-seismic events depends on the accuracy of the FBG-based micro-seismic sensor's coordinate, the location algorithm, the accuracy of first arrival, the installation location, the number of the micro-seismic sensor, and so on.

As the number of the micro-seismic sensors was too small, less effective micro-seismic events were obtained, and then the monitoring results needed further validation. In the future, we need to increase the measuring points to get more effective data to improve the monitoring accuracy. Then through the feedback information, we can monitor damage occurring in a rock and progressive failure process much earlier and timely give the early warning of instability of rock mass.

Acknowledgment

This work was partly supported by the Science and Technology Development Plan of Shandong Province (No. 2014GSF120017) and Science Fund for Young Scholars of SDAS (No. 2013QN005). This work was also partly supported by the Development Funds for SMEs (part of the European cooperation) (No. SQ2013ZOC600005).

References

[1] X. U. Bigen, P. W. Chunlai, S. H. Tang, and P. L. Cheng, "Study on large goaf management and its monitoring scheme design," *China Safety Science Journal*, 2007, 17(12): 147–152.

[2] H. Wang, W. M. Yang, B. Wang, and C. Z. Zhao, "Judgment on surface subsidence danger of mine goaf based on GIS technology," *Coal Engineering*, 2008, 9: 119–123.

[3] J. Wu, V. Masek, and M. Cada, "The possible use of fiber Bragg grating based accelerometers for seismic measurements," in *Conference on Electrical and Computer Engineering*, Canada, pp. 860–863, 2009.

[4] A. D. Kersey, T. A. Berkoff, and W. W. Morey, "Multiplexed fiber Bragg grating strain sensor with a fiber Fabry-Perot wavelength filter," *Optics Letters*, 1993, 18(16): 1370–1372.

[5] J. G. Liu, C. Schmidt-Hattenberger, and G. Borm, "Dynamic strain measurement with a fiber Bragg grating sensor system," *Measurement*, 2002, 32(2): 151–161.

[6] K. O. Lee, K. S. Chiang, and Z. Chen, "Temperature-insensitive fiber-Bragg-grating-based vibration sensor," *Optical Engineering*, 2001, 40(11): 2582–2585.

[7] J. Y. Wang, T. Y. Liu, C. Wang, X. H. Liu, D. H. Huo, and J. Chang, "A micro-seismic fiber Bragg grating (FBG) sensor system based on distributed feedback laser," *Measurement Science and Technology*, 2010, 21(9): 094012.

[8] J. Wang, B. Hu, W. Li, G. Song, L. Jiang, and T. Liu, "Design and application of fiber Bragg grating (FBG) geophone for higher sensitivity and wider frequency range," *Measurement*, 2016, 79: 228–235.

8

An Analytical Study on Bistability of Fabry-Perot Semiconductor Optical Amplifiers

Gang WANG[*], Shuqiang CHEN, and Huajun YANG

School of Physical Electronics, University of Electronic Science and Technology of China, Chengdu, 610054, China

[*]Corresponding author: Gang WANG   E-mail: buncan_wang@126.com

Abstract: Optical bistabilities have been considered to be useful for sensor applications. As a typical nonlinear device, Fabry-Perot semiconductor optical amplifiers (FPSOAs) exhibit bistability under certain conditions. In this paper, the bistable characteristics in FPSOAs are investigated theoretically. Based on Adams's relationship between the incident optical intensity I_{in} and the z-independent average intracavity intensity I_{av}, an analytical expression of the bistable loop width in SOAs is derived. Numerical simulations confirm the accuracy of the analytical result.

Keywords: Bistability; Fabry-Perot resonator; semiconductor optical amplifiers

1. Introduction

Optical bistability (OB) has been found to be a beneficial technology for designing optical sensors [1–2]. It is well known that Fabry-Perot semiconductor optical amplifiers (FPSOAs) can operate in a bistable regime with proper parameters. Therefore, further studies on bistability of FPSOAs are worthwhile from a practical point of view. In [3], the authors presented a commonly-used model for investigating the bistability in SOAs. By use of this model, the static and dynamic bistable characteristics in different types of SOAs have been investigated substantially [3–8]. However, these studies have been performed either numerically or experimentally. So far, few analytical results on the bistability in SOAs have been presented. In this paper, based on Adams's model [3], the necessary condition for bistability in SOAs is deduced. Moreover, analytical expressions of the switch-up

power, switch-down power of the incident beam, and the width of hysteresis loop in SOAs are derived. Numerical calculations are carried out to verify the analytical results.

The remainder of this paper is organized as follows. Section 2 gives detailed derivations of the necessary condition for bistability and the expression of hysteresis loop width in an FPSOA. Section 3 affords the numerical experiments to verify the correctness of the analytical results. Discussion and conclusions are provided in Section 4.

2. Theoretical analysis

Considering an SOA, it is assumed that in the cavity the optical intensity is uniform and the spontaneous emission is neglected. In the following, the analysis is carried out according to the relationships derived by Adams [3], in which the input intensity I_{in}, the output intensity I_{out}, and the

z-independent intracavity intensity I_{av} could be expressed as follows.

$$\frac{I_{out}}{I_s} = \frac{I_{in}}{I_s} \cdot \frac{(1-R_1)(1-R_2)e^{gL}}{(1-\sqrt{R_1R_2}e^{gL})^2 + 4\sqrt{R_1R_2}e^{gL}\sin^2\phi} \quad (1)$$

$$\frac{I_{av}}{I_s} = \frac{I_{out}}{I_s} \cdot \frac{(1-e^{-gL})(1+R_2e^{gL})}{(1-R_2)gL} \quad (2)$$

where R_1 and R_2 are the reflectivities of the two cavity mirrors, g is the net gain per unit length, ϕ is the single-pass phase change, and L is the cavity length. In the steady state, g and ϕ can be expressed as

$$g = \frac{\Gamma g_0}{1 + I_{av}/I_s} - \alpha \quad (3)$$

$$\phi = \phi_0 + \frac{g_0 Lb}{2}\left(\frac{I_{av}/I_s}{1 + I_{av}/I_s}\right) \quad (4)$$

$$g_0 = a(n - n_0) \quad (5)$$

where Γ is the optical confinement factor, α is the effective loss coefficient, g_0 is the unsaturated material gain, ϕ_0 is the initial phase detuning, I_s is the saturation optical intensity, b is the linewidth enhancement factor, n is the carrier density, a is the gain coefficient, and n_0 is the carrier density at transparency.

From (1) to (5), the input-output (IO) hysteresis loop can be observed by numerical simulation under certain parameter values [3]. In this paper, we analytically calculate the bistability in SOAs. From the numerical results and the definition of the bistability, it is obvious that in the bistable operation region, for a given input intensity I_{in}, there are three corresponding solutions of the output intensity I_{out} to (1)–(5). Namely, one is the unstable-state solution, and the other two are the stable-state solutions. Equation (2) shows that I_{out} is a single valued function of I_{av}. Thereby, it can be inferred that in the region where the hysteresis loop appears, there exist three values of I_{av} corresponding to one value of I_{in}. To ensure this feature, the function $I_{in}(I_{av})$ should have two extreme values in this region. Hence, from the mathematical point of view, the necessary condition for bistability of SOA is that the

first-order derivative of the function $I_{in}(I_{av})$ should have two zero points.

From (1) and (2), one can obtain the relationship between I_{in} and I_{av}:

$$\frac{I_{in}}{I_s} = \frac{(1-\sqrt{R_1R_2}e^{gL})^2 + 4\sqrt{R_1R_2}e^{gL}\sin^2\phi}{(1-R_1)(e^{gL}-1)(1+R_2e^{gL})} \cdot gL \cdot \frac{I_{av}}{I_s} \cdot \quad (6)$$

For convenience, let $x = I_{av}/I_s$, $y = I_{in}/I_s$. Then (6) can be rewritten as

$$y = \frac{\ln(G) \cdot [(1-\sqrt{R_1R_2}G)^2 + 4\sqrt{R_1R_2}G\sin^2\phi]}{(1-R_1)(G-1)(1+R_2G)} \cdot x \quad (7)$$

where $G=\exp(gL)$ represents the single pass gain in the cavity.

Clearly, (7) owns a complicated form, which makes it difficult to obtain the zero point of the first-order derivative of y. In order to theoretically find the necessary condition for the occurrence of hysteresis loop of SOAs and further achieve analytical results on the bistability, some approximations are demanded to simplify the form of (7). Naturally, the right hand of (7) can be expanded into Taylor series in terms of x. Since we try to examine whether the first-order derivative of y has two zero points, it is reasonable that three-order approximation of the Taylor series of y is adopted. Thus y can be rewritten as follows:

$$y \approx F_0 + F_1 \cdot (x - x_0) + \frac{1}{2!}F_2 \cdot (x - x_0)^2 + \frac{1}{3!}F_3 \cdot (x - x_0)^3 \quad (8)$$

where $F_0 = y(x_0)$ and $F_i = \left.\dfrac{d^i y}{dx^i}\right|_{x=x_0}$ $(1 \leq i \leq 3)$ are the ith order derivatives of y at the reference point $x=x_0$. In the following, (8) is a start point for further deduction and analysis. After straight and complex computation, the coefficients in (8) can be derived in the following form:

$$\frac{dy}{dx} = \frac{dT_1}{dx}x + T_1 + \frac{dT_2}{dx} \cdot \sin^2\phi \cdot x + T_2\left(\sin^2\phi + \sin 2\phi \cdot x \cdot \frac{d\phi}{dx}\right) \quad (9)$$

$$\frac{d^2 y}{dx^2} = \frac{d^2 T_1}{dx^2}\cdot x + 2\frac{dT_1}{dx} + \frac{d^2 T_2}{dx^2}\cdot \sin^2\phi\cdot x +$$

$$\frac{dT_2}{dx}(2\sin^2\phi + \sin 2\phi\cdot x) +$$

$$T_2\left[\frac{d^2\phi}{dx^2}\cdot \sin 2\phi\cdot x + 2\left(\frac{d\phi}{dx}\right)^2\cdot \cos 2\phi\cdot x +\right.$$

$$\left. 2\frac{d\phi}{dx}\cdot \sin 2\phi\right] \qquad (10)$$

$$\frac{d^3 y}{dx^3} = \frac{d^3 T_1}{dx^3}x + 3\frac{d^2 T_1}{dx^2} + \frac{d^3 T_2}{dx^3}\cdot \sin^2\phi\cdot x +$$

$$\frac{d^2 T_2}{dx^2}\left(3\sin^2\phi + \frac{d\phi}{dx}\cdot \sin 2\phi\cdot x + \sin 2\phi\cdot x\right) +$$

$$\frac{dT_2}{dx}\left[\frac{d^2\phi}{dx^2}\sin 2\phi\cdot x + 2\left(\frac{d\phi}{dx}\right)^2\cdot \cos 2\phi\cdot x +\right.$$

$$\left. 2\frac{d\phi}{dx}(2\sin 2\phi + \cos 2\phi\cdot x) + \sin 2\phi\right] \times$$

$$T_2\left[\frac{d^3\phi}{dx^3}\sin 2\phi\cdot x + 3\frac{d^2\phi}{dx^2}\sin 2\phi +\right.$$

$$6\frac{d^2\phi}{dx^2}\frac{d\phi}{dx}\cos 2\phi\cdot x - 4\left(\frac{d\phi}{dx}\right)^3\sin 2\phi\cdot x +$$

$$\left. 6\left(\frac{d\phi}{dx}\right)^2\cos 2\phi\right] \qquad (11)$$

where

$$T_1 = \frac{\ln(G)\cdot(1-\sqrt{R_1 R_2}G)^2}{(1-R_1)(G-1)(1+R_2 G)}$$

$$T_2 = \frac{4\ln(G)\cdot\sqrt{R_1 R_2}G}{(1-R_1)(G-1)(1+R_2 G)}$$

$$\frac{dT_1}{dx} = \frac{dT_1}{dG}\cdot\frac{dG}{dx}, \quad \frac{dT_2}{dx} = \frac{dT_2}{dG}\cdot\frac{dG}{dx}$$

$$\frac{dG}{dx} = -\frac{\Gamma\cdot g_0\cdot L}{\exp\left[\left(\alpha - \frac{\Gamma\cdot g_0}{x+1}\right)L\right](x+1)^2}$$

$$\frac{d\phi}{dx} = \frac{g_0 Lb}{2(x+1)^2}.$$

The selection of the reference point x_0 for the Taylor series expansion is of importance as well. Here more attention should be paid to the fact that the bistability in SOA occurs at the long wavelength side of the cavity resonance wavelength, as pointed out by aforementioned literatures [3, 6]. Therefore, it can be inferred that in the region that bistability

occurs the phase detuning factor ϕ is negative. This characterizes the interval of x in which we want to examine the feature of the first order derivative of y. Of course, the reference point should be chosen such that the phase detuning factor is less than zero. In this paper, x_0 is taken as half of the point at which ϕ is equal to 0. Using (4), it can be derived as

$$x_0 = \frac{-\phi_0}{\Gamma g_0 Lb + 2\phi_0}. \qquad (12)$$

Substitute (9)–(12) into (8), y is re-expressed as a cubic function of x. Differentiating both sides of (8), we can get

$$\frac{dy}{dx} = \frac{F_3}{2}x^2 + (F_2 - F_3\cdot x_0)x + \left(\frac{F_3}{2}\cdot x_0^2 - F_2\cdot x_0 + F_1\right). (13)$$

Let (13) equal to 0, then a quadratic equation is achieved in the form

$$\frac{F_3}{2}x^2 + (F_2 - F_3\cdot x_0)x + \left(\frac{F_3}{2}\cdot x_0^2 - F_2\cdot x_0 + F_1\right) = 0. \quad (14)$$

which owns the discriminant Δ as

$$\Delta = F_2^2 - F_1\cdot F_3. \qquad (15)$$

Equation (14) would own two real roots when Δ is greater than 0. Thus the necessary condition for bistability of SOAs can be described as

$$\Delta = F_2^2 - F_1\cdot F_3 > 0. \qquad (16)$$

When this condition is satisfied, an IO hysteresis loop can be acquired. Next, based on the above results, we would derive the analytical expression of the hysteresis loop width.

When $\Delta > 0$, the two real roots of (14) are

$$x_1 = \frac{F_3\cdot x_0 - F_2 + \sqrt{\Delta}}{F_3} \qquad (17)$$

$$x_2 = \frac{F_3\cdot x_0 - F_2 - \sqrt{\Delta}}{F_3}. \qquad (18)$$

Substitute (17) and (18) into (7), the switch-up power P_{up} and switch-down power P_{down} of the input beam for the hysteresis loop in the SOA can be easily obtained:

$$P_{up} = \frac{\ln(G_2)\cdot[(1-\sqrt{R_1 R_2}G_2)^2 + 4\sqrt{R_1 R_2}G_2\sin^2\phi_2]}{(1-R_1)(G_2-1)(1+R_2 G_2)}.$$

$$x_2\cdot I_s\cdot A_{in} \qquad (19)$$

$$P_{down} = \frac{\ln(G_1) \cdot [(1 - \sqrt{R_1 R_2} G_1)^2 + 4\sqrt{R_1 R_2} G_1 \sin^2 \phi_1]}{(1 - R_1)(G_1 - 1)(1 + R_2 G_1)} \cdot$$
$$x_1 \cdot I_s \cdot A_{in} \qquad (20)$$

where G_1 and ϕ_1 denote the values of G and ϕ at $x=x_1$, G_2 and ϕ_2 denote the values of G and ϕ at $x=x_2$, and A_{in} represents the area of the input beam. Combining (19) and (20), we can derive the bistable loop width H as

$$H = P_{up} - P_{down}. \qquad (21)$$

3. Numerical simulations

In the numerical simulation, the parameter values are taken as follows, $A_{in}=0.8\,\mu m^2$, $L=500\,\mu m$, $R_1=R_2=0.3$, $n_0=1.5\times10^{24}$ m^{-3}, $a=2.7\times10^{-20}$ m^2, $\alpha=1000\,m^{-1}$, $b=4.8$, and $\Gamma=0.1$. Figs. 1 and 2 plot the dependence of Δ on the initial phase factor and the normalized carrier concentration, respectively. According to the above derived criterion for the occurrence of hysteresis, when Δ is greater than 0, the bistable loop could be found. In Fig. 1, it can be seen that Δ is positive when ϕ_0 is in the interval $[-0.78\pi, -0.18\pi]$ as the normalized carrier concentration is 0.95. Further numerical simulations are carried out under the same parameters, which show that in this interval the SOA would appear the bistable behaviors. Similarly, Fig. 2 reveals that if the normalized carrier density is greater than 0.88, Δ is positive when the initial phase detuning factor ϕ_0 is -0.22π. Numerical calculations also exhibit that just in this region the bistable loop can be observed. All these confirm the effectiveness of the derived criterion (16) for the hysteresis in the SOAs.

Figure 3 displays the numerical and analytical results of the switch-up power as a function of the normalized carrier concentration for various initial phase detuning factors. Obviously, the discrepancy between the simulations and the derived formulae is negligible. Moreover, it can be seen that P_{up} dwindles as the carrier density increases, and with the enhancement of the initial phase detuning, P_{up}

rises up drastically. In Fig. 4, the tendency of the switch-down power P_{down} is plotted, as the normalized carrier concentration grows up for various initial phase detuning factors. The analytical results show a good agreement with the numerical ones. Furthermore, P_{down} possesses a similar tendency to P_{up}, when the carrier concentration or the initial phase detuning factor increases. In Figs. 5 and 6, the bistable loop width is shown as a function of the normalized carrier density and the initial phase detuning factors, respectively. It is observed that analytical calculations are slightly smaller than numerical simulations. Meanwhile, the hysteresis loop becomes wider with the increment of either the carrier density or the initial phase detuning factor.

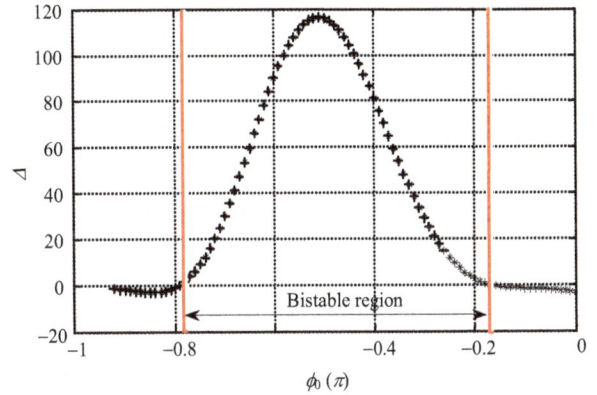

Fig. 1 Dependence of Δ on the initial phase detuning factor ϕ_0, $n/n_{th}=0.95$.

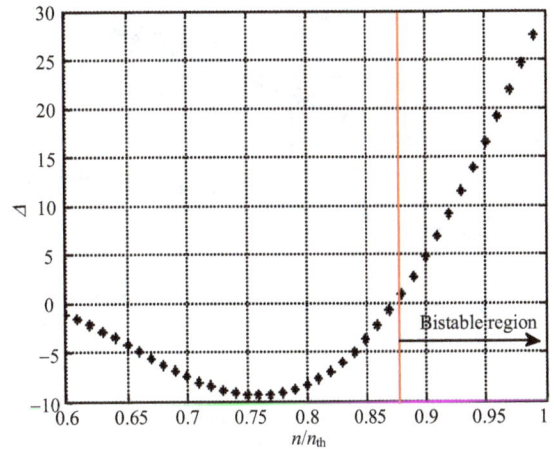

Fig. 2 Dependence of Δ on the normalized carrier concentration n/n_{th}, $\phi_0=-0.22\pi$.

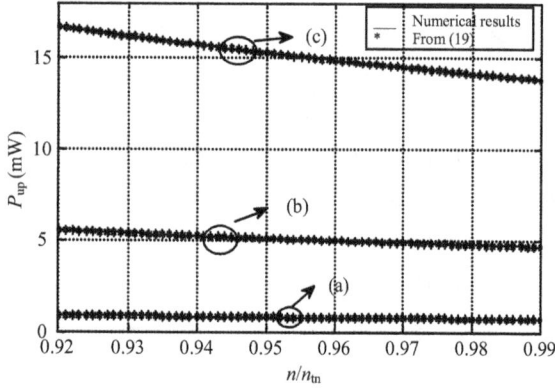

Fig.3 Jump-up power P_{up} versus normalized carrier density n/n_{th} for three cases: (a) $\phi_0 = -0.22\pi$, (b) $\phi_0 = -0.42\pi$, and (c) $\phi_0 = -0.62\pi$ (The solid lines and the symbols represent the numerical simulations and the analytical results, respectively).

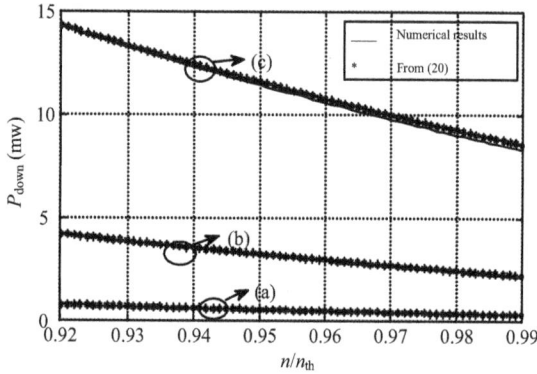

Fig.4 Comparison between analytical and numerical results of the jump-down input power P_{down} for three cases: (a) $\phi_0 = -0.22\pi$, (b) $\phi_0 = -0.42\pi$, and (c) $\phi_0 = -0.62\pi$ (Other parameters are the same as Fig.3).

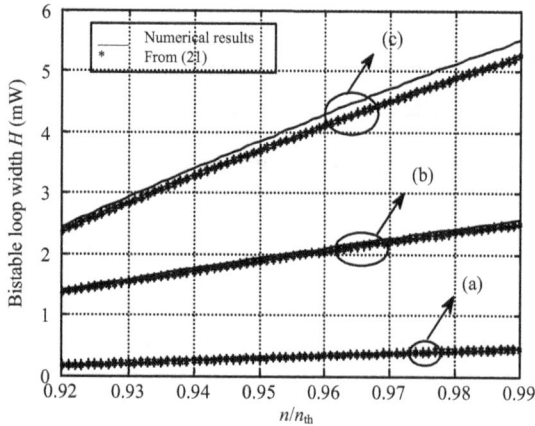

Fig.5 Bistable loop width as a function of the normalized carrier density n/n_{th} for three cases: (a) $\phi_0 = -0.22\pi$, (b) $\phi_0 = -0.42\pi$, and (c) $\phi_0 = -0.62\pi$ (The solid lines and the symbols represent the numerical simulations and the analytical results, respectively).

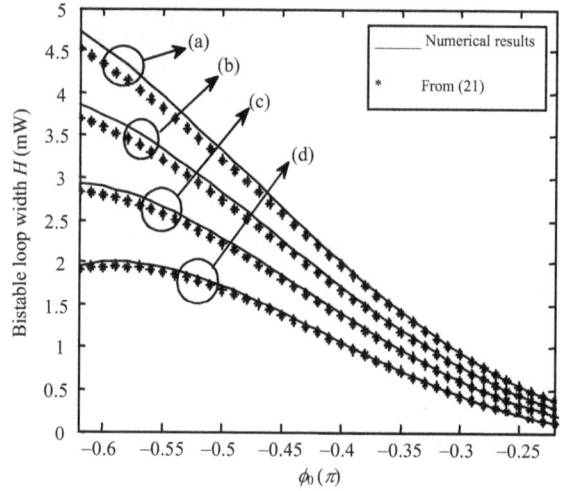

Fig.6 Analytical and numerical results of bistable loop width as a function of the initial phase detuning factor ϕ_0 for four cases: (a) $n/n_{th}=0.97$, (b) $n/n_{th}=0.95$, (c) $n/n_{th}=0.93$, and (d) $n/n_{th}=0.91$.

4. Conclusions

In summary, based on Adams's relationship between the z-independent intracavity intesnsity I_{av} and the incident intensity I_{in} in an SOA, we deduce the criterion for the hysteresis in SOAs by expanding I_{in} as a Taylor series in terms of I_{av}. As a result, the analytical expressions of the switch-up power, the switch-down power of the incident beam, and the width of the hysteresis loop in SOAs are derived for the first time. Numerical simulations are performed to test the effectiveness of the criterion and the accuracy of the expressions. The comparisons reveal that our analytical formulae are in good agreement with the numerical results. These conclusions would afford a valuable way for future investigations into the bistability in SOAs and other devices with a similar structure.

References

[1] P. M. Mejias and R. M. Herrero, "Bistable fiber-optics interferonetric sensor," *Applied Optics*, 1988, 27(5): 811–813.

[2] S. H. Shehadeh, D. Steup, and J. Weinzierl, "Cascaded linear and nonlinear optical resonators:

Towards a smart deflection sensor," in *Microsystems and Nanoelectronics Research Conference*, Canada, Oct. 13–14, 2009.

[3] P. Pakdeevanich and M. J. Adams, "Measurements and modeling of reflective bistability in 1.55-μm laser diode amplifiers," *IEEE Journal of Quantum Electronics*, 1999, 35(12): 1894–1903.

[4] N. F. Mitchell, J. O'Gorman, and J. Hegarty, "Optical bistability in asymmetric Fabry-Perot laser-diode amplifiers," *Optics Letters*, 1994, 19(4): 269–271.

[5] P. Wen, Michael S, M. Gross, and S. Esener, "Observation of bistability in vertical-cavity semiconductor optical amplifier (VCSOA)," *Optics Express*, 2002, 10(22): 1273–1278.

[6] A. Hurtado, A. Gonzalez-Marcos, and J. A. Martin-Pereda, "Modeling reflective bistability in vertical-cavity semiconductor optical amplifiers," *IEEE Journal of Quantum Electronics*, 2005, 41(3): 376–383.

[7] C. F. Marki, D. R. Jorgesen, H. J. Zhang, P. Wen, and S. C. Esener, "Observation of counterclockwise, clockwise and butterfly bistability in 1550 nm VCSOAs," *Optics Express*, 2007, 15(8): 4953–4959.

[8] W. L. Zhang and S. F. Yu, "Bistabilities of birefringent vertical-cavity semiconductor optical amplifiers with antiresonant reflecting optical waveguide," *IEEE Journal of Quantum Electronics*, 2010, 46(1): 11–18.

9

Fiber Fabry-Perot Interferometer for Curvature Sensing

Catarina S. MONTEIRO[1,2*], Marta S. FERREIRA[1,2], Susana O. SILVA[1,2], Jens KOBELKE[3], Kay SCHUSTER[3], Jörg BIERLICH[3], and Orlando FRAZÃO[1,2]

[1]*Department of Physics and Astronomy, Sciences Faculty, Porto University, Rua do Campo Alegre 687, 4169-007 Porto, Portugal*

[2]*INESC-TEC, Rua do Campo Alegre 687, 4169-007 Porto, Portugal*

[3]*IPHT Jena – Leibniz Institute of Photonic Technology, Albert-Einstein-Str. 9, 07745 Jena, Germany*

*Corresponding author: Catarina S. MONTEIRO   E-mail: catarina.s.monteiro@inesctec.pt

Abstract: A curvature sensor based on an Fabry-Perot (FP) interferometer was proposed. A capillary silica tube was fusion spliced between two single mode fibers, producing an FP cavity. Two FP sensors with different cavity lengths were developed and subjected to curvature and temperature. The FP sensor with longer cavity showed three distinct operating regions for the curvature measurement. Namely, a linear response was shown for an intermediate curvature radius range, presenting a maximum sensitivity of $68.52 \, \text{pm/m}^{-1}$. When subjected to temperature, the sensing head produced a similar response for different curvature radii, with a sensitivity varying from $0.84 \, \text{pm/}°C$ to $0.89 \, \text{pm/}°C$, which resulted in a small cross-sensitivity to temperature when the FP sensor was subjected to curvature. The FP cavity with shorter length presented low sensitivity to curvature.

Keywords: Curvature measurement; Fabry-Perot interferometer; optical fiber sensor

1. Introduction

Optical fiber sensors based on Fabry-Perot (FP) cavities are an interesting solution for engineering applications. Due to the small dimensions, they can be easily embedded in more advanced structures. The first FP cavities were fabricated in 1993, by splicing two standard single mode fibers (SMFs) with a silica capillary tube and tested for the measurement of dynamic strain [1]. Over the last decades, this type of interferometric sensor has been used for sensing a wide range of physical parameters such as temperature, vibration, pressure, and refractive index [2–5]. A temperature sensor was proposed consisting of a tip of a single mode fiber

coated with a thin film of polyvinyl alcohol [2]. The sensor provided a stable solution for temperature measurements, presenting a maximum sensitivity of $173.5 \, \text{pm/}°C$ above $80°C$. An all-fiber sensor based on FP interferometry was proposed for vibration measurements, showing insensitivity to environmental effects, along with the possibility of being embedded in composite material and used in harsh environments [3]. Refractive index was a key parameter for sensing purposes, and an FP-based sensing head formed by an air cavity was demonstrated [4]. In this case, the refractive index of gases was monitored as a function of pressure. An FP cavity based on a diaphragm-free hollow core silica tube was proposed [5]. The sensor head, composed by a hollow core silica tube spliced to a single mode fiber, was tested for pressure, refractive index, and temperature changes. The proposed sensor head was an interesting solution for both pressure and refractive index variations on fluids.

The optical fiber provides a great solution for curvature measurement purposes due to lightweight, compact dimensions, immunity to electromagnetic interference, and the resistance to corrosion. A curvature sensor based on a photonic crystal fiber was reported [6], showing a great sensitivity to the measurand and negligible sensitivity to temperature. Recently, a highly sensitive curvature sensor based on abrupt tapered fiber joined with a micro FP interferometer was presented [7]. The sensor head was sensitive only in visibility since the bending was applied before the FP cavity, not affecting the optical path length. Optical fiber sensors for curvature sensing were already tested for medical purposes [8], where an optical sensor was used to sense the radius of curvature of a rotary endodontic file inside an artificial tooth root canal.

In this work, an FP cavity for curvature sensing was proposed. The sensor was produced by fusion splicing a capillary silica tube between two SMFs. The capillary tube had an outer diameter similar to the one of the single mode fiber and an inner diameter of 60 µm. Two FP sensors with different cavity lengths were tested in curvature and temperature and characterized in visibility and wavelength shift. Finally, the cross-sensitivity for both measurands was also analyzed.

2. Experimental results

The experimental setup, shown in Fig. 1, is constituted by a broadband optical source, centered at 1550 nm and with a bandwidth of 100 nm, an optical circulator and an optical spectrum analyzer (OSA). The circulator provides interrogation of the sensing head in reflection and the OSA reads the spectral response of the sensing head with a resolution of 0.1 nm.

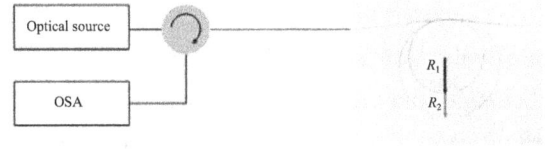

Fig. 1 Experimental setup, with photograph of the curvature sensor with different curvature radii.

Two FP cavities with different lengths were characterized. The first sensing head had a cavity length of 1140 µm and was called an FP-Long sensor, whereas the second one had a cavity length of 500 µm and is called FP-Short sensor. Both cavities were produced using a hollow core silica tube with an inner diameter of ~ 60 µm, spliced between two SMFs. The arc discharge was performed with a splice machine in manual operation, where an offset of the capillary tube was applied to prevent its collapse. In Fig. 2, a scheme of the sensors used and a microscope photograph of the hollow core silica tube cross-section are presented.

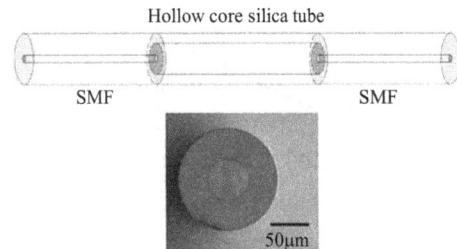

Fig. 2 Scheme of the sensor heads and a microscope image of the hollow core silica tube cross-section.

The spectrum of the two sensor heads used is shown in Fig. 3. The FP-Long sensor presented a visibility of 0.47 and had a free spectral range ($\Delta\lambda$) of 1.04 nm. Regarding the FP-Short sensor, it had a visibility of 0.41 and a free spectral range of 2.40 nm. The light that was transmitted to the capillary tube traveled mostly through the hollow core, which was filled with air. Refractive index of the optical cavity can be estimated using (1), where n is the refractive index and L the cavity length [9].

$$n = \frac{\lambda^2}{2L\Delta\lambda}. \tag{1}$$

Using a central operation wavelength λ of 1550 nm, refractive indices n of 1.0002 and 1.0009 were calculated for FP-Long and FP-Short, respectively. Refractive index for both sensing heads was close to the value for air, which went according to the expected. The difference between obtained results and theoretical value of refractive index of air can be explained by possible errors committed while determining the cavity length due to software limitations.

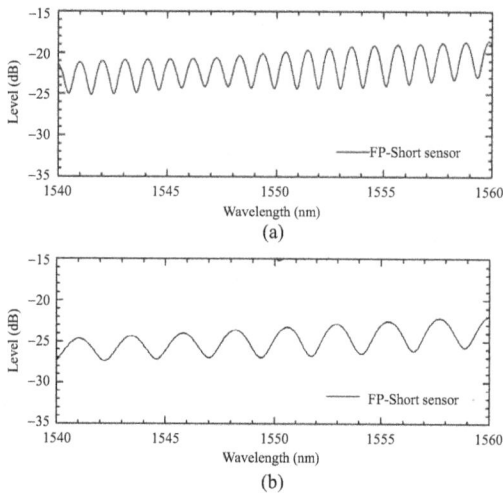

Fig. 3 Reflection spectrum of (a) FP-Long and (b) FP-Short without curvature applied.

A curvature study was carried out using an overhand knot configuration, as depicted in Fig. 1. The position of the sensor head was maintained parallel to the position of the knot. A variation on the relative position of the sensing head produced a variation in visibility, while maintaining the position of the fringes of the spectrum. Figure 4(a) shows the spectral behavior of the FP-Long when submitted to curvature. By stretching the fiber ends, the curvature radius diminished, resulting in the change in both visibility and wavelength as presented in Fig. 4(a). The curvature radius was maintained large enough to ensure that the losses were reduced. The visibility of the sensor head decreased 20% because the reduction of the curvature radius diminished the coupling efficiency of the FP mirrors, while the wavelength shift was due to the variation of the optical cavity length. In the case of FP-Short, even though the wavelength shift was negligible,

visibility decreased when curvature increased as depicted in Fig. 4(b). The visibility diminished 10% in the range of curvature studied, which was half of the value achieved in the FP-Long sensor.

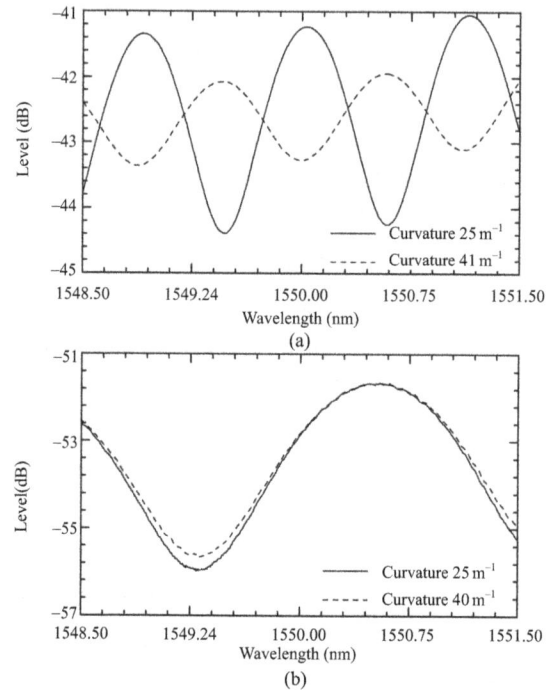

Fig. 4 Optical reflective spectrum for different curvature radius of (a) FP-Long and (b) FP-Short sensor.

Figure 5 shows the wavelength response of both sensing heads for different applied curvatures. The longer cavity, FP-Long, presented a non-linear response, but it was reasonable to divide the curvature range in three sub-regions where the response was linear. A sensitivity of 21.56 ± 0.68 pm/m^{-1} was obtained for a high curvature radius, in the range of $45\,\mathrm{m}^{-1}$ to $55\,\mathrm{m}^{-1}$. In the case of a lower curvature radius, between $25\,\mathrm{m}^{-1}$ and $35\,\mathrm{m}^{-1}$, a sensitivity of 17.27 ± 0.85 pm/m^{-1} was attained. A maximum sensitivity of 68.52 ± 1.65 pm/m^{-1} was achieved for a curvature radius comprised between $35\,\mathrm{m}^{-1}$ and $45\,\mathrm{m}^{-1}$.

The different regions of curvature response can be explained by the material behavior to the curvature applied. In the region of $25\,\mathrm{m}^{-1}$ to $35\,\mathrm{m}^{-1}$, the curvature radius applied to the sensor was not enough to curve the silica tube. Over this region, the spectral variation was due to compression over the interfaces caused by bending. On the other hand, over the region of $45 \mathrm{m}^{-1}$ to $55 \mathrm{m}^{-1}$ the bend reached a saturation point at which the silica tube stopped bending. For higher curvature radius, the sensing head broke at the splicing points.

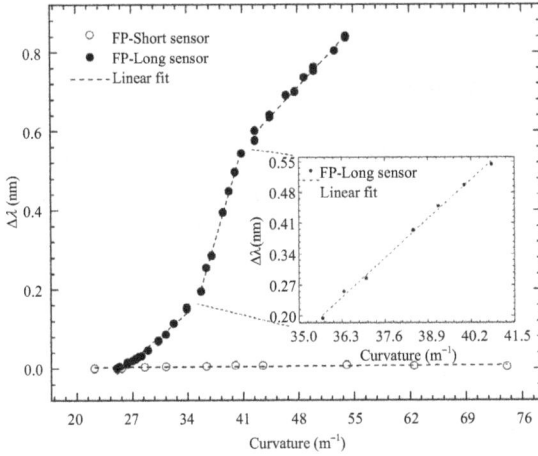

Fig. 5 Curvature response of FP-Long and FP-Short (inset: detail of the curvature response of FP-Long over the region of $35 \mathrm{m}^{-1}$ to $45 \mathrm{m}^{-1}$).

Table 1 summarizes the curvature sensitivities for the considered ranges. Curvature was also applied to FP-Short using the knot method, resulting in a negligible sensitivity to curvature, as observed in Fig. 5, with a corresponding sensitivity of $0.09 \mathrm{pm/m}^{-1}$. The result was somehow expected, since the geometry of the FP-Short sensor was unchanged when curvature was applied, due to its small size. This implied that the optical path was independent of curvature changes.

Table 1 Sensitivity for curvature of FP-Long sensor.

Cavity length (μm)	Range of measurement (m^{-1})	Sensitivity (pm/℃)
	25 − 35	17.27
1140	35 − 45	68.52
	45 − 55	21.56

Temperature responses of both sensing heads were also studied. Each sensor was placed in an oven, where temperature varied from 30℃ to 80℃, with a resolution of 0.1℃. The FP-Long was characterized in temperature considering two different curvatures, namely, $23.5 \mathrm{m}^{-1}$ and $50 \mathrm{m}^{-1}$, as presented in Fig. 6(a). The small wavelength change

for the two different curvatures confirms the low sensitivity of the sensor to temperature, with sensitivities of $0.84 \mathrm{pm/℃}$ and $0.89 \mathrm{pm/℃}$, for $23.5 \mathrm{m}^{-1}$ and $50 \mathrm{m}^{-1}$ curvature radii, respectively.

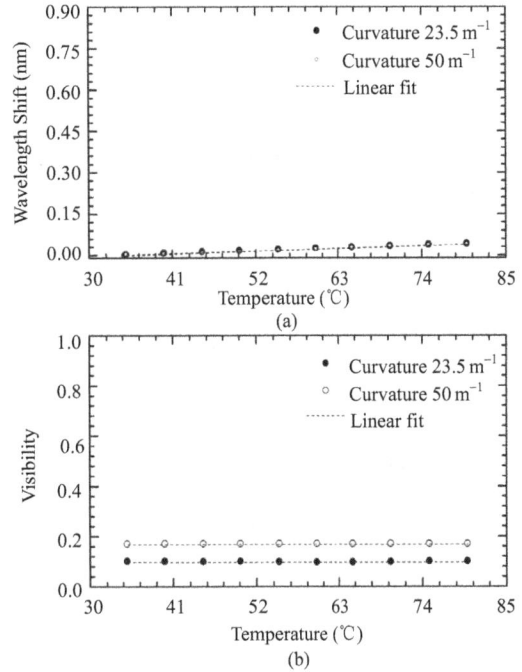

Fig. 6 Temperature characterization for FP-Long submitted to $23.5 \mathrm{m}^{-1}$ and $50 \mathrm{m}^{-1}$ of curvature applied (a) in wavelength and (b) in visibility.

The FP-Short was also tested without any curvature applied, presenting a sensitivity of $0.83 \mathrm{pm/℃}$, as shown in Fig. 7(a). Since the sensor head was only constituted by pure silica, thermal expansion of the capillary tube was the main behavior for the low temperature sensitivity. When compared with an ultra-long capillary tube (40 mm length) the temperature sensitivity was $20.6 \mathrm{pm/℃}$ [10]. The cross-sensitivity between curvature and temperature was also determined for the FP-Long sensor. Results of $0.052 \mathrm{m}^{-1}/℃$ and $0.041 \mathrm{m}^{-1}/℃$ for the lower curvature radius of $25 \mathrm{m}^{-1}$ to $35 \mathrm{m}^{-1}$ and higher curvature radius of $45 \mathrm{m}^{-1}$ to $55 \mathrm{m}^{-1}$, respectively, were achieved. In the range of higher sensitivity to curvature a cross-sensitivity of $0.013 \mathrm{m}^{-1}/℃$ was also attained. For both sensor heads, visibility was almost constant throughout temperature variations, as depicted in Figs. 6(b) and 7(b).

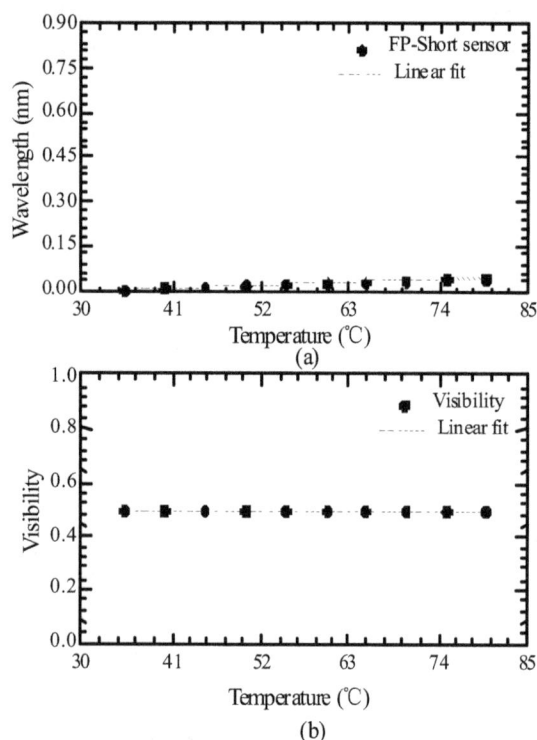

Fig. 7 Temperature characterization for FP-Short sensor not submitted to curvature (a) in wavelength and (b) in visibility.

3. Conclusions

A curvature sensor based on an Fabry-Perot interferometer was demonstrated. Curvature and temperature responses were compared for two different cavity lengths. The sensing head with a longer cavity length (FP-Long) showed a non-linear response to curvature, presenting a sensitivity of $21.56\,\text{pm/m}^{-1}$ at the low curvature region of $25\,\text{m}^{-1}$ to $35\,\text{m}^{-1}$, and $17.27\,\text{pm/m}^{-1}$ at the curvature range of $45\,\text{m}^{-1}$ to $55\,\text{m}^{-1}$. The FP-Long sensor presented a maximum curvature sensitivity of $68.52\,\text{pm/m}^{-1}$, in the range of $35\,\text{m}^{-1}$ to $45\,\text{m}^{-1}$. For the FP-Long sensing head, the temperature response was studied, considering two different curvature radii, which led to the conclusion that curvature had a low influence on the temperature response. Besides, the longer cavity exhibited negligible sensitivity to temperature, presenting a maximum cross-sensitivity of $0.052\,\text{m}^{-1}/\,^\circ\text{C}$. The smallest cavity (FP-Short) showed low sensitivity to the parameters studied,

providing an alternative solution to measure others physical parameters. The proposed configuration can be easily adapted in pipeline with curvature radii between 0.04 m (1.58 inches) and 0.02 m (0.79 inches).

Acknowledgment

Project NORTE-01-0145-FEDER-000036 is financed by the North Portugal Regional Operational Programme (NORTE 2020), under the PORTUGAL 2020 Partnership Agreement, and through the European Regional Development Fund (ERDF).

References

[1] J. S. Sirkis, D. D. Brennan, M. A. Putman, T. A. Berkoff, A. D. Kersey, and E. J. Friebele, "In-line fiber etalon for strain-measurement," *Optics Letters*, 1993, 18(22): 1973–1975.
[2] Q. Rong, H. Sun, X. Qiao, J. Zhang, M. Hu, and Z. Feng, "A miniature fiber-optic temperature sensor based on a Fabry-Perot interferometer," *Journal of Optics*, 2012, 045002(14): 059501–059501.
[3] Q. Zhang, T. Zhu, Y. Hou, and K. S. Chiang, "All-fiber vibration sensor based on a Fabry-Perot interferometer and a microstructure beam," *Journal of the Optical Society of America B*, 2013, 30(5): 1211–1215.
[4] G. Z. Xiao, A. Adnet, Z. Zhang, F. G. Sun, and C. P. Grover, "Monitoring changes in the refractive index of gases by means of a fiber optic Fabry-Perot interferometer sensor," *Sensors and Actuators A: Physical*, 2005, 118(2): 177–182.
[5] M. S. Ferreira, L. Coelho, K. Schuster, J. Kobelke, J. L. Santos, and O. Frazão, "Fabry-Perot cavity based on a diaphragm-free hollow-core silica tube," *Optics Letters*, 2011, 36(20): 4029–4031.
[6] H. Gong, H. Song, X. Li, J. Wang, and X. Dong, "An optical fiber curvature sensor based on photonic crystal fiber modal interferometer," *Sensors and*

Actuators A: Physical, 2013, 195(6): 139–141.

[7] M. Cano-Contreras, A. D. Guzman-Chavez, R. I. Mata-Chavez, E. Vargas-Rodriguez, D. Jauregui-Vazquez, D. Claudio-Gonzalez, *et al.*, "All-fiber curvature sensor based on an abrupt tapered fiber and a Fabry-Pérot interferometer," *IEEE Photonics Technology Letters.*, 2014, 26(22): 2213–2216.

[8] C. S. Shin and M. W. Lin, "An optical fiber-based curvature sensor for endodontic files inside a tooth root canal," *IEEE Sensors Journal*, 2010, 10(6):

1061–1065.

[9] M. S. Ferreira, P. Roriz, J. Bierlich, J. Kobelke, K. Wondraczek, C. Aichele, *et al.*, "Fabry-Perot cavity based on silica tube for strain sensing at high temperatures," *Optics Express*, 2015, 23(12): 16063–16070.

[10] M. S. Ferreira, K. Schuster, J. Kobelke, J. L. Santos, and O. Frazão, "Spatial optical filter sensor based on hollow-core silica tube," *Optics Letters*, 2012, 37(5): 890–892.

10

Study of Φ-OTDR Stability for Dynamic Strain Measurement in Piezoelectric Vibration

Meiqi REN, Ping LU, Liang CHEN, and Xiaoyi BAO[*]

Department of Physics, University of Ottawa, Ottawa, ON, K1S5G5, Canada

[*]Corresponding author: Xiaoyi BAO   E-mail: xbao@uottawa.ca

Abstract: In a phase-sensitive optical-time domain reflectometry (Φ-OTDR) system, the challenge for dynamic strain measurement lies in large intensity fluctuations from trace to trace. The intensity fluctuation caused by stochastic characteristics of Rayleigh backscattering sets detection limit for the minimum strength of vibration measurement and causes the large measurement uncertainty. Thus, a trace-to-trace correlation coefficient is introduced to quantify intensity fluctuation of Φ-OTDR traces and stability of the sensor system theoretically and experimentally. A novel approach of measuring dynamic strain induced by various driving voltages of lead zirconate titanate (PZT) in Φ-OTDR is also demonstrated. Piezoelectric vibration signals are evaluated through analyzing peak values of fast Fourier transform spectra at the fundamental frequency and high-order harmonics based on Bessel functions. High trace-to-trace correlation coefficients varying from 0.824 to 0.967 among 100 measurements are obtained in experimental results, showing the good stability of our sensor system, as well as small uncertainty of measured peak values.

Keywords: Optical fiber sensors; phase-sensitive optical time domain reflectometry (OTDR); vibration

1. Introduction

Distributed optical fiber sensors (DOFSs) based on detection and analysis of backscattered light have attracted significant research attention over the past two decades since they provide many advantages over conventional sensors. The distributed static measurements such as temperature or strain sensing have been extensively demonstrated with Brillouin-based DOFSs and optical frequency domain reflectometry (OFDR) [1–3]. However, as for distributed vibration sensing, the measurable frequency response is limited in above schemes because they require to measure the spectral shift continuously and average 10 times - 100 times in

order to increase signal to noise ratio [4, 5]. Phase-sensitive optical time domain reflectometry (Φ-OTDR) is one of promising DOFSs to detect the dynamic disturbances along the fiber. The principle of the Φ-OTDR is based on the interference between Rayleigh backscattered lights from different scattering centers along the fiber within a pulse width. The relative phase changes of the electric fields from scattering centers are highly sensitive to the disturbances at a certain position. Thus, the Φ-OTDR was successfully detection scheme [6, 7]. The long-range Φ-OTDR for intrusion sensing is demonstrated with the assistance of Raman amplification [8]. The Φ-OTDR was also demonstrated for vibration sensing by adopting heterodyne coherent detection [9]. The wavelet denoising method was also introduced in the Φ-OTDR to get a better performance with 50-cm spatial resolution [10]. The high frequency measurement was achieved by combining a Mach-Zehnder interferometer and a Φ-OTDR system [11].

The aforementioned works demonstrated the vibration measurements with frequencies ranging from Hz to kHz in terms of different applications. Although quantifying the vibration is important in some situations such as oil and gas exploration and crack damage detection in structural health monitoring, it was rarely discussed in literature. Several techniques have been demonstrated in Φ-OTDR for quantitative vibration [12] and dynamic strain measurement [13] based on measuring the phase differences between two adjacent sections. However, the challenge for dynamic strain measurements lies in the stochastic feature of Φ-OTDR traces, causing the signal fading and large measurement uncertainty [14–16]. Thus, minimizing intensity fluctuation by optimizing the Φ-OTDR system is necessary for precise dynamic strain measurements in Φ-OTDR.

In this paper, we demonstrate an approach for monitoring the dynamic strain induced by sinusoidal piezoelectric vibration signals. The trace-to-trace correlation coefficient is introduced to study the intensity fluctuation of Φ-OTDR traces, evaluate the stability of the sensor system and measurement uncertainty through standard deviation among 100 vibration measurements. Through the introduction to the trace to trace correlation coefficient, we find an important parameter F, which simulates the strength of phase fluctuation, and we link it to the repeatability of the dynamic strain measurement. With our proposed method, the quantitative vibration signal in a certain range determined by the first order Bessel function is identified by monitoring the peak value of fundamental frequency in the fast Fourier transform

(FFT) spectra.

2. Operation principle

The Φ-OTDR can be used in vibration measurements by distinguishing changes in speckle-like traces of Rayleigh backscattered light. Vibration positions along the sensing fiber can be extracted by subtracting the trace sequences, and vibration frequency can be obtained by performing the FFT on the time domain waveform at each vibration location. The Rayleigh backscattered light in a single-mode fiber can be modeled as a sequential series of spatially discrete reflectors. When a pulse of highly coherent light with a central frequency of f_0 is injected into the sensing fiber, the electric field of the Rayleigh backscattered light at the fiber incident end ($z = 0$) at time τ is the sum of the electric field of every Rayleigh scattering centers over half of the pulse duration from $\tau v_g/2-w$ to $\tau v_g/2$, which can be expressed as

$$E_{\tau,z=0} = E_0 \sum_{z_j=(\tau v_g-w)/2}^{\tau v_g/2} r_j e^{i(2\pi f_0\tau+\varphi_j+\delta\varphi_j)} e^{-2\alpha z_j} \quad (1)$$

where E_0 is the initial electric field amplitude, v_g is the group velocity, w is the spatial length of the pulse width, and α is the attenuation coefficient of the fiber. The subscript j represents the jth reflector, and r_j is the reflectance. $\varphi_j = 2\beta z_j$ is the phase where β is the propagation constant. $\delta\varphi_j$ is the time-dependent phase perturbation caused by internal instability of the Φ-OTDR system. External vibration will cause an extra phase change $\varphi_{vib}(t)$ of the detected signal, and the output intensity can be written as [17]

$$I = \sum_{j=1}^{N}(E_0 r_j)^2 e^{-4\alpha z_j} +$$
$$2\sum_{j=1}^{N-1}\sum_{k=j+1}^{N} E_0^2 r_j r_k e^{-2\alpha(z_j+z_k)}\cos(\Phi_{jk}+\delta\Phi_{jk}+\varphi_{vib}(t))$$
$$(2)$$

where $\Phi_{jk} = \varphi_j - \varphi_k$ and $\delta\Phi_{jk} = \delta\varphi_j - \delta\varphi_k$ are respectively the differences between phases of two backscattered signals from jth and kth reflectors, being part of a total number of N Rayleigh scattering centers within the spatial length of a short pulse.

As the cosine voltage is applied to the cylinder lead zirconate titanate (PZT) where a section of sensing fiber is wrapped around, the fiber length as well as the corresponding modal index and birefringence will experience periodic changes because of piezoelectric effect of piezo ceramics. So the phase change induced by the PZT can be written as

$$\varphi_{\text{vib}}(t) = \frac{2\pi nL}{\lambda}\eta V_m \sin(2\pi f_{\text{vib}}t) \qquad (3)$$

where λ is the wavelength of the light source, n is the refractive index of the optical fiber, L is the fiber length wrapped around the PZT, η is the sensitivity coefficient, and V_m is the vibration amplitude. It is indicated that the phase change $\varphi_{\text{vib}}(t)$ is proportional to $V_m \sin(2\pi f_{\text{vib}}t)$. By inserting $\varphi_{\text{vib}}(t)$ into (2) and ignoring the DC-term, the intensity of the Φ-OTDR is an interference signal determined by the AC-term:

$$I \approx 2\sum_{j=1}^{N-1}\sum_{k=j+1}^{N} E_0^2 r_j r_k e^{-2\alpha(z_j + z_k)} \cos\left(\Phi_0 + \varphi_{\text{vib}}(t)\right)$$

$$= 2\sum_{j=1}^{N-1}\sum_{k=j+1}^{N} \Gamma E_0^2 r_j r_k e^{-2\alpha(z_j + z_k)} \begin{bmatrix} \cos(\Phi_0)\cos\left(V_m \sin(2\pi f_{\text{vib}}t)\right) \\ -\sin(\Phi_0)\sin\left(V_m \sin(2\pi f_{\text{vib}}t)\right) \end{bmatrix} \qquad (4)$$

where $\Phi_0 = \Phi_{jk} + \delta\Phi_{jk}$ is a random variable, and Γ is the converting factor which is related to the material and size of PZT itself. Equation (4) can be further expanded by using an infinite series of Bessel functions:

$$I = 2\sum_{j=1}^{N-1}\sum_{k=j+1}^{N} \Gamma E_0^2 r_j r_k e^{-2\alpha(z_j + z_k)} \cdot$$

$$\left\{ \cos(\Phi_0)\left[J_0(V_m) + 2\sum_{n=0}^{\infty} J_{2n}(V_m)\cos(2n \cdot 2\pi f_{\text{vib}}t) \right] \atop -\sin(\Phi_0)\left[2\sum_{n=0}^{\infty} J_{2n+1}(V_m)\sin\left((2n+1)\cdot 2\pi f_{\text{vib}}t\right) \right] \right\}$$

$$= 2\sum_{j=1}^{N-1}\sum_{k=j+1}^{N} \Gamma E_0^2 r_j r_k e^{-2\alpha(z_j + z_k)} \cdot$$

$$\begin{bmatrix} J_0(V_m)\cos(\Phi_0) + \\ J_1(V_m)\cos(\Phi_0 + 2\pi f_{\text{vib}}t) - J_1(V_m)\cos(\Phi_0 - 2\pi f_{\text{vib}}t) + \\ J_2(V_m)\cos(\Phi_0 + 2\pi \cdot 2 f_{\text{vib}}t) + J_2(V_m)\cos(\Phi_0 - 2\pi \cdot 2 f_{\text{vib}}t) + \\ J_3(V_m)\cos(\Phi_0 + 2\pi \cdot 3 f_{\text{vib}}t) - J_3(V_m)\cos(\Phi_0 - 2\pi \cdot 3 f_{\text{vib}}t) + \dots \end{bmatrix} \qquad (5)$$

From (5), the output intensity contains sinusoidal signal components with frequencies $f_{\text{vib}}, 2f_{\text{vib}}, 3f_{\text{vib}}, \dots$ and amplitudes relative to $J_1(V_m), J_2(V_m), J_3(V_m), \dots$, and P_{vib} is the peak value of the FFT spectrum at the vibration frequency f_{vib}. P_{vib} is proportional to the vibration amplitude V_m in a certain range which is determined by the first order Bessel function, thus, V_m can be extracted from the P_{vib} measurement.

3. Φ-OTDR system stability analysis

In Φ-OTDR, system noise is attributed to underlying phase fluctuations originating from various noise sources such as laser frequency drift, phase noise, and thermal noise in electrical and polarization dependence of optical components, and thermal noise in the sensing fiber. The phase fluctuations due to the system noise contribute to intensity fluctuations of Φ-OTDR time domain traces, leading to instability of the sensor system, which can be quantized by an uncertainty of the measured P_{vib}. Thus, analyzing the system stability of Φ-OTDR is important for vibration measurements.

It is noted that the irregular vibrations often occur in nature where they can be decomposed into a multitude of frequency components with unequal cycles. For instance, a piezoelectric transducer tube is often used to generate a single-frequency vibration; however, an imperfect fiber wrapping process can induce non-uniform vibration responses of the sensing fiber. These additional random vibration components may cause an extra phase change $\delta\varphi_{\text{vib}}$ of the detected signal beyond the system noise, contributing to the uncertainty in P_{vib}.

In order to analyze the stability of Φ-OTDR, a trace-to-trace correlation coefficient $R(M)$ is introduced to quantify the intensity fluctuation of Φ-OTDR time domain traces as well as the standard deviation of the measured P_{vib}, which is defined as

$$R(M) = \frac{1}{M-1}\sum_{i=1}^{M-1} c\left[k_i(z), k_{i+1}(z)\right] \qquad (6)$$

where c is the correlation function between two adjacent Φ-OTDR traces, M is the total number of traces, and $k_i(z)$ is the normalized intensity of the ith trace at the position z. The correlation coefficient describes similarity between two time domain traces. If both traces are identical, the correlation coefficient is 1.

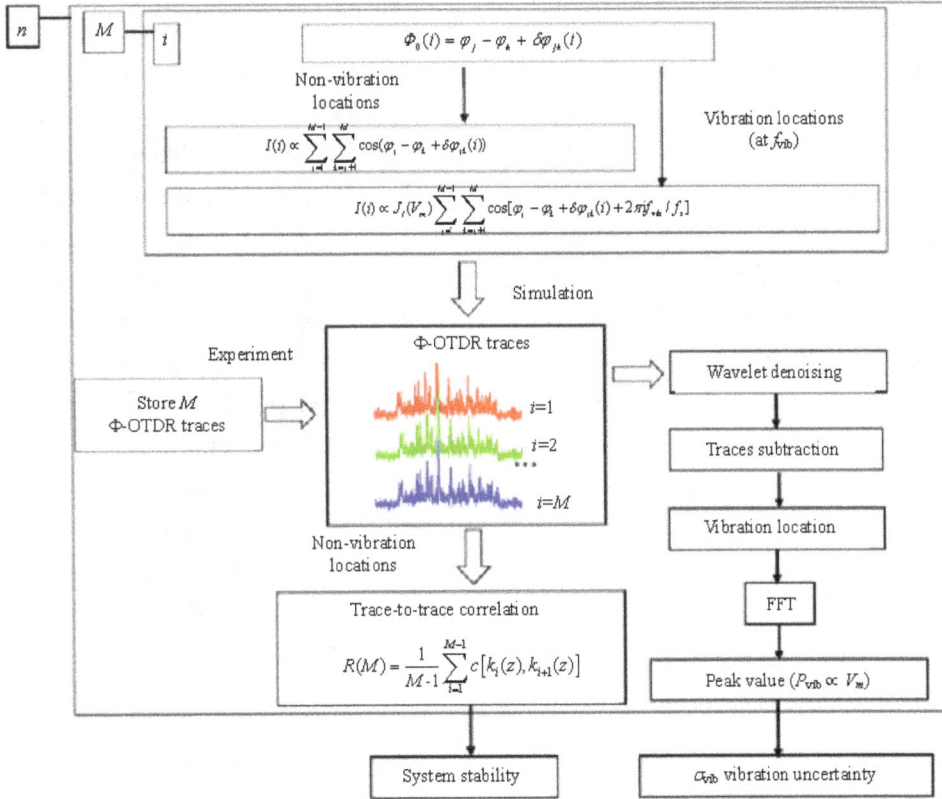

Fig. 1 Flow chart of vibration measurment by using Φ-OTDR.

The system stability is analyzed by calculating R in order to compare with experimental results. The simulation procedure is described in the flow chart of vibration measurement in Φ-OTDR as shown in Fig. 1. In simulation, phase φ of the individual reflector is a random quantity with a uniform probability distribution over the range of $[0, 2\pi]$. The phase fluctuation $\delta\Phi_{jk}$ has the uniform distribution on $F \cdot [0, 2\pi]$, where the coefficient F has a specific value in the range $[0, 1]$. F factor is used to simulate the strength of phase fluctuations which contribute to intensity fluctuations, and it depends on the noise in the Φ-OTDR system. Figure 2 shows simulated trace-to-trace correlation coefficients under different F values based on (6) by using 1000 traces. The simulation result illustrates that R decreases monotonically with increasing F. The larger F value is associated with larger phase fluctuation induced by (1) the laser phase noise; (2) local birefringence of the optical fiber, and both of factors will introduce the fluctuation between different traces. For a given fiber, the 2nd contribution is small and fixed; the only variable is the laser phase noise induced time dependent changes. If the laser is phase locked which has slow time variable in phase, the F factor tends to be small, and hence the trace to trace correlation R will remain high as expected, which will allow for the small error in dynamic strain measurement. To get the stable output for the vibration amplitude measurement, which is a relative measurement, it is important to keep low F value and high R value. The minimum F sets the limit for the measureable phase change due to the dynamic strain. This can only happen when the laser phase noise has slow time dependence and birefringence of the fiber is small, and this assumption is valid for all of the single mode fiber with weak Rayleigh backscattering.

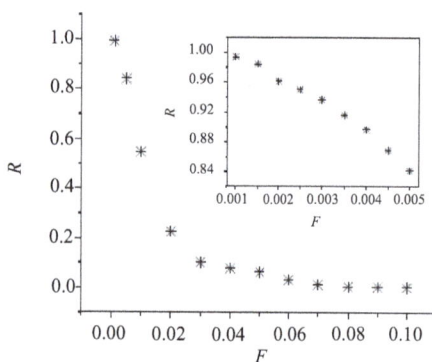

Fig. 2 Simulation results of R as a function of F.

In the experiment, Φ-OTDR traces are acquired and stored in the computer as illustrated in Fig. 1. In order to analyze the system stability, trace-to-trace correlation function is performed at non-vibration locations. The wavelet denoising technique is adopted in the signal post-processing to reduce the random noise, and then the vibration location can be identified by doing traces subtraction. After obtaining the vibration location, the FFT can be performed along traces at the vibration location, and P_{vib} will be recorded. The experiment will be carried out for n times, and the standard deviation σ_{vib} of the measured P_{vib} is also calculated to quantify the measurement uncertainty.

The experimental setup of the Φ-OTDR system is shown in Fig. 3. An external cavity laser source operated at 1550 nm with a narrow line width of 50 kHz is launched into an electro-optic modulator (EOM) controlled by a function generator to generate optical pulses with a repetition rate of 10 kHz and a pulse width of 50 ns. The EOM bias voltage is locked at the minimum point of the transfer function by utilizing a bias controller with a feedback circuit to maintain the high extinction ratio of the optical pulse. The modulated light is amplified by an erbium-doped fiber amplifier (EDFA1) and then sent into a 675 m sensing fiber (Corning, SMF-28e) via a circulator. The Rayleigh backscattered signals are further amplified through EDFA2, and the amplified spontaneous emission (ASE) noise is filtered out by an optical bandpass filter. Then the signals are measured by a photo-detector (PD) and processed by a data acquisition (DAQ) card.

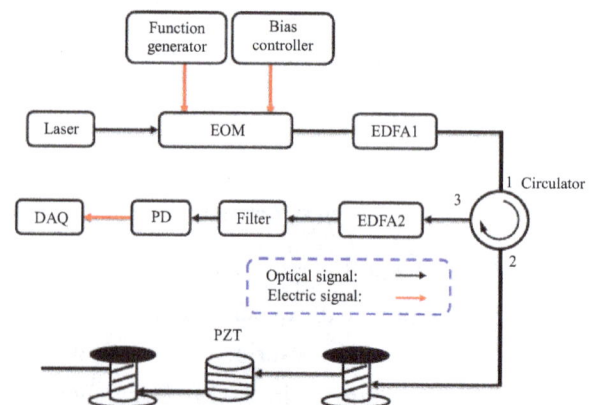

Fig. 3 Experimental setup of the Φ-OTDR system for vibration amplitude measurement.

In the vibration measurement experiment, a section of the sensing fiber is wrapped around a PZT cylinder, and this piezoelectric transducer is driven by a function generator with an output voltage ranging from 0 V to 20 V and a frequency ranging from several Hz up to several kHz.

100 vibration measurements are carried out (n=100) to test the system stability. Consecutive Φ-OTDR raw traces (M = 1000) in one measurement are experimentally obtained, and the first two adjacent traces as well as the last trace are shown in Fig. 4(a). The corresponding intensity fluctuation at non-vibration locations caused by the noise effects can be characterized by calculating R based on (6). Figure 4(b) shows two traces correlation coefficients $R(2)$ in one experiment. The values are calculated between the first trace and ith trace, where i is from 2 to 1000. The length of data used is from the starting point to the end point except for the vibration region, labeled in Fig. 4(a). The correlation coefficients $R(2)$ randomly varies from 0.824 to 0.967. The average correlation coefficient $R(1000)$ for one measurement is 0.902, corresponding to a simulated R value when $F \approx 0.004$, as indicated in the inset of Fig. 2. Figure 4(c) shows a histogram distribution with a bin size of 0.01 of calculated correlation coefficients at non-vibration locations for a set of 100 measurements where the highest probability occurs around $R(1000)$ = 0.903. The high correlation coefficients obtained in the experiment show small intensity fluctuations of Φ-OTDR traces and high stability of sensor system, which is able to get vibration dynamic strain measurement with small uncertainty.

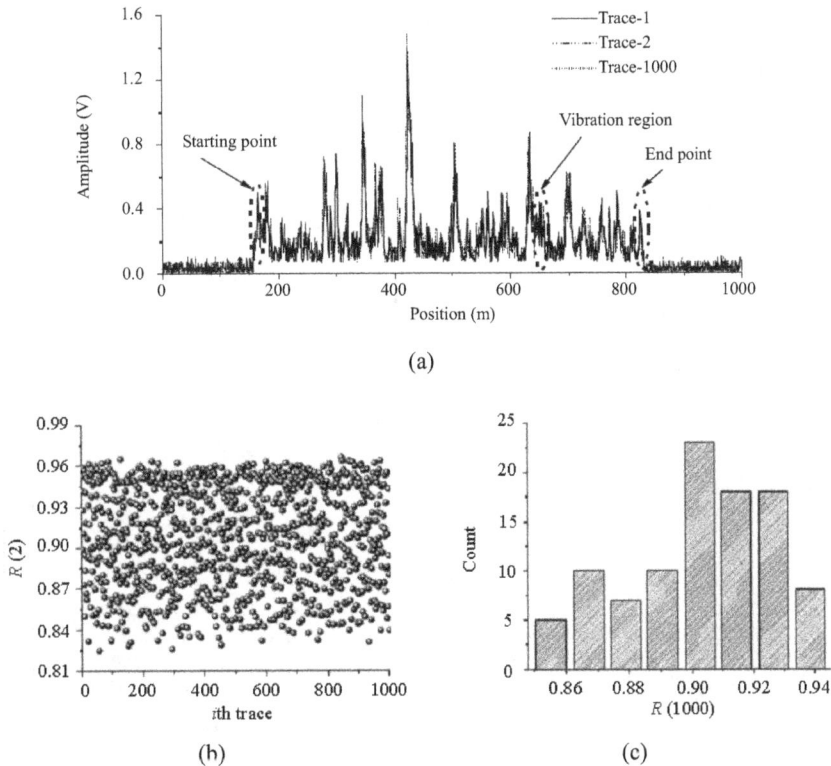

(a)

(b)

(c)

Fig. 4 Correlation coefficient analysis of Φ-OTDR traces: (a) the measured Φ-OTDR traces including a vibration region, (b) the distribution of $R(2)$ in one measurement with 1000 traces, and (c) the histogram distribution of calculated $R(1000)$ at non-vibration locations.

4. Vibration measurement

Φ-OTDR for vibration sensing can be simulated by calculating intensity at the vibration location based on (5), and its procedure is shown in Fig. 1. Figure 5(a) exhibits the first order and second order Bessel functions, $J_1(V_m)$ and $J_2(V_m)$, respectively. The data points of A, B, and C on the curve of the first order Bessel function correspond to those in Figs. 5(b), 5(c), and 5(d) which are simulated power spectra with $V_m = 0.25$ V, 1 V, and 2 V. Since the output of any order Bessel functions would change with different V_m, the vibration dynamic strain can thus be acquired by observing P_{vib} of the corresponding frequency in the power spectra. In Fig. 5(b), there is one peak at the vibration frequency $f_{vib} = 500$ Hz under the driving voltage of 0.25 V. According to (5), harmonics of higher orders in terms of Bessel functions remain and become more and more noticeable for larger V_m values. Another peak at $2f_{vib}$ in Fig. 5(c) and even three other peaks at harmonic frequencies in Fig. 5(d) appear in the power spectra when V_m is further increased. The amplitudes of these peaks are determined by the magnitude of the second order and third order Bessel functions.

The above theoretical and simulation analysis indicates that the detection of V_m becomes available by monitoring relative peak magnitudes at the fundamental frequency and high-order harmonics, however, this work only demonstrates the case that the first order Bessel function plays a dominant role. The 500-Hz vibration event is identified at 490 m as shown in Fig. 6(a). Figures 6(b), 6(c), and 6(d) show the measured power spectra with different driving voltages of $V_m = 1.8$ V, 3 V, and 4 V.

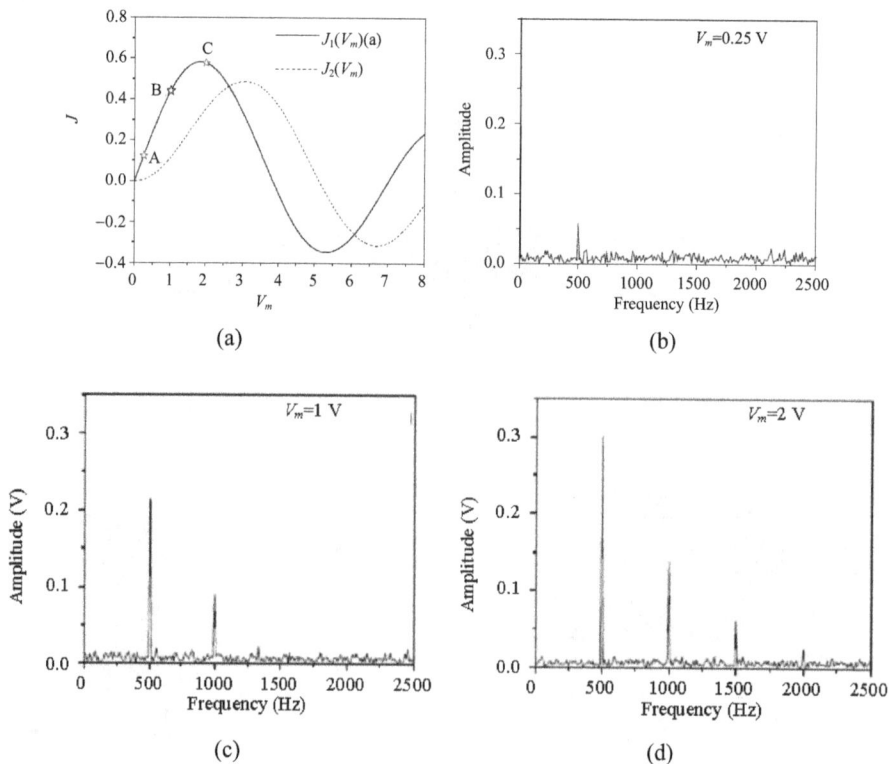

Fig. 5 Simulation results of vibration analysis: (a) the first order and second order Bessel functions. A, B, and C corresponds to those in Figs. 5(b), 5(c), and 5(d), respectively; simulation results of power spectra under different V_m: (b) $V_m = 0.25$ V, (c) $V_m = 1$ V, and (d) $V_m = 2$ V.

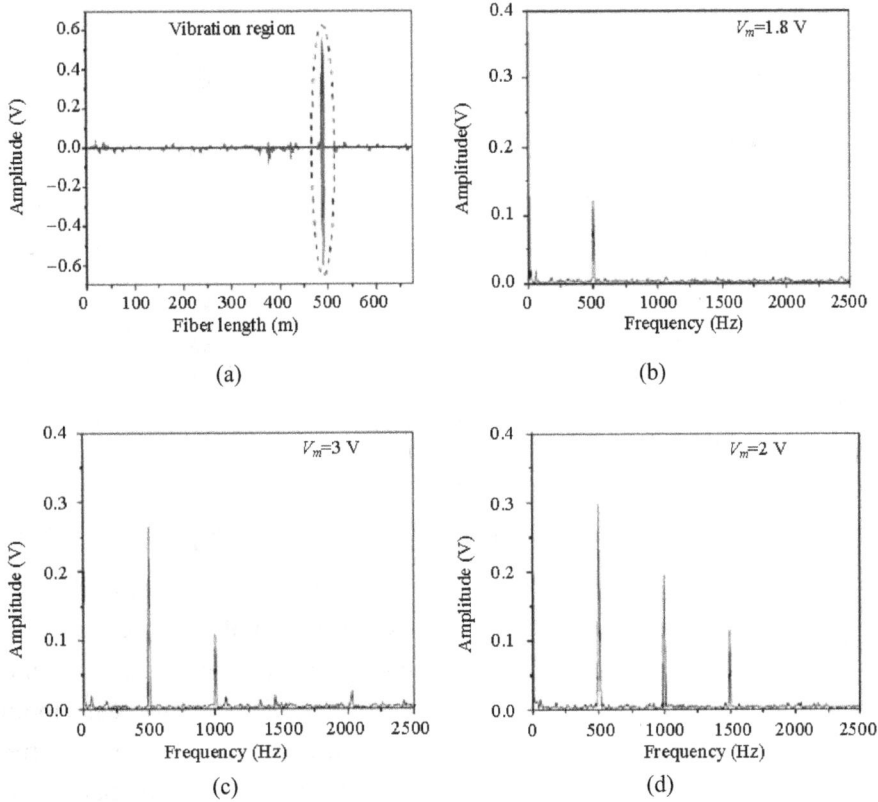

Fig. 6 Experimental results of vibration analysis: (a) vibration location information; experiment results of power spectra under different V_m: (b) $V_m = 1.8$ V, (c) $V_m = 3$ V, and (d) $V_m = 4$ V.

5. Measurement uncertainty analysis

In simulation, the applied vibration signal is modeled as a perfectly sinusoidal signal, and the system noise at the vibration region is assumed to be the same as that at other positions along the sensing fiber. Thus, the standard deviation of peak value σ_{vib} at the vibration frequency depends only on the system noise that can be characterized by the correlation coefficient $R(1000)$. Figure 7(a) shows the simulated σ_{vib} with varying R under $V_m = 0.25$ V, for 100 measurements. Figures 7(b) and 7(c) show the simulation and experimental results of σ_{vib} with different M under $V_m = 0.25$ V and $V_m = 2.2$ V, respectively. The Φ-OTDR system stability analysis at non-vibration locations has indicated that R_{nv} is about 0.9 corresponding to $F_{nv} = 0.004$. However,

the actual measured σ_{vib} [star dot in Fig. 7(c)] is much larger than the simulation result in Fig. 7(b1) and is comparable to the simulation result in Fig. 7(b2) when $F_v = 0.055$ and $R_v = 0.03$. An increase in F and a decrease in R imply that the fiber-wrapped PZT induces imperfect vibration signals and causes undesired phase changes which could further increase the measurement uncertainty of P_{vib}. In addition, σ_{vib} becomes smaller when a larger M is used for the spectrum analysis as shown in Fig. 7(c), and σ_{vib} is 0.0067 when $M = 1000$. This is because that the average random noise is approaching to the same level in each measurement when more traces are used. Thus, 1000 traces are selected in the experiment to achieve a small measurement uncertainty and a short measurement time.

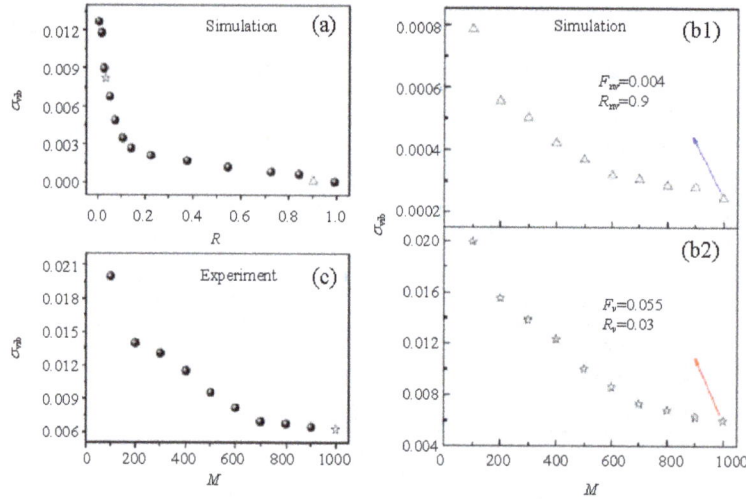

Fig. 7 Simulation result of σ_{vib} as a function of R :(a) and σ_{vib} as a function of M: (b) simulation and (c) experiment.

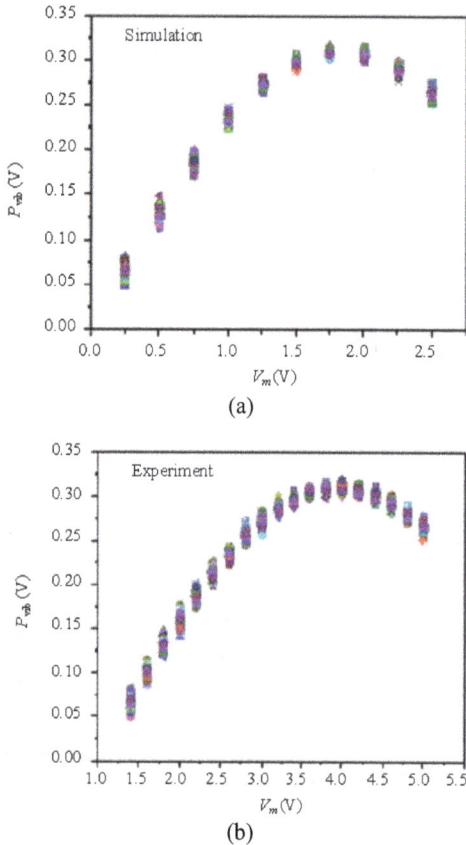

Fig. 8 Change in P_{vib} with different V_m: (a) simulation and (b) experiment.

Figure 8(a) shows the simulation result of P_{vib} with different V_m by using $F_v = 0.055$ and $M = 1000$ to match the experimental result as shown in Fig. 8(b). Both results are obtained with 100 measurements to guarantee reliable and repeatable measurements. The dynamic strain induced by various driving voltages of PZT can be statistically extracted from the average value of P_{vib}. It should be noted that the measured V_m is the actual vibration amplitude of the sinusoidal vibration signal in the experimental result. The corresponding phase change induced by external vibration depends not only on V_m but also on the sensitivity coefficient η of PZT and fiber length wrapped on the PZT. Since our interest is the vibration strength, the simplified model is simulated by only considering V_m as the modulation depth to quantify the vibration amplitude which is an average value rather than a time dependent variable. This results in the different values of V_m between the simulation and experimental results. The experimental results show that the Φ-OTDR system has a high stability and low phase fluctuation because of high correlation coefficients and relative small measurement uncertainty obtained in the experimental results.

6. Conclusions

In conclusion, the high stability Φ-OTDR system has been successfully demonstrated for quantitative vibration measurements. The PZT vibration dynamic strain can be characterized by monitoring the peak values in the power spectra obtained by using the FFT of the time-domain waveform at the vibration location. The trace-to-trace correlation coefficient is introduced to analyze the system noise, and it is also beneficial to the measurement uncertainty analysis. The experimental results demonstrate a good agreement with the simulation results.

Acknowledgment

M. Ren would also like to acknowledge the support from the NSERC CREATE: Sustainable Engineering in Remote Areas (SERA) program, and suggestion from Dr. D. P. Zhou. We like to acknowledge the supports from Canada Research Chairs; Natural Sciences and Engineering Research Council of Canada (NSERC).

References

[1] X. Bao and L. Chen, "Recent progress in Brillouin scattering based fiber sensors," *Sensors*, 2011, 11(4): 4152–4187.

[2] M. A. Soto, G. Bolognini, and F. D. Pasquale, "Enhanced simultaneous distributed strain and temperature fiber sensor employing spontaneous Brillouin scattering and optical pulse coding," *IEEE Photonics Technology Letter*, 2009, 21(7): 450–452.

[3] B. Soller, D. Gifford, M. Wolfe, and M. Foggatt, "High resolution optical frequency domain reflectometry for characterization of components and assemblies," *Optics Express*, 2005, 13(2): 666–674.

[4] D. Zhou, Z. Qin, W. Li, L. Chen, and X. Bao, "Distributed vibration sensing with time-resolved optical frequency-domain reflectometry," *Optics Express*, 2012, 20(12): 13138–13145.

[5] Q. Cui, S. Pamukcu, W. Xiao, and M. Pervizpour, "Truly distributed fiber vibration sensor using pulse base BOTDA with wide dynamic range," *IEEE Photonics Technology Letter*, 2011, 23(24): 1887–1889.

[6] J. C. Juarez, E. W. Maier, K. N. Choi, and H. F. Taylor, "Distributed fiber-optic intrusion sensor system," *Journal of Lightwave Technology*, 2005, 23(6): 2081–2087.

[7] J. C. Juarez and H. F. Taylor, "Field test of a distributed fiber-optic intrusion sensor system for long perimeters," *Applied Optics*, 2007, 46(11): 1968–1971.

[8] F. Peng, H. Wu, X. Jia, Y. Rao, Z. Wang, and Z. Peng, "Ultra-long high-sensitivity Φ-OTDR for high spatial resolution intrusion detection of pipelines," *Optics Express*, 2014, 22(11): 13804–13810.

[9] Y. Lu, T. Zhu, L. Chen, and X. Bao, "Distributed vibration sensor based on coherent detection of phase-OTDR," *Journal of Lightwave Technology*, 2010, 28(22): 3243–3249.

[10] Z. Qin, L. Chen, and X. Bao, "Wavelet denoising method for improving detection performance of distributed vibration sensor," *IEEE Photonics Technology Letter*, 2012, 24(7): 542–544.

[11] T. Zhu, Q. He, X. Xiao, and X. Bao, "Modulated pulses based distributed vibration sensing with high frequency response and spatial resolution," *Optics Express*, 2013, 21(3): 2953–2963.

[12] G. Tu, X. Zhang, Y. Zhang, F. Zhu, L. Xia, and B. Nakarmi, "The development of Φ-OTDR system for quantitative vibration measurement," *IEEE Photonics Technology Letter*, 2015, 27(12): 1349–1352.

[13] A. Masoudi, M. Belal, and T. P. Newson, "A distributed optical fibre dynamic strain sensor based on phase-OTDR," *Measurement Science and Technology*, 2013, 24(8): 085204.

[14] P. Healey, "Fading in heterodyne OTDR," *Electronics Letters*, 1984, 20(1): 30–32.

[15] H. Izumita, S. Furukawa, Y. Koyamada, and I. Sankawa, "Fading noise reduction in coherent OTDR," *IEEE Photonics Technology Letter*, 1992, 4(2): 201–203.

[16] H. Izumita, Y. Koyamada, S. Furukawa, and I. Sankawa, "Stochastic amplitude fluctuation in coherent OTDR and a new technique for its reduction by stimulating synchronous optical frequency hopping," *Journal of Lightwave Technology*, 1997, 15(2): 267–278.

[17] A. K. Wojcik, "Signal statistics of phase dependent optical time domain reflectometry," Ph.D. dissertation, Texas A & M University, College Station, TX, USA, 2006.

11

Numerical Investigation into a Surface Plasmon Resonance Sensor Based on Optical Fiber Microring

Chunliu ZHAO[*], Yanru WANG, Dongning WANG, and Zhewen DING

Institute of Optoelectronic Technology, China Jiliang University, Hangzhou, 310018, China

[*]Corresponding author: Chunliu ZHAO   E-mail: clzhao@cjlu.edu.cn

Abstract: A reflective surface plasmon resonance (SPR) sensor based on optical fiber microring is proposed. In such a sensor, plasmons on the outer surface of the metallized channels containing analyte can be excited by a fundamental mode of a thin-core fiber (TCF). The refractive index (RI) sensing can be achieved as the surface plasmons are sensitive to changes in the refractive index of the analyte. Numerical simulation results show that the resonance spectrum shifts toward the shorter wavelength gradually when the analyte refractive index increases from 1.0 to 1.33, whereas it shifts toward the longer wavelength gradually when the analyte refractive index increases from 1.33 to 1.43, and there is a turning point at the refractive index value of 1.33. The highest sensitivity achieved is up to 2.30×10^3 nm/RIU near the refractive index value of 1.0. Such a compact sensor has potential for gaseous substance monitoring.

Keywords: Fiber optics; surface plasmons; sensors; microstructure

1. Introduction

Surface plasmons propagating at the metal-dielectric interface are extremely sensitive to changes in the refractive index of the dielectric medium. This feature constitutes the core of many surface plasmon resonance (SPR) sensors [1]. Typically, these sensors are implemented in the Kretschmann configuration [2] to direct *p*-polarized light through a glass prism and reflect it from a thin metal (Au and Ag) film deposited on the prism facet. However, this method remains a gold standard for commercial SPR systems, which is bulky, expensive, and not suitable for some applications such as remote sensing. To overcome these limitations, using optical fiber instead of a prism has attracted a lot of research attention due to its advantages of miniaturization, label-free sensing, remote sensing capabilities, and real-time monitoring. The SPR optical fiber sensors have board applications in biology, environment, chemistry, medicine, etc. [3–8]. Over the past decade, many fiber-based SPR sensors have been reported [9–14], including sensors based on side-polished fiber, D-type fiber, tapered fiber, and fiber Bragg gratings. These sensors are based on Kretschmann [2] configuration, to mentallize fiber surfaces and enable evanescent coupling with a plasmon, and one has to first strip the fiber jacket and then etch or taper fiber cladding almost to the core. This laborious procedure compromises fiber integrity, making the resultant sensor be prone to mechanical failures. In recent years, with the rapid development of the micromachining technology, many microstructures on optical fiber have been fabricated and used as microfluidic channels for the analyte measurement. SPR sensors based on microstructured optical fibers (MOFs) have attracted immense research interest worldwide [15–20]. Deposition of metal layers inside of the MOFs can be performed either with the high-pressure chemical vapor deposition technique [21] or with electroless plating techniques used for the fabrication of metallized hollow waveguides and microstructures [22, 23]. Microfluidics on microstructured fibers is enabled by passing analyte though the fiber porous cladding, thus, partially solving the packaging problem.

In this paper, we present a reflective SPR sensor based on an MOF by introducing a microring as the microfluidic channel into the cladding of the thin-core fiber (TCF). A gold layer is deposited on the outer surface of the microring to form the Otto configuration [2]. As far as we know, no experimental observations and theoretical analysis on optical fiber SPR sensors based on the Otto configuration have been reported. The excitation of plasmon resonances in the proposed sensor is investigated numerically using the finite element method (FEM). Numerical simulation results show that such a compact structure can excite SPR effectively. The resonance spectrum shifts toward the shorter wavelength gradually when the refractive index (RI) of the analyte increases from 1.0 to 1.33 and shifts toward the longer wavelength gradually when the RI of the analyte increases from 1.33 to 1.43, and there is a turning point at the RI value of 1.33, which is a phenomenon not observed in other structures. The highest sensitivity achieved is 2.30×10^3 nm/RIU near the RI value of 1.0. Such a fiber microring SPR sensor has a large measurement range, compact size, and a durable structure which can be used in gaseous material detections.

2. Geometry and theory

Figure 1 shows the cross-sections of the proposed SPR sensor based on optical fiber microring. This kind of structure design is based on

[24, 25], in which an air ring is introduced as the microfluidic channel into the cladding of the TCF. A gold layer with thickness 50 nm is desposited on the outer surface of the microring. The reflective type sensor is formed by a layer of gold film with thickness 50 nm coated on the optical fiber end. The sensing head is composed of a core layer, a residual cladding layer, an air layer, a metal layer, and a cladding layer, going from inside to outside. The thickness is 50 nm for both the residual cladding and the air ring. The air ring can be used as a microfluidic channel, in which the tested samples with different RIs can be injected directly. In the simulation, the diameter and the RI of the fiber core are selected as 7 μm and 1.46, respectively, and the cladding RI is 1.45.

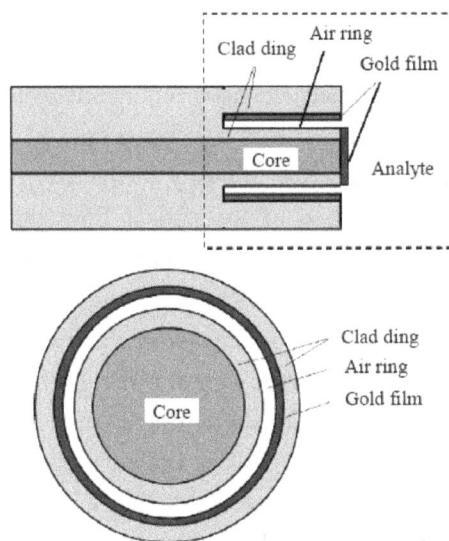

Fig. 1 Cross-sectional structure of the SPR sensor head based on optic fiber microring.

When light propagates in the fiber, due to the reduction of the fiber cladding, the fundamental mode spreads out of the residual cladding region and forms a dynamic evanescent field. In the SPR setup based on the Otto configuration, the surface plasma wave (SPW) is supported by the tail of the evanescent wave at the metal-microring (analyte) interface. Owing to the strong absorption of the metal film, there are lots of plasmon charges at the interface, and the SPW propagates along the metal-microring (analyte) boundary. Furthermore, the SPR can be excited when the light wave vector matches the SPW vector at the plane of $z=0$ [2], leading to local field reinforcement at the interface between the gold film and the sensing sample. When light propagates to the fiber end, the light is reflected due to the presence of the gold layer coated on the fiber end. The reflected light passes through the mental-microring interface again to reinforce the SPR effect. The SPR effect in the surface of the metal film is easy to be influenced by the surrounding medium, and the position of the SPR spectrum is changed with the RI of the analyte injected into the microring. Thus, by detecting the wavelength shift, the RI of the analyte can be measured.

3. Numerical results and discussion

Commercial software COMSOL Multiphysics with perfectly matched layer boundaries based on FEM was used to analyze the proposed sensor. Figure 2 illustrates the intensity of the electric field in the cross section of the optical fiber microring structure using the study "mode analyses" from COMSOL Multiphysics. The wavelength of the light is 460 nm. As shown in Fig. 2(a), the electric field intensity at the interface between the gold film and the microring is obviously reinforced when the gold film is deposited and SPR is excited (the thickness of Au film=50 nm, analyte RI n_a=1.33), while in Fig. 2(b), no SPR is excited as the absence of the gold film. It means that the proposed sensing structure is advisable to excite the SPR effect.

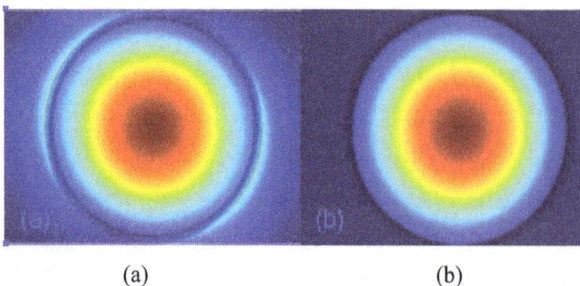

(a)   (b)

Fig. 2 Electric field distribution in the cross section of the fiber microring at the wavelength λ=460 nm: (a) when the SPR is excited with the gold film thickness of 50 nm and analyte RI is 1.33 and (b) no SPR is excited in the absence of the gold film.

The performance of the sensor depends on its structural parameters. According to the SPR theory based on the Kretschmann configuration, due to the radiation damping, if the thickness of the metal is too small, the SPW at the metal surface will be strongly damped. While the thickness of the metal is too large, the SPW can no more be efficiently excited because of the strong absorption of the metal [2]. Thus, the thickness of the Au film has significant effect on the surface plasmon wave excitation. To observe the effect of Au film thickness on the sensor performance, the Au film thickness is varied from 40 nm to 70 nm while other parameters remain as constant (residual cladding thickness= 50 nm, air ring thickness=50 nm, and n_a=1.0) in the simulation. When using COMSOL Multiphysics software, the electric field intensity (E) at different wavelengths can be obtained. The corresponding intensity (I) is obtained by the following equation

$$I(w) = \frac{1}{2} c \varepsilon_0 E^2 \qquad (1)$$

where c is the speed of light; ε_0 is the permittivity of vacuum; E is the electric field intensity. And then the intensity in dBm is transformed form the following formula

$$I(dBm) = 10 \cdot \lg \left[\frac{I(w)}{1mw} \right]. \qquad (2)$$

All calculations and SPR spectra are performed by using Matlab. The SPR spectra corresponding to different Au film thicknesses are shown in Fig. 3, where it can be observed that the resonance wavelength shifts toward longer wavelength with an increase in thickness of the Au film. This indicates that due to an increase in Au film thickness, light penetration through the cladding decreases. The gold film thickness of 50 nm is selected for the RI sensor simulation.

Besides the Au film thickness, effects of the microring thickness on the SPR spectrum has also been studied (residual cladding thickness=50 nm, Au film thickness=50 nm, and $n_a = 1.0$), and the relevant graphs obtained are shown in Fig. 4, where it can be seen that the resonance peak shifts toward longer wavelength when the microring thickness increases from 20 nm to 60 nm. The air ring thickness of 50 nm is selected for the RI sensor simulation.

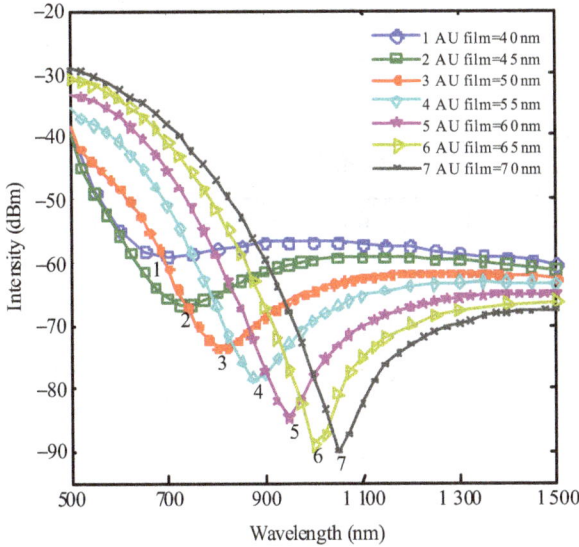

Fig. 3 SPR spectra with different thicknesses of gold film (residual cladding thickness=50 nm, air ring thickness =50 nm, and n_a=1.0).

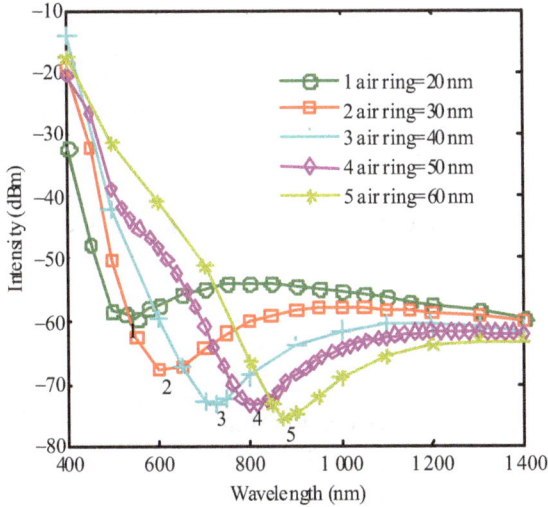

Fig. 4 SPR spectra with different thicknesses of air ring (residual cladding=50 nm, Au film=50 nm, and n_a=1.0).

In addition, the influence of the thickness of the remaining cladding near the fiber core (Au film thickness=50 nm, air ring thickness=50 nm, and $n_a = 1.0$) on the SPR resonance wavelength is demonstrated in Fig. 5. It can be seen from Fig. 5 that the thickness of the remaining cladding near the fiber core has little effect on the SPR spectrum.

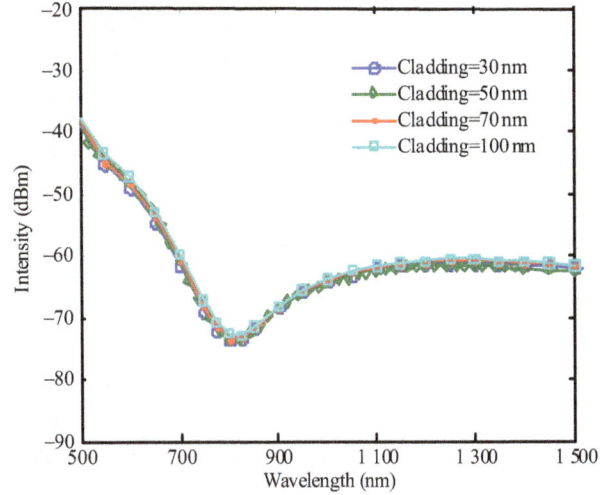

Fig. 5 SPR spectra with different thicknesses of the remaining cladding near the fiber core (Au film thickness =50 nm, air ring thickness =50 nm, and n_a=1.0).

The results of the resonant wavelength shift obtained by varying the analyte RI from 1.0 to 1.43 are revealed in Fig. 6, where the thicknesses of the residual cladding, gold film, and the microring are all equal to 50 nm. It can be observed from Fig. 6 that there is a turning point in the change of resonance wavelength with the RI at the RI value of 1.33. In Fig. 6(a), the resonance spectrum shifts toward the shorter wavelength gradually when the analyte RI increases from 1.0 to 1.33. The relationship between the resonant wavelength and the analyte RI in the range of 1.0–1.3 is shown in Fig. 6(b). The corresponding regression equation is

$$y = 3.179\times10^3 x^2 - 8.486\times10^3 x + 6.117\times10^3$$

where y is the resonant wavelength in nm, and x is the analyte RI (n_a). The resonant wavelength shifts from 815 nm to 700 nm when the n_a varies from 1.0 to 1.05. The positive RI sensitivity (S) obtained is 2.30×10^3 nm/RIU, determined by the following equation

$$S = \left|\frac{d\lambda_R}{dn_a}\right| \qquad (3)$$

where $d\lambda_R$ is the resonant peak shift, and dn_a is the analyte RI variation. The proposed sensor shows the sensitivity of 2300 nm/RIU, 1600 nm/RIU, 1200 nm/RIU, and 600 nm/RIU for analyte RI variation ranges of 1.0 to 1.05, 1.05 to 1.1, 1.1 to 1.2, and 1.2 to 1.33, respectively.

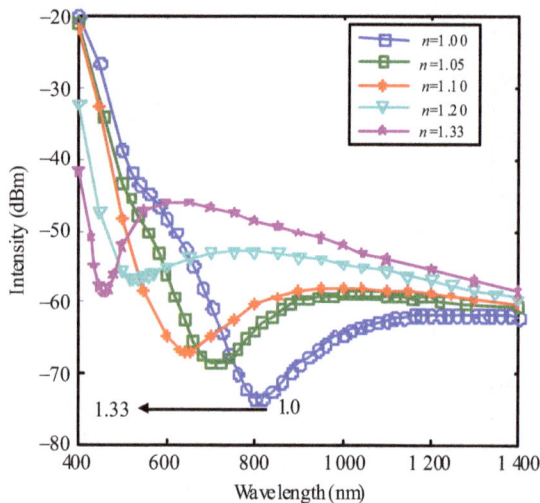

(a)

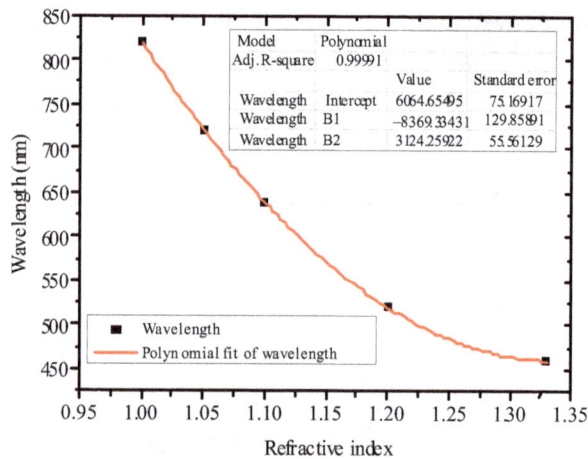

(b)

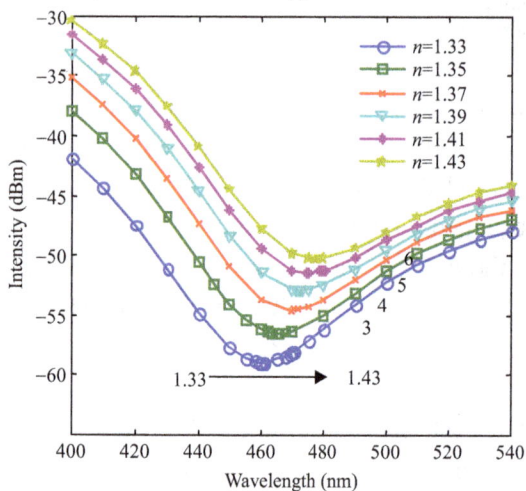

(c)

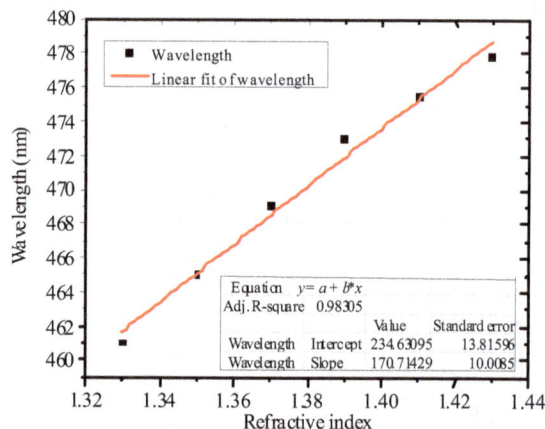

(d)

Fig. 6 SPR spectra with analyte RI variation: (a) from 1.0 to 1.33, (c) from 1.33 to 1.43, and (b) and (d) the relationship between resonant wavelength and analyte RI.

It can be seen from Fig. 6(c) that the resonance spectrum shifts toward the longer wavelength gradually when the analyte RI increases from 1.33 to 1.43. Figure 6(d) shows the linear line fitting of the resonant wavelength with respect to the analyte RI, and the regression equation is

$$y = 170.71x + 234.63, \ 1.33 \le x \le 1.43$$

where y is the resonant wavelength of the analyte in nm, and x is the analyte RI. The average sensitivity is 170 nm/RIU, and R^2 is 0.9830, indicating a good fitting of the sensor response.

Theoretical investigation and numerical analysis have been carried out to explain the existence of the turning point in the resonance wavelength variation with the RI change. According to the basic theory, the SPR effect in the interface occurs as the light wave vector matches the SPW vector at the plane of $z=0$. The relationship between the real part of effective RI (n_{eff}) of the core-guided mode and wavelength with the RI in the range of 1.31 to 1.35 is shown in Fig. 7, where it can be seen that the effective RI decreases with an increase in wavelength. The effective RI increases with an increase in analyte RI from 1.31 to 1.33 and decreases with an increase in analyte RI from 1.33 to 1.35 when the wavelength is in the range of

450 nm to 480 nm. Due to the small change in analyte RI, the real part of the effective RI of core-guided mode changes, which causes the change in the phase matching wavelength between the core guided mode and the SPW mode. An increase in the effective RI of the core indicates that less light is coupled to the air layer and the metal interface, which makes the resonant wavelength shift to the shorter wavelength. The reduction of the effective RI of the core-guided mode indicates that more light is coupled to the air layer and the metal interface, which makes the resonant wavelength shift to the longer wavelength. Thus, an abrupt change appears in the effective RI and results in a turning point in the resonant wavelength variation with analyte RI. The electric field intensity distributions with the RI values of 1.31, 1.33, and 1.35 are shown in Fig. 8, the red checkmark represents the wavelength in which the electric field intensity of the core reaches

the minimum value, and the strongest resonance effect is achieved. It can be seen from Fig. 8 that the resonance wavelengths at the RI values of 1.31, 1.33, and 1.35 are 470 nm, 460 nm, and 465 nm, respectively, and there is a turning point of the resonance wavelength at the RI value of 1.33.

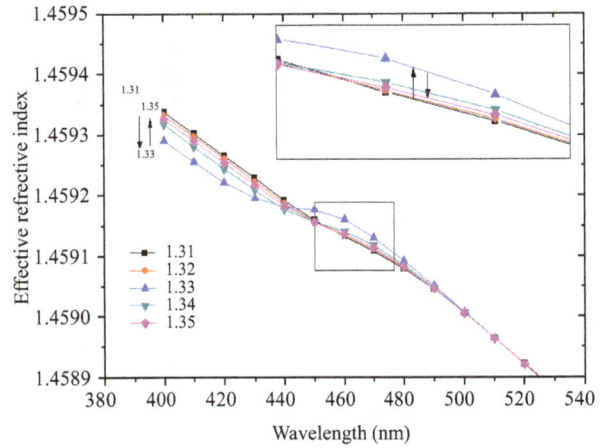

Fig. 7 Effective RI with analyte RI variation from 1.31 to 1.35.

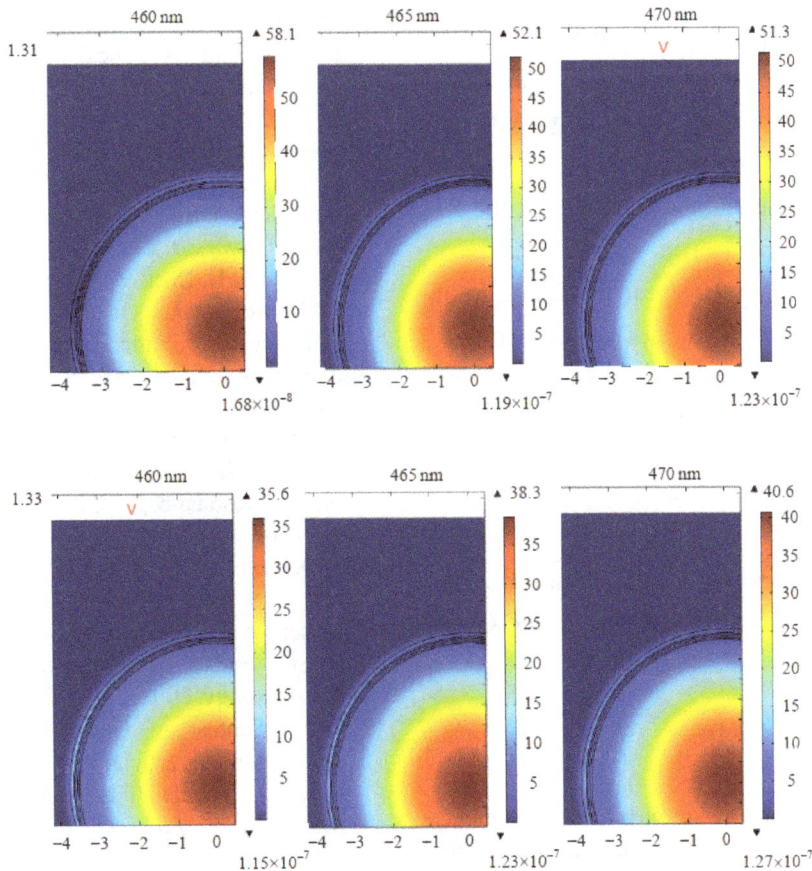

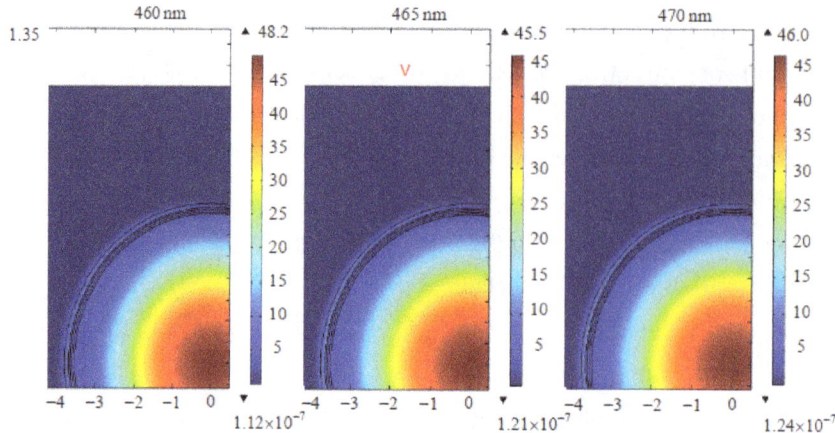

Fig. 8 Electric field intensity distributions with the refractive index are 1.31, 1.33, and 1.35.

4. Conclusions

In this paper, we propose a reflective type SPR RI sensor based on optical fiber microring. The sensing performance and the effects of the geometrical parameters on the resonant spectrum are investigated by use of the FEM method. Simulation results show that the monitoring range of analyte RI is from 1.0 to 1.43. The resonance spectrum shifts toward the shorter wavelength gradually with an increase in analyte RI from 1.0 to 1.33 and shifts toward the longer wavelength gradually with an increase in analyte RI from 1.33 to 1.43, and a turning point in the resonance wavelength at the RI value of 1.33 is observed. The highest sensitivity can be up to 2.30×10^3 nm/RIU near the RI value of 1.0. The sensor has a large measurement range, compact size, and durable structure, and it can be used in gaseous material detections.

Acknowledgment

This work was supported by the Natural Science Foundation of Zhejiang Province China under Grant No.LY17F050010.

References

[1] D. R. Tilley, "Surface polaritons: electromagnetic waves at surfaces and interfaces," *Journal of Modern Optics*, 1983, 30(11): 1501–1506.

[2] J. Homola, S. S. Yee, and G. Gauglitz, "Surface plasmon resonance sensors: review," *Sensors and Actuators B: Chemical*, 1999, 54(1): 3–15.

[3] Y. Wang, S. Meng, Y. Liang, L. Li, and W. Peng, "Fiber-optic surface plasmon resonance sensor with multi-alternating metal layers for biological measurement," *Photonic Sensors*, 2013, 3(3): 202–207.

[4] C. Perrotton, R. J. Westerwaal, N. Javahiraly, M. Slaman, H. Schreuders, B. Dam, *et al.*, "A reliable, sensitive and fast optical fiber hydrogen sensor based on surface plasmon resonance," *Optics Express*, 2013, 21(1): 382–390.

[5] S. K. Srivastava, R. Verma, and B. D. Gupta, "Surface plasmon resonance based fiber optic sensor for the detection of low water content in ethanol," *Sensors and Actuators B: Chemical*, 2011, 153(1): 194–198.

[6] A. Nooke, U. Beck, A. Hertwig, A. Krause, H. Krüger, V. Lohse, *et al.*, "On the application of gold based SPR sensors for the detection of hazardous gases," *Sensors and Actuators B: Chemical*, 2010, 149(1): 194–198.

[7] M. Xue, Q. Jiang, C Zhang, and J. Lin, "A kind of biomolecular probe sensor based on TFBG surface plasma resonance," *Photonic Sensors*, 2015, 5(2): 102–108.

[8] J. Homola, J. Dostalek, S. Chen, A. Rasooly, S. Jiang, and S. S. Yee, "Spectral surface plasmon resonance biosensor for detection of staphylococcal enterotoxin B in milk," *International Journal of Food Microbiology*, 2002, 75(1): 61–69.

[9] D. F. Santos, A. Guerreiro, and J. M. Baptista, "Numerical investigation of a refractive index SPR D-type optical fiber sensor using COMSOL Multiphysics," *Photonic Sensors*, 2013, 3(1): 61–66.

[10] H. Y. Lin, C. H. Huang, G. L. Cheng, N. K. Chen, and H. C. Chui, "Tapered optical fiber sensor based

on localized surface plasmon resonance," *Optics Express*, 2012, 20(19): 21693–21701.

[11] L. C. C. Coelho, J. M. M. M. D Almeida, H. Moayyed, J. L. Santos, and D. Viegas, "Multiplexing of surface plasmon resonance sensing devices on etched single-mode fiber," *Journal of Lightwave Technology*, 2015, 33(2): 432–438.

[12] J. Zhu, L Qin, S. Song, J. Zhong, and S. Lin, "Design of a surface plasmon resonance sensor based on grating connection," *Photonic Sensors*, 2015, 5(2): 159–165.

[13] R. Verma, A. Sharma, and B. Gupta, "Modeling of tapered fiber-optic surface plasmon resonance sensor with enhanced sensitivity," *IEEE Photonics Technology Letters*, 2007, 19(22): 1786–1788.

[14] A. Jian, L. Den, S. Sang, Q. Duan, X. Zhang, and W. Zhang, "Surface plasmon resonance sensor based on an angled optical fiber," *IEEE Sensors Journal*, 2014, 14(9): 3229–3235.

[15] A. Hassani and M. Skorobogatiy, "Design of the microstructured optical fiber-based surface plasmon resonance sensors with enhanced microfluidics," *Optics Express*, 2006, 14(24): 11616–11621.

[16] M. Hautakorpi, M. Mattinen, and H. Ludvigsen, "Surface-plasmon-resonance sensor based on three-hole microstructured optical fiber," *Optics Express*, 2008, 16(12): 8427–8432.

[17] M. Erdmanis, D. Viegas, M. Hautakorpi, S. Novotny, J. L. Santos, and H. Ludvigsen, "Comprehensive numerical analysis of a surface-plasmon-resonance sensor based on an H-shaped optical fiber," *Optics Express*, 2011, 19(15): 13980–13988.

[18] X. Yu, Y. Zhang, S. Pan, P. Shum, M. Yan, Y. Leviatan, *et al.*, "A selectively coated photonic crystal fiber based surface plasmon resonance sensor," *Journal of Optics*, 2009, 12(1): 74–77.

[19] Y. Lu, C. J. Hao, B. Q. Wu, M. Musideke, L. C. Duan, W. Q. Wen, *et al.*, "Surface plasmon resonance sensor based on polymer photonic crystal fibers with metal nanolayers," *Sensors*, 2013, 13(1): 956–965.

[20] A. K. Mishra, S. K. Mishra, and B. D. Gupta, "Gas-clad two-way fiber optic SPR sensor: a novel approach for refractive index sensing," *Plasmonics*, 2015, 10(5): 1071–1076.

[21] P. J. A. Sazio, A. Amezcua-Correa, C. E. Finlayson, J. R. Hayes, T. J. Scheidemantel, N. F. Baril, *et al.*, "Microstructured optical fibers as high-pressure microfluidic reactors," *Science*, 2006, 311(5767): 1583–1586.

[22] J. A. Harrington, "A review of IR transmitting, hollow waveguides," *Fiber & Integrated Optics*, 2000, 19(3): 211–227.

[23] N. Takeyasu, T. Tanaka, and S. Kawata, "Metal deposition deep into microstructure by electroless plating," *Japanese Journal of Applied Physics*, 2005, 44(35): L1134–L1137.

[24] Y. Zhu, M. Cheng, H. Wang, Y. Zhang, and J. Yang, "Design of a surface-plasmon-resonance sensor based on a microstructured optical fiber with annular-shaped holes," *Plasma Science and Technology*, 2014, 16(9): 867–872.

[25] A. Hassani and M. Skorobogatiy, "Design criteria for microstructured-optical-fiber based surface plasmon resonance sensors," *JOSA B*, 2007, 24(6): 1423–1429.

12

Experimental Analysis of Beam Pointing System Based on Liquid Crystal Optical Phase Array

Yubin SHI[*], Jianmin ZHANG, and Zhen ZHANG

State Key Laboratory of Laser Interaction with Matter, Northwest Institute of Nuclear Technology, Xi'an, 710024, China

[*]Corresponding author: Yubin SHI   E-mail: shiyubin@nint.ac.cn

Abstract: In this paper, we propose and demonstrate an elementary non-mechanical beam aiming and steering system with a single liquid crystal optical phase array (LC-OPA) and charge-coupled device (CCD). With the conventional method of beam steering control, the LC-OPA device can realize one dimensional beam steering continuously. An improved beam steering strategy is applied to realize two dimensional beam steering with a single LC-OPA. The whole beam aiming and steering system, including an LC-OPA and a retroreflective target, is controlled by the monitor. We test the feasibility of beam steering strategy both in one dimension and in two dimension at first, then the whole system is build up based on the improved strategy. The experimental results show that the max experimental pointing error is 56 μrad, and the average pointing error of the system is 19 μrad.

Keywords: Lasers and laser optics; laser beam shaping; liquid crystals; optical engineering

1. Introduction

Laser acquisition, tracking, pointing (ATP) system has played an important role in free space optical communication, laser radar, and other applications for a long while [1–3]. It is well known that traditional ATP systems based on the mechanical mirrors and gimbals are complex and expensive, with relatively large volumes and high weight. For the traditional ATP systems, beam steering and stabilizing are still major limitations [4]. However, optical phased array techniques can avoid these problems effectively. In recent years, liquid crystal optical phased array (LC-OPA) techniques have been widely developed and have been considered to be potential in several applications such as laser steering, tracking, and optical tweezers [5–8].

Methods of improving LC-OPA beam steering efficiency, steering angle, and pointing (steering) accuracy have been reported [9–11], however, little work has been done to test and verify the steering performances of LC-OPA in a beam steering system.

In this paper, we present a beam steering system with a single LC-OPA device, which can realize two-dimensional continuous beam steering within a max steering angle. This paper is organized as follows. Firstly, the experimental setup is introduced. Secondly, the conventional and improved beam steering strategy of LC-OPA is described. Then, a numerical simulation of two dimensional beam steering and a beam pointing experiment based on LC-OPA are reported, and the steering error is analyzed. Finally, the whole system based on the steering strategy is established, and the performance of the system is measured.

The schematic diagram for beam aiming and pointing system is shown in Fig. 1. The essential instruments are LC-OPA (BNS Company, LC-OPA with 256×256 24 μm×24 μm pixels). The light source used is a 1064-nm Nd: YVO$_4$ polarized laser beam. A telescope is employed for beam expanding and collimating. A polarized beam splitter (PBS1), a half-wave plate, and a polarized beam splitter (PBS2) are used together to provide the LC-OPA with s-linearly polarized laser beam, which makes it possible to adjust energy of the laser. Beam splitter (BS1) is utilized for the beam transmitted from the LC-OPA vertically. The emission beam is separated by the beam splitter (BS2) in two parts. Most of the beam is reflected to the target in far field. The last part is incident on a retro-reflector and then focused on the charge-coupled device (CCD) by a lens with 12.5-mm focal length. The CCD camera (Dolphin F-145B, 15 Hz) with 1280×960 6.45 μm×6.45 μm pixels is used to diagnose the far-field intensity distribution of the output beam and the far-field target. Since we hope that the beam efficiency is high in our system, the splitting ratio (index of transmission: index of reflection) of BS1 should be 5:5 to guarantee the maximum energy of beam, and the reflection of the BS2 should be close to 1 (the beam still need to be transmitted as a small signal). In the experiment, the splitting ratio (index of transmission: index of reflection) of the BS1 is 44:56, and the splitting ratio of the BS2 is 84:16.

The CCD camera is directly interfaced by using an IEEE 1394 (Firewire) connection to computer. With the CCD detector and the retro-reflector, the position information between the target and the laser is acquired, and then the tilt phase distributions which loaded on LC-OPA can be calculated by the computer. This experiment is done at an indoor laboratory facility, and the distance between the BS2 and the target is approximately 2 m.

The beam steering method is verified, and the beam steering accuracy is measured with CCD image system in front of the BS2. On the basis of the results, the beam pointing experiment is done further.

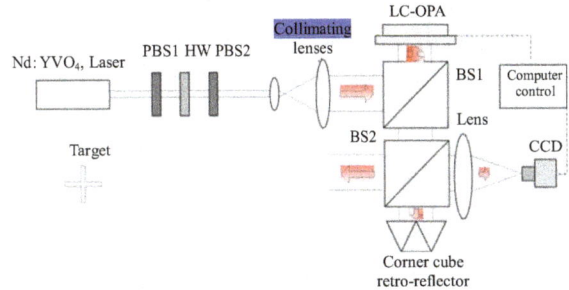

Fig. 1. Schematic diagram of laser aiming and pointing experiment system based on LC-OPA. Both BS1 and BS2 are the non-polarization beam splitters, and BS1 is used to ensure that the transmitted beam falls on LC-OPA perpendicularly, and BS2 is used to ensure that part of the beam enters into CCD.

According to the conventional method, when the phase modulation of LC-OPA is periodic, the steering angle of the incident beam can be given by [5]

$$\theta = \arcsin(\lambda / Nd) \qquad (1)$$

where θ is the deflection angle, λ is the working wavelength, N is the number of phase shifters (or called pixels) in one period, and d is the size of a single phase shifter. In addition, the largest realizable deflection angle θ_{max} is $\arcsin(\lambda/2d)$. However, periodic phase modulation of LCOPA cannot realize continuous beam deflection angle within $\pm \theta_{max}$. Based on the theory above, Engström provided a non-periodic method to realize one dimensional continuous beam steering strategy in the first part [9], which can be summarized as:

$$\phi_j(\theta) = \text{round}\left[\phi_{j,\,\text{ideal}}(\theta)\cdot M/2\pi\right]\cdot 2\pi/M \qquad (2)$$

where $\phi_{j,\,\text{ideal}}$ is the ideal staircase phase related to the deflection angle of j pixel, $j=1, 2, \cdots, N$, and N is the total number of pixels. M is the equidistant phase level between 0 and 2π, and round simply round the value to the closest integer value. From (2), two dimensional (2D) beam steering method can be derived further:

$$\phi_{j,\,\text{final}} = \text{imrotate}\left[\phi_j(\theta), \Theta\right] \qquad (3)$$

where imrotate represents rotate ϕ_j with angle Θ conter-clockwise, Θ is arbitrary number between 0 and 2π, and $\phi_{j, \text{final}}$ is the modulated phase for j pixel for 2D beam steering.

Compared with the conventional 2D beam steering strategy, elevation angle α and azimuth angle β are always used to describe the deflection angle. Here the relation between (θ, Θ) and (α, β) is summarized as follows:

$$\theta = \arctan\left[\sqrt{\tan^2(\alpha) + \tan^2(\beta)}\right] \qquad (4)$$

and

$$\Theta = \text{arc} \cot\left[\tan(\beta)/\tan(\alpha)\right]. \qquad (5)$$

The numerical simulation results are done with pixel size 24 µm and wavelength of 1064 nm. The waist of the Gauss beam is 0.3 mm, and the focus of the lens is 0.1 m. We assume there are no fringing fields degrading the performance in the numerical simulation.

The numerical results show the relationship between the aiming angle and the realized angle in Fig. 2. There are some fluctuations in the linear relationship between the aiming angle and steering angle, and this phenomenon is inevitable due to the quantized staircase phase of an LC-OPA. We call this error as a theoretical error. The zeroth diffraction order is due to the unity fill factor of pixels in an LC-OPA device (active area of the pixel is smaller than the separation of pixels). Furthermore, the normalized steering error is conventionally defined as

$$\varepsilon = |\theta - \theta_{\text{ideal}}|/\theta_{\text{spot}} \qquad (6)$$

where θ_{ideal} is the aiming angle, θ is the realized steering angle, and θ_{spot} is the diffraction limited spot size of the beam [6]. The experimental steering error at first is given in Fig. 3. As shown in Fig. 3, the steering error is symmetric around 0 mrad, and it appears that the steering error is periodic within the max steering angle. Also, it is obvious that the fluctuation of the error become bigger as the steering angle becomes larger. It suggests that we had better use the beam steering strategy in middle of the field of view (FOV) in a beam steering system.

Theoretically, the pointing error can be explained as the results of quantized staircase phase of an LC-OPA and fringe effect. However, phase aberration in the system influences the steering accuracy of the system, especially phase tilt. The results also infer that the strategy is relatively stable in a small FOV.

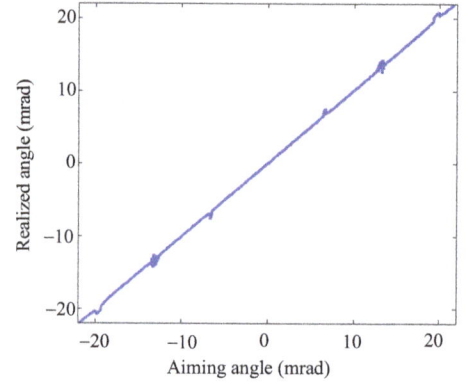

Fig. 2 Numerical simulation result of relationship between realized angle and steering angle.

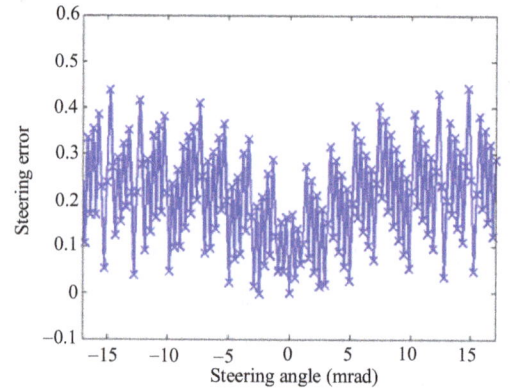

Fig. 3 Experimental result of relationship between beam steering error and the steering angle.

Figure 4 shows the results of two-dimensional beam steering. Four different values of Θ are given, when θ equals 10 mrad. The original beam is deflected to different directions. The results show that the 2D beam strategy is feasible. However, there are several diffraction orders. It can be attributed to the non-unity fill factor of spatial light modulator (SLM).

We also define the beam steering errors of Θ as

$$\gamma = (\Theta - \Theta_{\text{ideal}})/\theta_{\text{spot}} \qquad (7)$$

where Θ_{ideal} is the desired steering angles, and Θ is the realized steering angle. The beam steering error of Θ is shown in Fig. 5. It shows that the steering error γ is bigger as the steering angle becomes bigger.

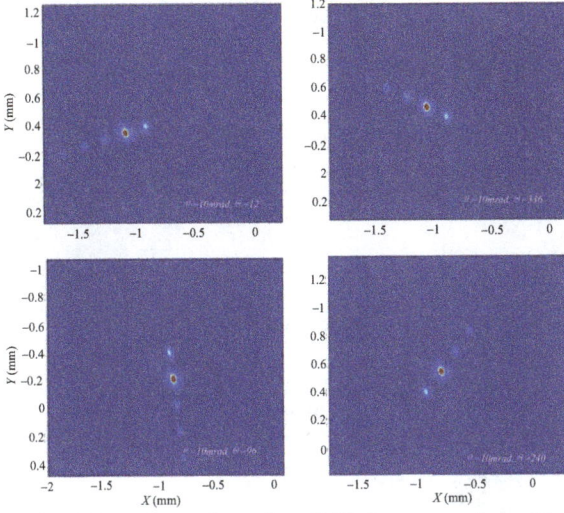

Fig. 4 Experimental results of 2D beam steering with $\theta=$ 10 mrad.

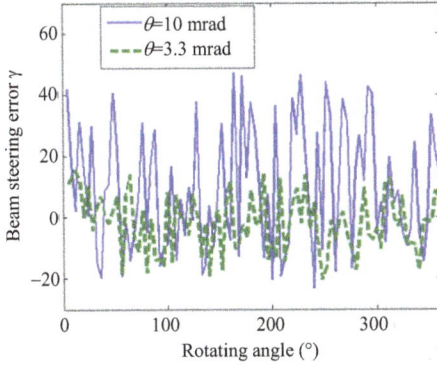

Fig. 5 Steering error of 2D beam steering with $\theta=$10 mrad and $\theta=$3.3 mrad.

After verifying the feasibility of the 2D beam steering method, we combine the 2D beam steering strategy and the LC-OPA together in order to realize beam aiming and steering in an electro-optical system.

The images of retroreflective target are shown in Fig. 6. In order to evaluate the aiming and pointing error, a cross target about 5 mm wide and 5 mm high is used. When the target is illumined by light, the target can be acquired by the CCD camera and the image is shown in Fig. 6(a). When it is illumined by the transmitted laser, the image of the target is shown in Fig. 6(b).

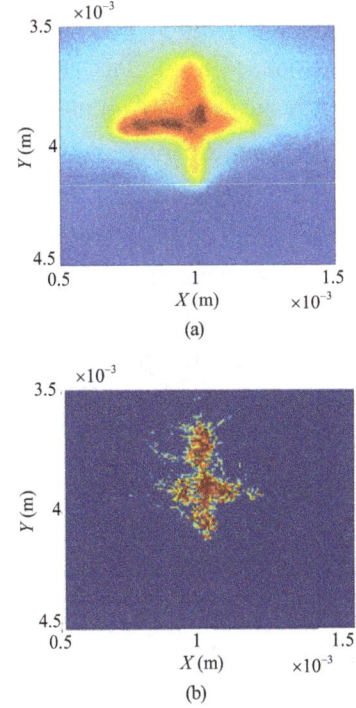

Fig. 6 Images of cross target: (a) image of target without lasers illuminating on it and (b) image of target with lasers illuminating on it (the background noise has been removed).

When the target is moved by the stepper motor in two directions, the centroid of target can be calculated from the target image. If we adjust the corner cube retro-reflector properly, the beam which is incident on it can be regarded as a beacon to calculate the steering angle. Then the phase distribution can be calculated further and can be loaded on the LC-OPA to realize beam steering. Here the pointing error is defined as

$$\text{error}_p = \left\{\left[(x_{\text{retro}} - x_{\text{tar}})^2 + (y_{\text{retro}} - y_{\text{tar}})^2\right]^{0.5} \cdot d_{\text{CCD}}\right\}/L \quad (8)$$

where x_{tar} and y_{tar} are the centroid coordinates of the target, x_{retro} and y_{retro} are the centroid coordinates of the retro-reflect beam, L is the distance between the target and BS2, and d_{CCD} is the pixel size of the CCD. When the target is moving mainly along horizontal direction and vertical direction, the results are recorded and shown in Figs. 7 and 8.

Compare Fig. 7(a) with Fig. 7(b), we can see that the target is moving along X axis (horizontal) from the variation of vertical coordinate. Even though the target is moving along one single direction, another direction will produce jitter in fact. The reason why this happens can be ascribed to two main points. Firstly, the 1D beam steering method has errors between the aiming angle and the realized angle in theory. Secondly, influence factors cannot be avoided in the experiment, such as the target position, target shape, *et al*. As a result of what have been discussed above, changes in both directions should be taken into consideration, and both will cause pointing error in the system. From these result, if the centroid of the retro-reflect beam is not zero, it shows that the beam steers to the target. But there is difference between the centroid of the target and the retro-reflect beam. In order to minimize the pointing error in this kind of evaluation criterion, we try our best to make the size of the target close to the far field spot size.

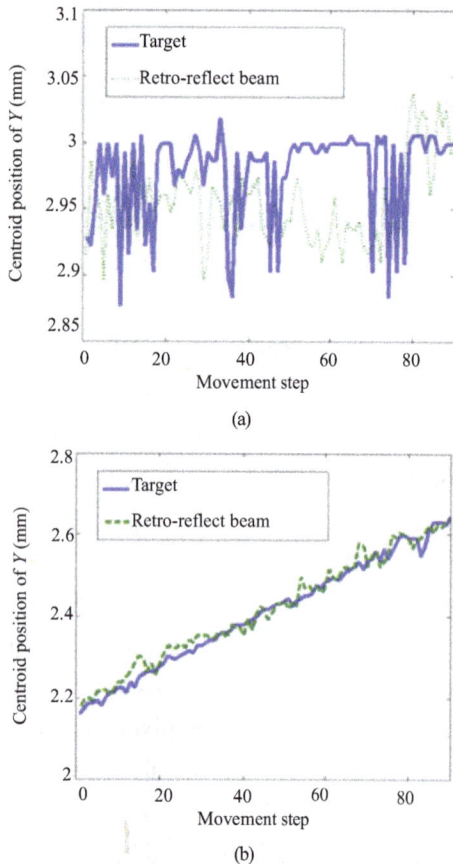

Fig. 7 Results of target moving along horizontal: (a) relationship between movement step and centroid position of *Y* and (b) relationship between movement step and centroid position of *X*.

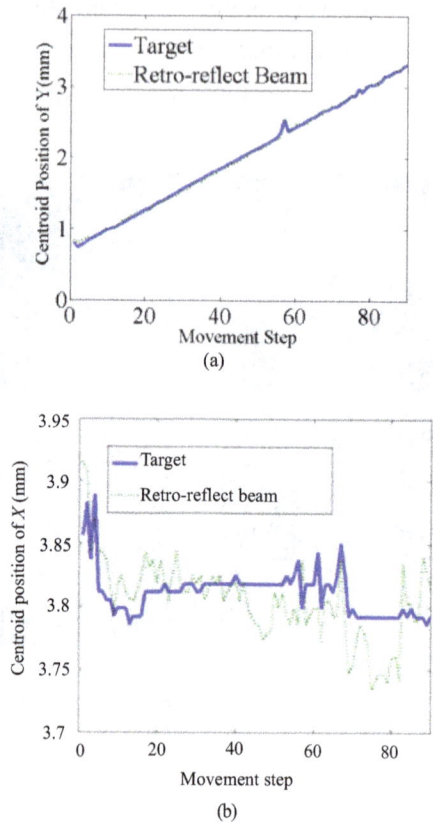

Fig. 8 Results of target moving along vertical: (a) relationship between movement step and centroid position of *Y* and (b) relationship between movement step and centroid position of *X*.

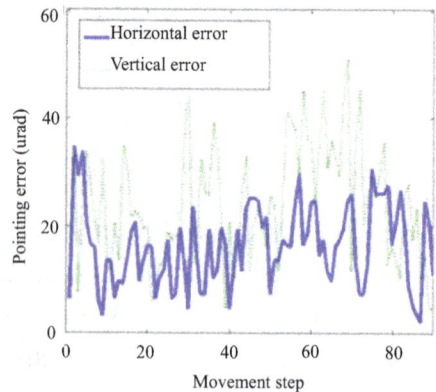

Fig. 9 Results of total pointing error.

Considering the composition of total pointing error, the beam steering method will cause pointing error, which is shown in Fig. 3, and it is mainly caused by the method itself and LC-OPA devices. On the other hand, errors coming from outer of the system will also contribute to the total pointing error. Factors, such as shape of the target, non-uniformity of the target surface reflectivity, spot size of the beam, background noise, and phase aberration, will cause the error of target location. So the total pointing error is shown in Fig. 9. After further calculation, an average pointing error of the LC-OPA aiming and pointing system is 19 μrad, and the max pointing error is 56 μrad.

In conclusion, a non-mechanical target aiming and pointing system with a single LC-OPA is demonstrated, and the LC-OPA based system has combined with an improved 2D beam steering method, and the feasibility of this method has been proved both in theory and experiment. A retro-reflect target is used to evaluate the aiming and pointing system. The system has a maximum FOV of ±22 mrad on the theory, and the average pointing error of the system is 19 μrad. The maximum pointing error is 56 μrad. Detailed error analysis of this system and tracking performance would be performed in the future work.

References

[1] C. R. Cooke, "Automatic laser tracking and ranging system," *Applied Optics*, 1972, 11(2): 277–284.

[2] P. M. Livingston, J. L. Jacoby, and W. S. Tierney, "Laser beam active tracking for specular objects to fractions of Lambda/D," *Applied Optics*, 1985, 24(13): 1919–1925.

[3] F. E. Hoge, "Integrated laser/radar satellite ranging and tracking system," *Applied Optics*, 1974, 13(10): 2352–2358.

[4] J. F. Riker, 'Active tracking lasers for precision target stabilization," *Proc. SPIE*, 2003, 5087: 1–12.

[5] J. E. Curtis, B. A. Koss, and D. G. Grier, "Dynamic holographic optical tweezers," *Optics Communications*, 2002, 207(1–6): 169–175.

[6] P. F. Mcmanamon, P. J. Bos, M. J. Escuti, J. Heikenfeld, S. Serati, H. Xie, *et al.*, "A review of phased array steering for narrow-band electrooptical systems," *Proceeding of the IEEE*, 2009, 97(6): 1078–1096.

[7] E. Ha, L. Allard, L. Sjo, D. Engstro, S. Ha, Q. Wang, *et al.*, "Retrocommunication utilizing electroabsorption modulators and nonmechanical beam steering," *Optical Engineering*, 2005, 44(4): 45001-1–45001-8.

[8] E. Haellstig, J. Stigwall, M. Lindgren, and L. Sjoqvist, "Laser beam steering and tracking using a liquid crystal spatial light modulator," *Proc. SPIE*, 2003, 5087: 13–23.

[9] D. Engström, J. Bengtsson, E. Eriksson, and M. Goksör, "Improved beam steering accuracy of a single beam with a 1D phase-only spatial light modulator," *Optics Express*, 2008, 16(22): 18275–18287.

[10] A. Linnenberger, S. Serati, and J. Stockley, "Advances in optical phased array technology," *Proc. SPIE*, 2006, 6304: 1–9.

[11] L. Kong, Y. Zhu, Y. Song, and J. Yang, "Beam steering approach for high-precision spatial light modulators," *Chinese Optics Letters*, 2010, 8(11): 1085–1089.

13

Design and Optimization of Photonic Crystal Fiber Based Sensor for Gas Condensate and Air Pollution Monitoring

Md. Ibadul ISLAM[1], Kawsar AHMED[1,2*], Shuvo SEN[1], Sawrab CHOWDHURY[1], Bikash Kumar PAUL[1,2], Md. Shadidul ISLAM[1], Mohammad Badrul Alam MIAH[1], and Sayed ASADUZZAMAN[1,2,3]

[1]*Department of Information and Communication Technology (ICT), Mawlana Bhashani Science and Technology University, Santosh, Tangail 1902, Bangladesh*

[2]*Group of Bio-photomatiχ, Tangail, 1902, Bangladesh*

[3]*Department of Software Engineering (SWE), Daffodil International University, Sukrabad, Dhaka, 1207, Bangladesh*

*Corresponding author: Kawsar AHMED   Email: kawsar.ict@mbstu.ac.bd and k.ahmed.bd@ieee.org

Abstract: In this paper, a hexagonal shape photonic crystal fiber (H-PCF) has been proposed as a gas sensor of which both micro-structured core and cladding are organized by circular air cavities. The reported H-PCF has a single layer circular core which is surrounded by a five-layer hexagonal cladding. The overall pretending process of the H-PCF is completed by using a full vectorial finite element method (FEM) with perfectly matched layer (PML) boundary condition. All geometrical parameters like diameters and pitches of both core and cladding regions have fluctuated with an optimized structure. After completing the numerical analysis, it is clearly visualized that the proposed H-PCF exhibits high sensitivity with low confinement loss. The investigated results reveal the relative sensitivity of 56.65% and confinement loss of 2.31×10^{-5} dB/m at the 1.33-μm wavelength. Moreover, effective area, nonlinearity, and *V*-parameter of the suggested PCF are also briefly described.

Keywords: Relative sensitivity; confinement loss; gas sensor; effective area; nonlinearity and photonic crystal fiber

1. Introduction

Photonic crystal fibers (PCFs) have ensured novel modification in the present fiber-optic communication system. It is the perforated fiber that goes through the entire length of fiber. The PCF can be designed for many significant applications in intelligent systems, and it has attracted too much concentration of researchers. The requirement of PCF is exponentially escalated for flexibility and variation of geometrical structures like D-shape [1], pentafgonal [2], tapered [3] panda-shaped [4], colloidal [5], spherical shaped [6], rhombic-shaped core [7], and spiral [8] shaped architecture. High nonlinearity [9], large mode field area [10], highly birefringence [2], and zero-flattened dispersion [11] characteristics have gained by modifying the design of PCF. Distinctive background materials such as Tellurite [12], Topas [13], and Silica [14] are also applied to improve the guiding properties of PCF. As a result, PCF can be utilized in the nanoparticle detection [15], humidity measurement [16], atom-atom interaction [17], micro-organism growth detection [18], bacteria detection [19], air pollution monitoring [14], gas leak detection systems with high accuracy and safety [20], environmental studies [21], sensing [22], and communication systems [23] for its unique optical properties. In recent years, PCFs have established various branches of sensing technology with the promotion of fabrication techniques.

The PCF based sensors are mainly utilized in plasmonic [24], remote irradiation [25], surface, temperature, tension, magnetic field, and gas sensing applications [22]. Nowadays, gas sensing capability of PCF is enhanced rapidly due to its several attractive features like the high relative sensitivity and low confinement loss. The performances of gas sensors are enlarged by changing the size or number of air cavities [26]. The global warming problem is very acute day by day due to various types of toxic gases including C_2H_2, CH_4, NO_2, SO_2, NH_3, H_2S, CO_2, and CO. These combustible and explosive gases are emitted from industries and vehicles. The monitoring and controlling of gas emission are needed to protect our environment from greenhouse effect and global climate change. The PCF-based gas sensors have hastily popular because of low-cost, small footprint, and long interaction length. As a result, it is used in areas such as civil engineering, environment protection, homeland security, and military defense [27].

In previous and current period, the researchers tried their best to upgrade the performance of PCF-based gas sensing technique. Cordeiro et al. [28] proposed a microstructure PCF that increased the energy of gas filling air holes. In 2003, Hoo et al. [29] reported an all-fiber gas sensor and obtained the relative sensitivity of 10.15% at the 1.4 μm wavelength. The improvement of relative sensitivity of 16.88% and the reduced confinement loss of 1.765×10^{-8} dB/m have been done by Morshed et al.

[30]. In [31], it suggested an index-guiding PCF and achieved the sensitivity and leakage loss of 29.80% and 4.50×10^{-1} dB/m, respectively. In 2013, Olyaee et al. [32] proposed a micro structured PCF and enhanced the sensitivity of 32.99% as well as minimizing the leakage loss of 2.59×10^{-5} dB/m at the 1.33-μm wavelength. Asaduzzaman and Ahmed [14] reported a micro structured elliptical cored photonic crystal fiber (E-PCF) for air pollution monitoring purpose where the core region was constituted with elliptical air holes and exhibited high relative sensitivity, birefringence, and nonlinearity.

In this research paper, we have suggested and successfully scrutinized a hexagonal photonic crystal fiber (H-PCF) where five-layer hexagonal cladding and single layer circular core territories are formed by circular air holes. The proposed H-PCF is mentioned as a gas sensor that improves the relative sensitivity of 56.65% and reduces the confinement loss of 2.31×10^{-5} dB/m at $\lambda = 1.33$ μm. In addition, the effective area and nonlinearity of 6.44 μm² and 22.73 W^{-1}·km^{-1} respectively are gained at the same wavelength. So, the reported PCF is massively efficient for pollution monitoring applications and it ensures the industrial safety.

2. Geometries of the proposed H-PCF

Figure 1 depicts the transverse cross-sectional view of the proposed H-PCF based on micro-structured core and cladding in circular and hexagonal manner, respectively. This technique was applied in [33] where the cladding region was hexagonal with six type's holes in the edges of the outermost cladding. In the work, the diameters of two outermost rings as well as three innermost rings were different. In our proposed H-PCF, the diameters of all five rings of cladding have been kept the same to interact the proper fabrication tolerance, which were defined as d [14]. The idea of the micro-structured core was introduced in [34]. In the proposed design, the angular displacement of two adjacent air holes in the cladding region is kept

60° to form a hexagonal structure. The dimensions of air holes in the core are smaller than that of cladding holes to satisfy the effective index guiding criterion. In our H-PCF, the micro-structured core is formed with an array of eight tiny air holes in circular manner, the diameter is defined as d_c, and the diameter of supplementary center air hole in the core region is d_{c0}. The distance between center to center of holes is called pitch of cladding denoted as Λ. The distance from center to the first air hole ring in the core region is defined as Λ_c. Due to the central hole, the refractive indices between core and cladding dropped and more light would confine into the core, thereby the sensitivity increases. To gain more relative sensitivity, the innermost ring is formed with six air holes in hexagonal manner closer to the core region [35]. The effect of the size, position, and shape of those holes on the sensitivity and confinement loss are numerically investigated by applying different numerical methods.

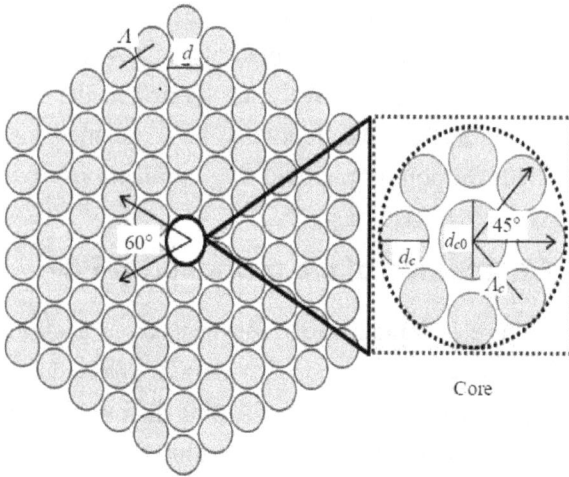

Fig. 1 Transverse cross-sectional view of the proposed H-PCF.

3. Synopsis of numerical method

The finite element method (FEM) is applied for solving Maxwell's equations due to correctly pretending the optical properties of the suggested H-PCF. Moreover, it has capacity to solve the complex architectural design of PCF and also contributes full vector analysis of distinctive PCF structures [36]. As a result, it provides a good approximation to prove the reliability of PCF structures. It is known that due to the finite number of air holes in the cladding, light can leak among the air holes called the leakage loss or confinement loss. So, to compute the leakage loss, an effective boundary condition which produces no reflection at the boundary is needed. In this case, perfectly matched layers (PMLs) are the most efficient absorption boundary condition for this purpose [37–38]. The confinement loss or leakage loss can be calculated as [14]

$$L_C = 8.868 \times K_0 \, \text{Im}[n_{\text{eff}}] (\text{dB/m}) \quad (1)$$

where $k_0 = 2\pi/\lambda$; λ is the wavelength of light, and $\text{Im}[n_{\text{eff}}]$ is the imaginary part of the refractive index.

Beer-Lambert law is used to evaluate the absorbance of evanescent field by the gas samples. The relationship between gas concentration and optical intensity is evaluated as [14]

$$I(\lambda) = I_0(\lambda) \exp[-r\alpha_m l_c] \quad (2)$$

where I and I_0 are the output and input light intensities, respectively, α_m is the absorption coefficient, l is the length of the H-PCF used for chemical detection, c is the concentration of the target samples, and r is a relative sensitivity coefficient defined as [34]

$$r = \frac{n_r}{n_{\text{eff}}} \times f \quad (3)$$

where n_r are the refractive index of gas, n_{eff} is the real part of the effective mode index, and f is the fraction of the total power located in the holes that can be expressed as [34]

$$f = \frac{\int_{\text{sample}} \text{Re}(E_x H_y - E_y H_x) dx dy}{\int_{\text{total}} \text{Re}(E_x H_y - E_y H_x) dx dy} \times 100 \quad (4)$$

where E_x, E_y, H_x, and H_y are the transverse electric and magnetic fields of the guided mode, respectively.

Utilizing the Sellmeier's equation [14], pure silica is set as the background material of the fiber of which the refractive index changes with the modification of the wavelength.

$$n(\lambda) = \sqrt{\left(1 + \frac{B_1\lambda^2}{\lambda^2 - c_1} + \frac{B_2\lambda^2}{\lambda^2 - c_2} + \frac{B_3\lambda^2}{\lambda^2 - c_3}\right)} \quad (5)$$

where B_1, B_2, B_3, C_1, C_2, and C_3 are the Sellmeier's coefficients. The refractive index of silica is $n(\lambda)$ which alters with the operating wavelength.

The V-parameter [39] is used to compute the single mode response of a fiber which can be defined as

$$V_{\text{eff}} = \frac{2\pi\Lambda}{\lambda}\sqrt{n_{co}^2 - n_{cl}^2} \quad (6)$$

where the refractive indices of core and cladding regions are denoted by n_{c0} and n_{cl}, respectively. To fulfill the single mode condition, the value of V_{eff} must be not greater than 2.405. Another important parameter is effective area (A_{eff}) which can be determined by the following equation [14]:

$$A_{\text{eff}} = \frac{\left(\iint |E|^2 \, dxdy^2\right)}{\iint |E|^4 \, dxdy} \quad (7)$$

where optical power is denoted by E. Equation (8) is applied to determine the nonlinear coefficient (γ) [14]

$$\gamma = \left(\frac{2\pi}{\lambda}\right)\left(\frac{n_2}{A_{\text{eff}}}\right) \quad (8)$$

where n_2 is called the nonlinear refractive index.

In this research, the main focus is on the methane (CH_4) and hydrogen fluoride (HF) gasses to have the absorption line at the wavelength of 1.33 µm. We have tried to improve the relative sensitivity and confinement loss at this wavelength. Furthermore, for the other wavelength, our proposed H-PCF has shown the better results.

4. Numerical results and discussion

FEM for disposing Maxwell's equations has been used for its reliability as well as high accuracy to investigate the optical properties of the proposed H-PCF [40–41]. Full vectorial FEM with the perfectly match layer (PML) boundary condition is one of the most potential numerical approaches. It is the most obtainable approach to engineer for

structuring and developing photonic elements and devices [42]. To investigate propagation properties of leaky modes, the PML as boundary conditions is an essential approach in PCFs and by employing these layers, all optical propagation characteristics can be appraised in a single run [41, 43, 44]. Figure 2 exhibits the fundamental mode field distributions along x-polarization [Fig. 2(a)] and y-polarization [Fig. 2(b)] for the proposed H-PCF at the operating wavelength of 1.33 µm. From this figure, it is evaluated that the Gaussian distribution of the fields is uniform as well as the fundamental modes which are strongly bounded by the core region of circular air holes. It also illustrates that the sample air holes close to the core region strongly interacts with the evanescent field that causes the high sensitivity as well as low confinement loss.

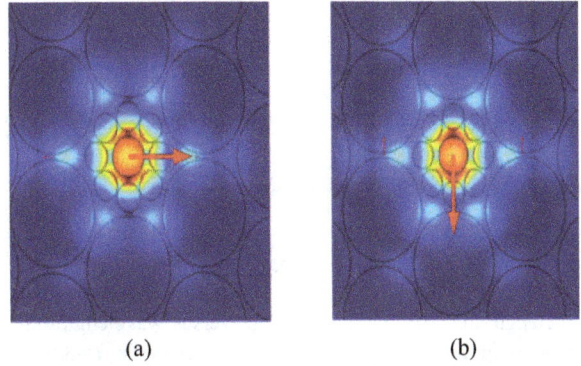

(a) (b)

Fig. 2 Modal intensity distribution of the proposed H-PCF at the wavelength of 1.33 µm: (a) x-polarization and (b) y-polarization.

Due to the increment of air hole diameter of innermost ring of cladding, the fraction of evanescent field penetrates into core that improves relative sensitivity [33]. Figure 3 represents the relative sensitivity versus wavelength curve [Fig. 3(a)] and confinement loss versus wavelength curve [Fig. 3(b)] for the optimized parameters of the proposed structure and investigates the behavior of guiding properties. A simple technique is used to optimize the parameters and figure out x-polarized mode to acquire optimum results. By selecting a proper mode of mesh size, the simulation task has been completed by applying 4.2 version COMSOL Multiphysics [14].

(a)

(b)

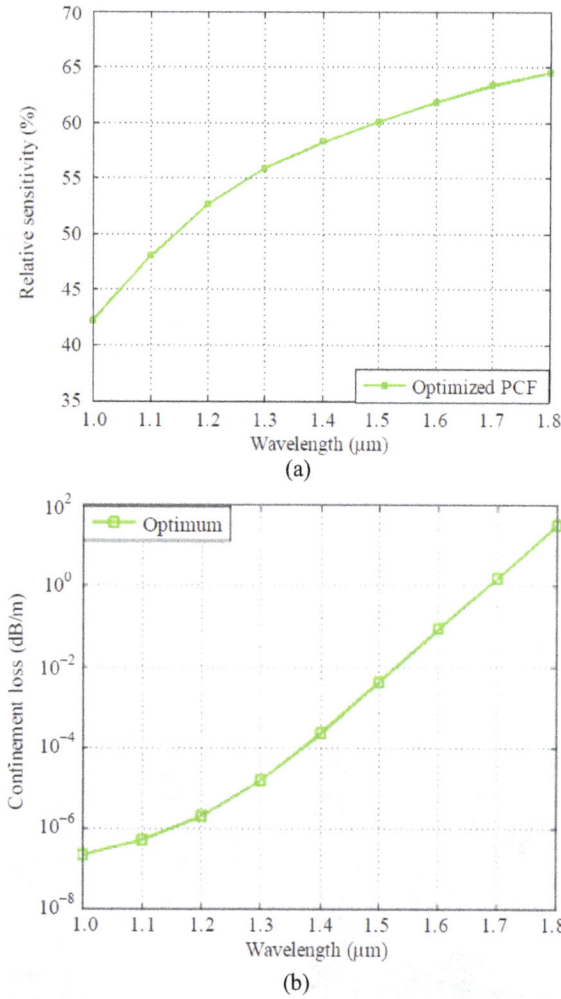

Fig. 3 Relationship between: (a) relative sensitivity versus wavelength and (b) confinement loss versus wavelength curve for the proposed H-PCF for optimized parameters: $\Lambda = 3.00\,\mu m$, $\Lambda_c = 1.08\,\mu m$, $d = 2.85\,\mu m$, $d_c = 0.774\,\mu m$, and $d_{c0} = 1.08\,\mu m$.

As the penetrating field in the cladding is unloading, inner rings have higher impact on sensitivity; so it is enhanced for the diameter of air holes in the innermost ring as much as possible. It can be seen that 56.65% relative sensitivity as well as 2.31×10^{-5} dB/m confinement loss can be obtained at the operating wavelength of 1.33 μm. From the observation, it can be seen that the proposed H-PCF has a higher relative sensitivity compared with prior PCFs [14, 30–32, 34].

For the central air holes, when enough light would be confined into the cladding region and interaction happens between light and gases, then sensitivity will be increased [31]. The air filling ratio (d/Λ) of 0.95 is acceptable for fabrication [45], and

we have carefully optimized all these parameters. Figure 4 describes the variations of air filling ration (d/Λ) in cladding by keeping other parameters fixed to $\Lambda = 3.00\,\mu m$, $\Lambda_c = 1.08\,\mu m$, $d_c = 0.774\,\mu m$, and $d_{c0} = 1.08\,\mu m$. At the operating wavelength 1.33 μm, the air filling ratio d/Λ is defined as 0.94, 0.95, and 0.96, the considered relative sensitivities are 55.21%, 56.65%, and 57.86%, and confinement losses are 5.08×10^{-4} dB/m, 2.31×10^{-5} dB/m, and 3.41×10^{-6} dB/m, respectively.

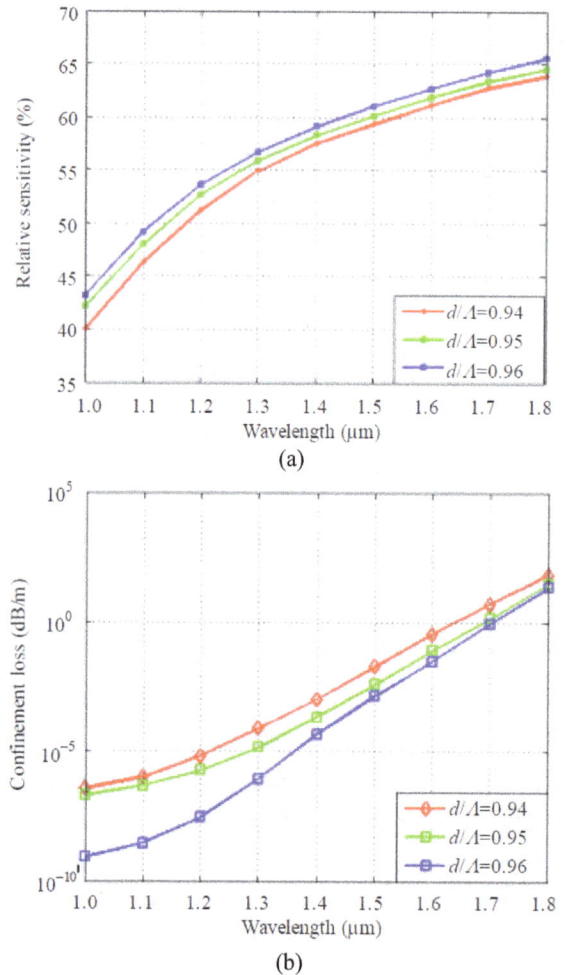

(a)

(b)

Fig. 4 Comparison between: (a) relative sensitivity versus wavelength and (b) confinement loss versus wavelength curve of the proposed H-PCF for optimized parameters: $\Lambda = 3.00\,\mu m$; $\Lambda_c = 1.08\,\mu m$; $d_c = 0.774\,\mu m$ and $d_{c0} = 1.08\,\mu m$; $d/\Lambda = 0.94$, 0.95, and 0.96.

By the reduction of air filling ratio d/Λ, the evanescent field penetrated into cladding causes the fraction power of holes alleviates and leads to a decrease in the absorption power by core region which causes the relative sensitivity to decrease. On the other hand, if the air filling ratio (d/Λ) is promoted, the evanescent field penetrating into core region causes an increase in the fraction power of holes which leads to an increase in the absorption power by core region so as to enhance the relative sensitivity. The higher sensitivity makes a fiber as an active candidate to detect noxious gas, colorless gas, and monitor environment pollution [14]. Now $d/\Lambda=$ 0.95 is selected as an optimized value for the further investigation process.

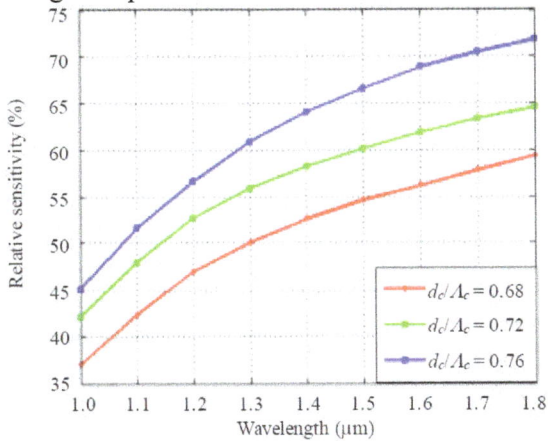

(a)

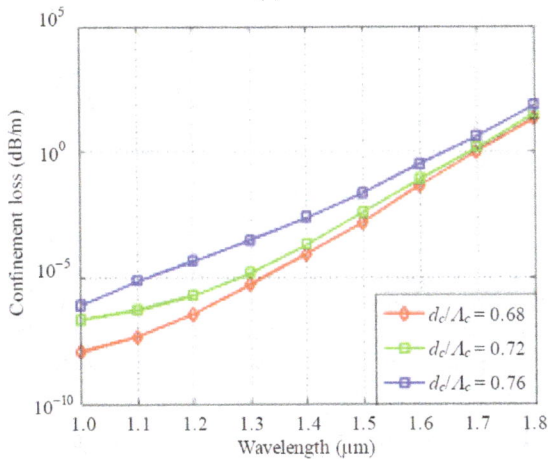

(b)

Fig. 5 Comparison between: (a) relative sensitivity versus wavelength and (b) confinement loss versus wavelength curve for the proposed H-PCF for optimized parameters: $\Lambda=3.00\,\mu m$; $\Lambda_c=1.08\,\mu m$; $d=2.85\,\mu m$ and $d_{c0}=1.08\,\mu m$; $d_c/\Lambda_c=0.68$, 0.72, and 0.76.

The micro-structured core PCFs help gain higher relative sensitivity and alleviate confinement loss [36]. Besides, it can be analyzed that the relative sensitivity is promoted with the enhancement of air holes diameter in the core region and decrement of

pitch Λ_c [30]. Figure 5 shows the effect of air filling ratio d_c/Λ_c of core region when other parameters are kept fixed. At the wavelength 1.33 μm, the variations of d_c/Λ_c are defined as 0.68, 0.72, and 0.76, and the calculated relative sensitivities are 51.28%, 56.65%, and 62.47%, respectively. At the same wavelength, the considered confinement losses are 8.53×10^{-6} dB/m, 2.31×10^{-5} dB/m, and 7.56×10^{-4} dB/m, respectively. It can be also mentioned that the evanescent field can be absorbed into the core region by promoting air filling ratio (d_c/Λ_c) which leads to enhancing the absorption power and causes high sensitivity as well as low confinement loss. Besides, the capability of core to confine the power fraction is low when the air filling ration d_c/Λ_c decreases.

Now to obtain the expected result, the pitch is chosen as $\Lambda=2.18$ μm for optimizing the investigation process. The variation of pitch (Λ) has great effect on relative sensitivity. High sensitivity makes the fiber more practical in the sensing applications like toxic and colorless gas detection [30]. The relative sensitivity versus wavelength curve for variations of pitch (Λ) of cladding is depicted in Fig. 6(a) when other parameters ($\Lambda_c=$ 1.08 μm; $d=2.85$ μm, and $d_c=0.774$ μm, and $d_{c0}=$ 1.08 μm) are kept constant. According to the variation of Λ as 2.96 μm, 3.00 μm, and 3.04 μm, the calculated relative sensitivities are 57.11%, 56.65%, and 54.32% and confinement losses [Fig. 6(b)] are 3.23×10^{-6} dB/m, 2.31×10^{-5} dB/m, and 9.18×10^{-4} dB/m, respectively. From the above observation, it can be mentioned that the higher relative sensitivity can be gained by reducing pitch of cladding Λ as much as possible [33]. In another sense, it can be mentioned that if Λ is enhanced, the evanescent field would not be confined properly into core. So the power fraction of holes alleviates which governs to decrease the absorption power by core region and the relative sensitivity decays. From Fig. 6(b), it is visualized that the confinement loss is decreases by enhancing pitch [34].

(a)

(b)

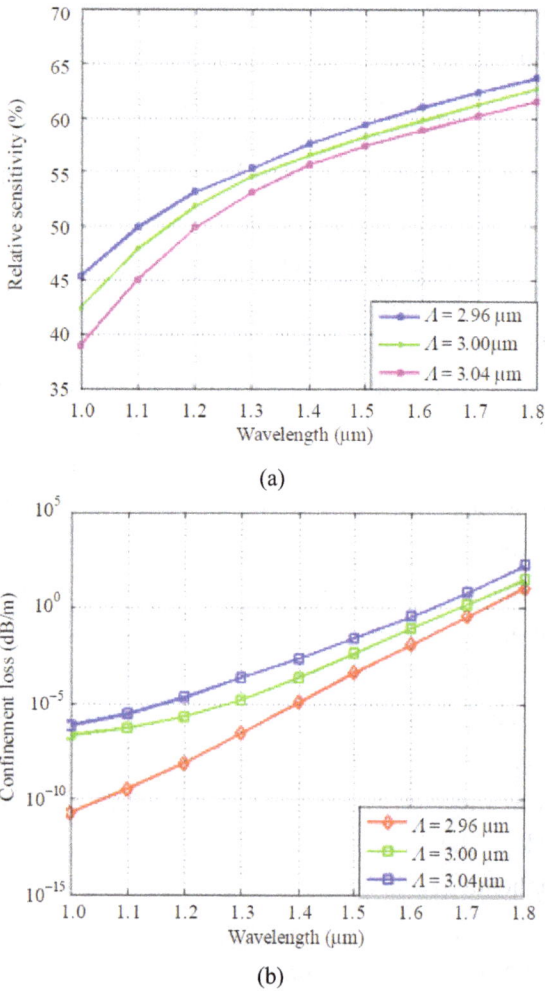

Fig. 6 Comparison between: (a) relative sensitivity versus wavelength and (b) confinement loss versus wavelength curve for the proposed H-PCF for optimized parameters: $\Lambda=2.96\,\mu m$, $3.00\,\mu m$ and $3.04\,\mu m$; $\Lambda_c=1.08\,\mu m$; $d=2.85\,\mu m$, $d_c=0.774\,\mu m$, and $d_{c0}=1.08\,\mu m$.

Figure 7 exhibits the effect of pitch variations of core on relative sensitivity and confinement loss by keeping other parameters fixed. From this figure, it is observed that the relative sensitivity enhances and the confinement loss alleviates when the pitch of core decreases and vice versa. From observations (Fig. 7), $\Lambda_c=1.08\,\mu m$ is selected for further processing. To improve the relative sensitivity and confinement loss simultaneously, the diameter of central air hole should be increased as much as possible [45]. Figure 8 describes the effect of diameter of central air hole on relative sensitivity and the confinement loss. At the operating wavelength 1.33 μm, the considered relative

sensitivities are 53.35%, 56.65%, and 59.68%, and confinement losses are 8.73×10^{-6} dB/m, 2.31×10^{-5} dB/m, and 6.75×10^{-4} dB/m, respectively.

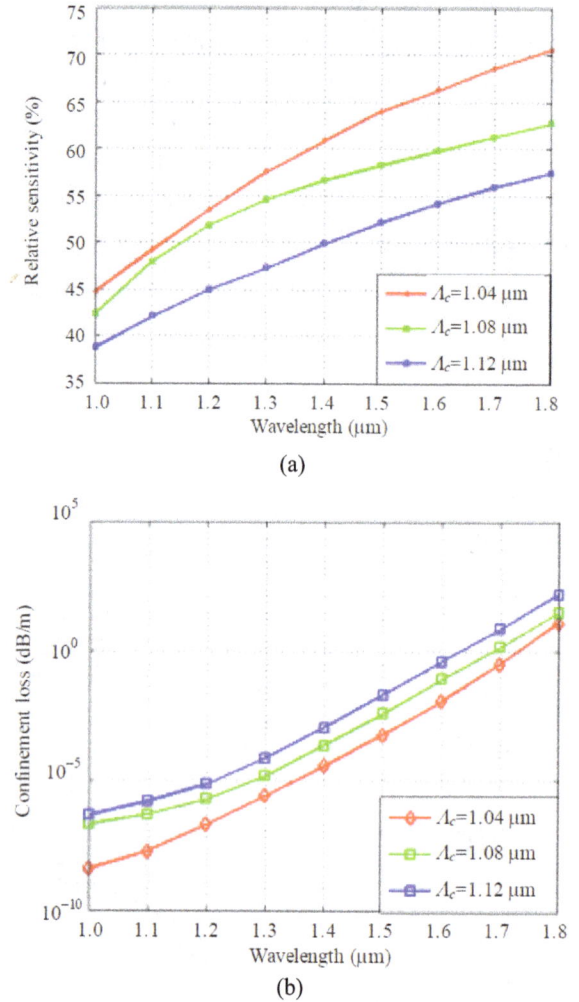

(a)

(b)

Fig. 7 Comparison between: (a) relative sensitivity versus wavelength and (b) confinement loss versus wavelength curve for the proposed H-PCF for optimized parameters: $\Lambda=3.00\,\mu m$; $\Lambda_c=1.04\,\mu m$, $1.08\,\mu m$, and $1.12\,\mu m$; $d=2.85\,\mu m$, $d_c=0.774\,\mu m$, and $d_{c0}=1.08\,\mu m$.

Figure 9 illustrates the effective area and nonlinear coefficient of the proposed H-PCF as a function of wavelength for the optimized parameters as $\Lambda=3.00\,\mu m$, $\Lambda_c=1.08\,\mu m$, $d=2.85\,\mu m$, $d_c=0.774\,\mu m$, and $d_{c0}=1.08\,\mu m$. It is noticed that effective area enhances according to the increment of wavelength. The mode power is narrowly demarked in the core region at longer wavelength, so the guiding waves diverse largely. As a result, the propagating modes hold a larger effective area [46]. A reverse relationship can be seen for the nonlinear coefficient, which is also depicted in Fig. 9. The effective area and nonlinear coefficient of $6.44\,\mu m^2$ and $22.73\,W^{-1}\cdot km^{-1}$ have been achieved respectively at $\lambda = 1.33\,\mu m$.

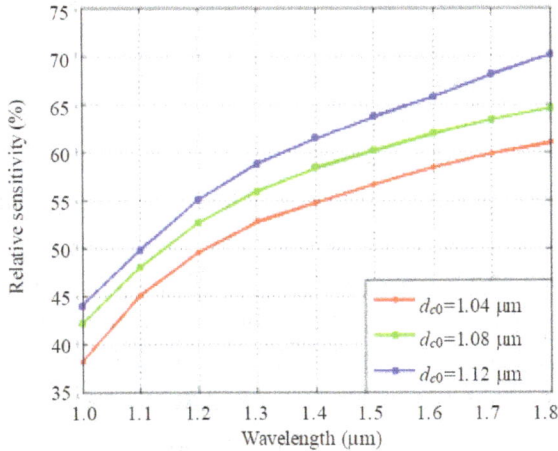

(a)

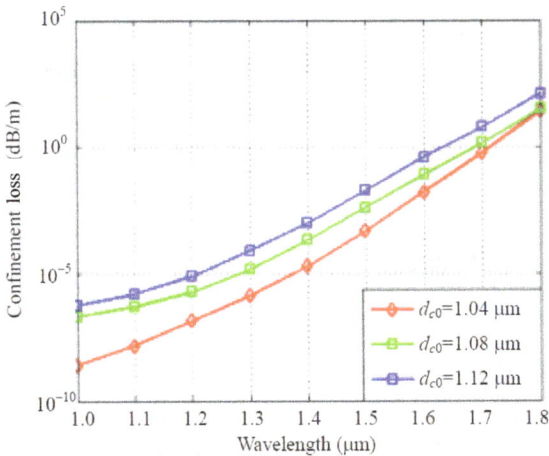

(b)

Fig. 8 Comparison between: (a) relative sensitivity versus wavelength and (b) confinement loss versus wavelength curve for the proposed H-PCF for optimized parameters: $\Lambda = 3.00\,\mu m$; $\Lambda_c = 1.08\,\mu m$; $d = 2.85\,\mu m$, $d_c = 0.774\,\mu m$, and $d_{c0} = 1.04\,\mu m$, $1.08\,\mu m$, and $1.12\,\mu m$.

The single-mode operation of the proposed H-PCF using V effective (V_{eff}) parameter can be investigated. Figure 10 shows the V_{eff} parameter as a function of operating wavelength with optimized design parameters for $\Lambda = 3.00\,\mu m$; $\Lambda_c = 1.08\,\mu m$; $d = 2.85\,\mu m$, $d_c = 0.774\,\mu m$, and $d_{c0} = 1.08\,\mu m$. The V_{eff} parameter can be expressed by (6). With approximate perfect electric and perfect magnetic conductor boundary condition, FEM is used at the outer enclosure to achieve the index of space-filling

mode [39]. At operating wavelength $1.33\,\mu m$, the desired agreement for operating as an SMF is $V_{eff} \leq 2.405$ [14]. From Fig. 10, it is visualized that the gained V_{eff} value is 0.95 at wavelength $1.33\,\mu m$ which is less than 2.405. From above observations, it can clearly be mentioned that the fiber remains constant to single mode over the entire bands.

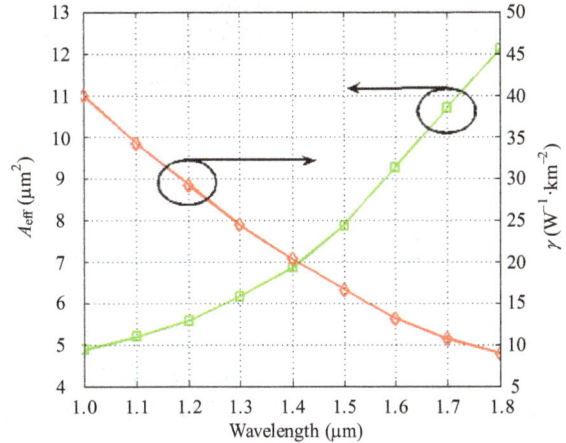

Fig. 9 Effective area and nonlinear coefficient of the x-polarization mode of the proposed H-PCF as a function of wavelength for $\Lambda = 3.00\,\mu m$; $\Lambda_c = 1.08\,\mu m$; $d = 2.85\,\mu m$, $d_c = 0.774\,\mu m$, and $d_{c0} = 1.08\,\mu m$.

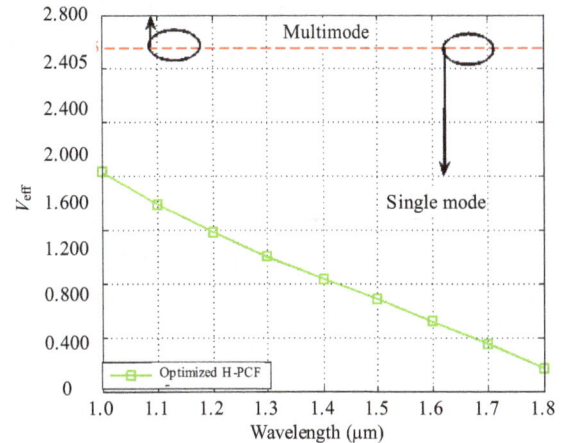

Fig. 10 V parameter of the proposed H-PCF as a function of wavelength for $\Lambda = 3.00\,\mu m$, $\Lambda_c = 1.08\,\mu m$, $d = 2.85\,\mu m$, $d_c = 0.774\,\mu m$, and $d_{c0} = 1.08\,\mu m$.

After completing numerical investigation, the effects of distinguished parameters on the proposed hexagonal photonic H-PCF are observed. It is evidently reported that the relative sensitivity can be promoted due to the decrement of global diameter shown in Table 1. High sensitivity leads to confine more light into core region as well as better electromagnetic power interaction with gasses.

Table 1 also exhibits the effect of changing global parameters of ±1% and ±2%, on confinement loss, effective area, and nonlinearity. It is noticed that the effective area promotes with an increase in air holes diameter as well as a decrease in the non-linear coefficient.

Table 1 Comparison of different index-guiding properties for optimum design parameters and also for fiber's global diameter variations of order ± 1 % – ± 2 % around the optimum value.

Change in diameters (%)	Relative sensitivity (%)	Confinement loss (dB/m)	A_{eff} (μm^2)	γ ($W^{-1} \cdot km^{-1}$)
+2	55.44	1.71×10^{-5}	7.46	19.63
+1	56.37	2.10×10^{-5}	6.47	22.63
Optimum	56.65	2.31×10^{-5}	6.44	22.73
−1	56.93	6.43×10^{-5}	6.29	23.28
−2	57.21	6.83×10^{-5}	6.25	23.43

Table 2 demonstrates the comparison among relative sensitivity, confinement loss, effective area, and nonlinearity of the proposed H-PCF with other prior designs in [14, 30, 34, 40, 45]. The proposed H-PCF provides about 56.65% sensitivity as well as low confinement loss at the 1.33 μm wavelength. Besides, the reported H-PCF shows higher effective area and nonlinear coefficient at the same wavelength compare to prior PCFs.

Table 2 Comparison of simulated results among proposed PCF and prior PCFs at 1.33 μm wavelength.

PCFs	Relative sensitivity (%)	Confinement loss (dB/m)	A_{eff} (μm^2)	γ ($W^{-1} \cdot km^{-1}$)
Ref. [34]	15.67	1.12×10^{-7}		
Ref. [30]	16.88	1.765×10^{-8}		
Ref. [40]	20.10	1.09×10^{-3}	-	-
Ref. [45]	48.26	1.26×10^{-5}	4.02	
Ref. [14]	53.07	3.21×10^{-6}	3.88	15.67
Proposed PCF	56.65	2.31×10^{-5}	6.44	22.73

Finally, the fabrication process is one of the fundamental issues in PCFs. In the proposed H-PCF three mixing air-holes are considered and shown in Fig. 1. Different methods have been offered for the fabrication of micro-structured fibers such as stack and draw [47], drilling [48], sol-gel casting [49], and extrusion [50] methods. Triangular or honeycomb lattices can be fabricated by stack and draw methods but cannot easily generate circular patterns. The drilling methods offer adjustment of both holes size and spacing but are usually limited to a few number of holes and bounded to circular shapes. Extrusion techniques permit design freedom, but are typically restricted for soft glasses because those cause the material loss with severely high values. However, the proposed structure can be fabricated by currently developed technology for technological advancement in the fabrication of PCFs [51]. El *et al.* [49] provided a sol-gel technique to fabricate the PCFs with all structures, and they provide the freedom to adjust air-hole size, shape, and spacing. In such situation, our proposed H-PCF can be fabricated by the sol-gel casting method. After drawing the relative sensitivity curve to the expected level in the way just illustrated, we then verify the accuracy of the property of the proposed design. A standard fiber draw, ±1% variations in fiber global diameter may occur during the fabrication process.

5. Conclusions

Two major propagation characteristics like the relative sensitivity and confinement loss of an optical sensor have been numerically examined in this paper. The reported optical sensor acts as a gas sensor which has five layers hexagonal cladding with circular shape core. Compared with the prior reported PCFs, our proposed H-PCF exposes high relative sensitivity with low confinement loss. After finishing all investigations, the relative sensitivity and confinement loss of 56.65% and 2.31×10^{-5} dB/m correspondingly are gained at the 1.33 μm wavelength. Moreover, the effective area of 6.44 μm^2 and nonlinearity of 22.73 $W^{-1} \cdot km^{-1}$ are achieved at the same wavelength. So, it is expected that the reported H-PCF will be efficient in gas sensing applications as well as greatly beneficial for the gas detection in industrial sectors and safety purpose with high accuracy.

References

[1] H. J. Kim, O. Kown, S. B. Lee, and Y. Han, "Measurement of temperature and refractive index based on surface long-period gratings deposited onto a d-shaped photonic crystal fiber," *Applied Physics B: Chemical,* 2010, 102(1): 81–85.

[2] F. Yu, Z. P. Wang, W. H. Yang, and C. Y. Lv, "Characteristics of highly birefringent photonic crystal fiber with defected core and equilateral pentagon architecture," *Advances in OptoElectronics,* 2016, 2016(6): 1–8.

[3] E. C. Mägi, P. Steinvurzel, and B. J. Eggleton, "Tapered photonic crystal fibers," *Optics Express,* 2004, 12(5): 776.

[4] S. L. Mousavi and M. Sabaeian, "Thermal stress-induced depolarization loss in conventional and panda-shaped photonic crystal fiber lasers," *Brazilian Journal of Physics,* 2016, 46(5): 481–488.

[5] S. H. Kim and G. R. Yi, "Colloidal photonic crystals for sensor applications," *Photonic Materials for Sensing, Biosensing and Display Devices,* 2016, 229: 51–78.

[6] H. P. Gong, M. L. Xiong, Z. H. Qian, C. L. Zhao, and X. Y. Dong, "Simultaneous measurement of curvature and temperature based on Mach-Zehnder interferometer comprising core-offset and spherical-shape structures," *IEEE Photonics Journal,* 2016, 8(1): 1–9.

[7] H. L. Bao, K. Nielsen, H. K. Rasmussen, P. U. Jepsen, and O. Bang, "Fabrication and characterization of porous-core honeycomb bandgap THz fibers," *Optics Express,* 2012, 20(28): 29507–29517.

[8] M. I. Islam, M. Khatun, S. Sen, K. Ahmed, and S. Asaduzzaman, "Spiral photonic crystal fiber for gas sensing application," in *Proceeding of IEEE 9th International Conference on Electrical and Computer Engineering,* Dhaka, Bangladesh, 2016, pp. 20–22.

[9] J. Han, S. Y. Li, and T. Zhang, "Design on a novel hybrid-core photonic crystal fiber with large birefringence and high nonlinearity," *Optical and Quantum Electronics,* 2016, 48(371): 1–11.

[10] B. K. Paul, K. Ahmed, S. Asaduzzaman, and M. S. Islam, "Folded cladding porous shaped photonic crystal fiber with high sensitivity in optical sensing applications: design and analysis," *Sensing and Bio-Sensing Research,* 2017, 12(1): 36–42.

[11] W. C. Cai, E. Liu, B. Feng, H. F. Liu, Z. M. Wang, W. Xiao, *et al.,* "Dispersion properties of a photonic quasi-crystal fiber with double cladding air holes," *Optik – International Journal for Light and Electron Optics,* 2016, 127(10): 4438–4442.

[12] Z. L. Liu, J. An, J. W. Xing, and H. L. Du, "Polarization rotator based on liquid crystal infiltrated tellurite photonic crystal fiber," *Optik – International Journal for Light and Electron Optics,* 2016, 127(10): 4391–4395.

[13] A. Argyros, "Microstructures in polymer fibers for optical fibers, THz waveguides, and fiber-based metamaterials," *ISRN Optics,* 2013, 2013(7): 1–22.

[14] S. Asaduzzaman and K. Ahmed, "Proposal of a gas sensor with high sensitivity, birefringence and nonlinearity for air pollution monitoring," *Sensing and Bio-Sensing Research,* 2016, 10: 20–26.

[15] D. Q. Yang, W. Yuan, and Y. F. Ji, "Nanoparticle detection using fano-resonance photonic crystal on optical fiber-tip," *SPIE,* 2016, 10158: 1015811–1015818.

[16] S. Rota-Rodrigo, A. Lopez-Aldaba, R. A. Perez-Herrera, M. D. L. Bautista, O. Esteban, and M. Lopez-Amo, "Simultaneous measurement of humidity and vibration based on a microwire sensor system using fast Fourier transform technique," *Journal of Lightwave Technology,* 2016, 34(19): 4525–4530.

[17] J. D. Hood, A. Goban, A. Asenjo-Garcia, M. Lu, S. P. Yu, D. E. Chang, *et al.,* "Atom-atom interactions around the band edge of a photonic crystal waveguide," *Proceedings of the National Academy of Sciences,* 2016, 113(38): 10507–10512.

[18] X. H. Liu, M. S. Jiang, Q. M. Sui, S. Y. Luo, and X. Y. Geng, "Optical fiber Fabry-Perot interferometer for microorganism growth detection," *Optical Fiber Technology,* 2016, 30: 32–37.

[19] E. Brzozowska, M. Koba, M. Smietana, S. Gorska, M. Janik, A. Gamian, *et al.,* "Label-free gram-negative bacteria detection using bacteriophage-adhesin-coated long-period gratings," *Biomedical Optics Express,* 2016, 7(3): 829.

[20] Z. Yang, M. L. Liu, M. Shao, and Y. J. Ji, "Research on leakage detection and analysis of leakage point in the gas pipeline system," *Open Journal of Safety Science and Technology,* 2011, 01(03): 94–100.

[21] M. F. H. Arif, S. Asaduzzaman, M. J. H. Biddut, and K. Ahmed, "Design and optimization of highly sensitive photonic crystal fiber with low confinement loss for ethanol detection," *International Journal of Technology,* 2016, 6: 1068–1076.

[22] K. Ahmed and M. Morshed, "Design and numerical analysis of microstructured-core octagonal photonic crystal fiber for sensing applications," *Sensing and Bio-Sensing Research,* 2016, 7(1): 1–6.

[23] F. Du, Y. Q. Lu, and S. T. Wu, "Electrically tunable liquid-crystal photonic crystal fiber," *Applied Physics Letters,* 2004, 85(12): 2181.

[24] A. A. Rifat, R. Ahmed, A. K. Yetisen, H. Butt, A. Sabouri, G. A. Mahdiraji, *et al.,* "Photonic crystal fiber based plasmonic sensors," *Sensors and*

Actuators B: Chemical, 2017, 243: 311–325.

[25] R. Zeltner, D. S. Bykov, S. Xie, T. G. Euser, and P. S. J. Russell, "Fluorescence-based remote irradiation sensor in liquid-filled hollow-core photonic crystal fiber," *Applied Physics Letters*, 2016, 108(23): 231107.

[26] S. H. Kassani, R. Khazaeinezhad, Y. Jung, J. Kobelke, and K. Oh, "Suspended ring-core photonic crystal fiber gas sensor with high sensitivity and fast response," *IEEE Photonics Journal*, 2015, 7(1): 1–9.

[27] S. J. Zheng, M. Ghandehari, and J. P. Ou, "Photonic crystal fiber long-period grating absorption gas sensor based on a tunable erbium-doped fiber ring laser," *Sensors and Actuators B: Chemical*, 2016, 223: 324–332.

[28] C. M. B. Cordeiro, M. A. Franco, G. Chesini, E. C. Barretto, R. Lwin, C. B. Cruz, *et al.*, "Microstructured-core optical fibre for evanescent sensing applications," *Optics Express*, 2006, 14(26): 13056.

[29] Y. L. Hoo, W. Jin, C. Shi, H. L. Ho, D. N. Wang, and S. C. Ruan, "Design and modeling of a photonic crystal fiber gas sensor," *Applied Optics*, 2003, 42(18): 3509.

[30] M. Morshed, M. I. Hasan, and S. M. A. Razzak, "Enhancement of the sensitivity of gas sensor based on microstructure optical fiber," *Photonic Sensors*, 2015, 5(4): 312–320.

[31] Z. G. Zhi, F. D. Zhang, M. Zhang, and P. D. Ye, "Gas sensing properties of index-guided PCF with air-core," *Optics & Laser Technology*, 2008, 40(1): 167–174.

[32] S. Olyaee and A. Naraghi, "Design and optimization of index-guiding photonic crystal fiber gas sensor," *Photonic Sensors*, 2013, 3(2): 131–136.

[33] S. Olyaee, A. Naraghi, and V. Ahmadi, "High sensitivity evanescent-field gas sensor based on modified photonic crystal fiber for gas condensate and air pollution monitoring," *Optik－International Journal for Light and Electron Optics*, 2014, 125(1): 596–600.

[34] S. Asaduzzaman, K. Ahmed, T. Bhuiyan, and M. F. H. Arif, "Design of simple structure gas sensor Based on hybrid photonic crystal fiber," *Cumhuriyet Science Journal*, 2016, 37(3): 187–196.

[35] X. Yu, Y. Zhang, Y. C. Kwok, and P. Shum, "Highly sensitive photonic crystal fiber based absorption spectroscopy," *Sensors and Actuators: B Chemical*, 2010, 145(1): 110–113.

[36] T. M. Monro, W. Belardi, K. Furusawa, J. C. Baggett, N. G. R. Broderick, and D. J. Richardson, "Sensing with microstructured optical fibres," *Measurement Science and Technology*, 2001, 12(7): 854–858.

[37] M. Morshed, S. Asaduzzaman, and K. Ahmed, "Proposal of simple gas sensor based on micro structure optical fiber," in *IEEE International Conference on Electrical Engineering and Information Communication Technology*, Dhaka, Bangladesh, pp. 1–5, 2015.

[38] J. W. Wang, C. Jiang, W. S. Hu, and M. Y. Gao, "Properties of index-guided PCF with air-core," *Optics & Laser Technology*, 2007, 39(2): 317–321.

[39] N. A. Mortensen, J. R. Folkenberg, M. D. Nielsen, and K. P. Hansen, "Modal cutoff and the V parameter in photonic crystal fibers," *Optics Letters*, 2003, 28(20): 1879.

[40] M. Morshed, M. F. H. Airf, S. Asaduzzaman, and K. Ahmed, "Design and characterization of photonic crystal fiber for sensing applications," *European Journal of Scientific Research*, 2016, 11(12): 228–235.

[41] S. Olyaee and F. Taghipour, "Doped-core octagonal Photonic crystal fiber with ultra-flattened nearly zero dispersion and low confinement loss in a wide wavelength range," *Fiber and Integrated Optics*, 2012, 31(3): 178–185.

[42] S. E. Kim, B. H. Kim, C. G. Lee, S. Lee, K. Oh, and C. S. Kee, "Elliptical defected core photonic crystal fiber with high birefringence and negative flattened dispersion," *Optics Express*, 2012, 20(2): 1385.

[43] P. Ma, N. F. Song, J. Jin, J. M. Song, and X. B. Xu, "Birefringence sensitivity to temperature of polarization maintaining photonic crystal fibers," *Optics & Laser Technology*, 2012, 44(6): 1829–1833.

[44] S. Sen, S. Chowdhury, K. Ahmed, and S. Asaduzzaman, "Design of a porous cored hexagonal photonic crystal fiber based optical sensor with high relative sensitivity for lower operating wavelength," *Photonic Sensors*, 2017, 7(1): 55–65.

[45] S. Asaduzzaman, K. Ahmed, and B. K. Paul, "Slotted-core photonic crystal fiber in gas-sensing application," *SPIE*, 2016, 10025: 100250O1–100250O9.

[46] M. Napierala, T. Nasilowski, E. Beres-Pawlik, F. Berghmans, J. Wojcik, and H. Thienpont, "Extremely large-mode-area photonic crystal fiber with low bending loss," *Optics express*, 2010, 18(15): 15408–15418.

[47] J. Broeng, D. Mogilevstev, S. E. Barkou, and A. Bjarklev, "Photonic crystal fibers: a new class of optical waveguides," *Optical Fiber Technology*, 1999, 5(3): 305–330.

[48] M. N. Petrovich, A. Brakel, F. Poletti, K. Mukasa, E.

Austin, V. Finazzi, *et al.*, "Microstructured fibers for sensing applications," *SPIE*, 2005, 6005: 60050E1–60050E15.

[49] H. H. El, Y. Ouerdane, L. Bigot, G. Bouwmans, B. Capoen, A. Boukenter, *et al.*, "Sol-gel derived ionic copper-doped microstructured optical fiber: a potential selective ultraviolet radiation dosimeter," *Optics Express*, 2012, 20(28): 29751.

[50] H. Ebendorff-Heidepriem, P. Petropoulos, S. Asimakis, V. Finazzi, R. Moore, K. Frampton, *et al.*, "Bismuth glass holey fibers with high nonlinearity," *Optics Express*, 2004, 12(21): 5082.

[51] K. M. Kiang, K. Frampton, T. M. Monro, R. Moore, J. Tucknott, D. W. Hewak, *et al.*, "Extruded single-mode non-silica glass holey optical fibers," *Electronics Letters*, 2002, 38(12): 546–547.

14

Fiber Optic Gyroscope Dynamic North-Finder Algorithm Modeling and Analysis Based on Simulink

Zhengyi ZHANG* and Chuntong LIU

Department Two, Rocket Force University of Engineering , Xi'an, 710025, China

*Corresponding author: Zhengyi ZHANG   E-mail: 18809231139@163.com

Abstract: In view of the problems such as the lower automation level and the insufficient precision of the traditional fiber optic gyroscope (FOG) static north-finder, this paper focuses on the in-depth analysis of the FOG dynamic north-finder principle and algorithm. The simulation model of the FOG dynamic north found algorithm with the least square method by points is established using Simulink toolbox, and then the platform rotation speed and sampling frequency, which affect FOG dynamic north found precision obviously, are simulated and calculated, and the optimization analysis is carried out as a key consideration. The simulation results show that, when the platform rotation speed is between 4.5°/s and 8.5°/s and the sampling frequency is at about 50 Hz in the case of using the parameters of this paper, the FOG dynamic north finding system can reach the higher precision. And the conclusions can provide the reference and validation for the engineering and practical of FOG dynamic north-finder.

Keywords: Fiber optics; optical fiber gyroscope; dynamic north found; modeling

1. Introduction

With the in-depth development of equipment information, modern warfare has been gradually changed to all-weather, all-direction, rapid maneuver, and precision strike direction, and this will require the weapon systems having the abilities of fast and precise positioning and directing [1, 2]. For the inertial north finding technology is not dependent on the outside world, it has become one of the important guarantees for the weapon system realizing autonomic, fast, mobile, and accurate targeting goals [3, 4]. Fiber optic gyroscope (FOG) north-finder is one of new inertial north-finding technologies, which is increasingly used in defense and civil applications. It not only provides ideal

directional information for the weapon system and equipment, but also provides precise orientation and attitude control reference for oil drilling, robots, and other civilian areas [5, 6].

The principle of the FOG north-finding mainly relies on its sensitiveness of the horizontal component of the earth angular rate. For this value is very small, the north-finder resolution mainly depends on the precision of the FOG. However, it will pay a higher cost for improving the level of hardware resources on the basis of manufacturing processes and technology at present. On the other hand, the north-finding algorithm affects the FOG north-finding precision and rapidity obviously, so the north-finding algorithm can be considered to be improved in order to better meet the practical requirements for the gyroscope north-finding system. Among the FOG north-finding algorithm, static north-finding is a traditional method and has been widely used, but the operation is complex, and north-finding precision is limited [7, 8]. The FOG dynamic north-finding plan is a new inertial positioning method, which is being explored and taken much of the focus of researchers at home and abroad. It refers to a new method that the FOG north seeker rotates with the turntable around its central axis in continuous constant-speed in the process, and it can calculate the initial orientation according to the FOG output signal [9]. Compared with the FOG static north finding plan, the FOG output signal can be modulated periodically by continuous constant-speed mechanical rotation. By using the appropriate solution, it can inhibit the constant drift and random drift of the FOG effectively, shorten the north finding time, and improve the accuracy of north finder solution. However, for the reason of structure and limited process level, there are few examples of the FOG dynamic north finding used in engineering [10].

According to the above background, this paper studies mainly on the FOG dynamic north finding principle and algorithm by using Simulink simulation toolbox of MATLAB, and focuses on the simulation model building and analysis of the least square method by points of the FOG dynamic north finding algorithm, which is used to realize the parameters optimization design of the rotation speed and sampling frequency for the FOG north finder system, and provide reference for engineering and applications of the FOG dynamic north finder.

2. FOG and its dynamic north finding principle

The FOG is a new all-solid-state angular rate sensor based on the Sagnac effect. Compared with the conventional electro-mechanical gyroscope, it not only has the advantages of quick start, wide dynamic range, and high overload, but also has

better zero deviation repeatability, and it is not sensitive to the movement and noise of the intersecting axes [11, 12]. In the practical engineering application, although the north finding ways are different, its basic principle is to measure the horizontal component (ω^b_{ie}) of the earth rotation angular rate (ω_{ie}) in the geographic coordinate system and calculate the included angle between the carrier ordinate axis direction and geographic north direction or give the azimuth of true north to realize autonomy orientation. Its principle is shown in Fig. 1(a).

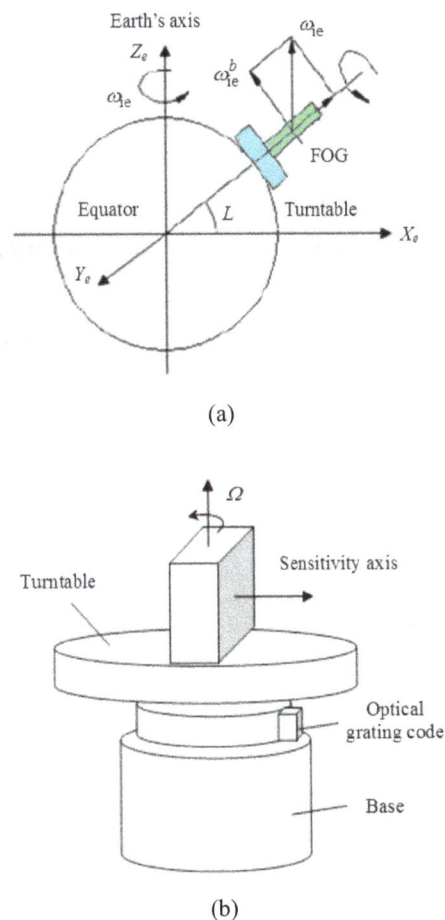

(a)

(b)

Fig. 1 Fiber optic gyro and its dynamic north-seeking principle: (a) principle of the fiber-optic gyro north-finding and (b) principle of the fiber optical gyro dynamic north seeker system.

The FOG dynamic north finding algorithm refers to a way in the north finding process where gyro rotates with the turntable from the initial location around the vertical axis at a constant angular rate Ω

continuously, and real-time sampling of gyro output value is carried out, then we can calculate the azimuth directly or included angle between the carrier ordinate axis direction and geographic north direction according to different algorithms. Its principle is shown in Fig. 1(b).

Ideally, when FOG rotates with the turntable at a constant angular rate ω, the projection of the horizontal component of the earth rotation angular rate on the carrier ordinate axis direction can be expressed as [13]

$$\omega_{ie}^b = \omega_{ie}\left[\cos L\cos\theta\cos(2\pi ft_i + \Psi_0) + \sin\theta\sin L\right] \quad (1)$$

where ω_{ie} is the earth rotation angular rate and its value is $15.04108°$ per hour, L is the latitude value of the carrier, Ψ_0 is the included angle between the carrier ordinate axis direction and geographic north direction, θ is the pitch angle, f is the rotation frequency of the turntable, and obviously ω is $2\pi f$.

Theoretically, when the input state of the fiber-optic gyroscope is constant, its output is a constant proportional to ω_{ie}^b. But the actual measurement of output includes several random error terms because of the existence of the gyro drift of itself, thermal noise of data acquisition circuit, variation of the rotation rate of the turntable, and low frequency noise (ground shaking and wind) in the ambient environment. Therefore, the practical dynamic output model of the fiber optic gyroscope can be expressed as [14]

$$\omega_{outi}(t) = K\omega_{ie}\times$$
$$\left[\cos L\cos\theta\sin\left(\frac{\pi}{2} - (2\pi ft_i + \Psi_0)\right) + \sin\theta\sin L\right] + \varepsilon_0 + \varepsilon_i \quad (2)$$

where K is the scale factor of the FOG, ε_0 is the constant zero deviation of the FOG, and ε_i is the random drift of the gyro including white noise.

In ideal conditions, the FOG can be fixed on the leveling turntable, then θ is zero. When the turntable rotates at a constant angular rate (Ω), the output model of the fiber optic gyro can be simplified as

$$\omega_{outi}(t_i) = K\omega_{ie}\cos L\cos(\Omega t_i + \Psi_0) + \varepsilon_0 + \varepsilon_i \quad (3)$$

where Ψ_0 can be calculated according to the above equation, then the azimuth of the true north direction is known, and the task of north finding is completed. Overall, the dynamic north-seeking scheme has higher requirements on the overall performance of the system, and its outstanding advantages include better precision and short north-finding time, which meet the requirements of the development direction toward integration, automation, high precision, and short time.

3. Modeling and simulation analysis of dynamic north-finding algorithm

3.1 Modeling and simulation analysis of dynamic north-finding algorithm

The above parts are basic principles of the FOG dynamic north finding. When FOG components rotate with the turntable at a constant speed, the dynamic output at a particular sampling frequency in real time can be measured. Then according to total sampling points of a full period or several full periods, least squares parameter estimation is used to calculate the initial included angle (Ψ_0) between the carrier ordinate axis direction and geographic north direction.

According to the fiber optic gyroscope dynamic output model in (3), it can be broken down into

$$\omega_{outi}(t_i) = K\omega_{ie}\cos L\cos\Omega t_i\cos\Psi_0 - K\omega_{ie}\cos L\sin\Omega t_i\sin\Psi_0 + \varepsilon_0 + \varepsilon_i \quad (4)$$

Let $A = K\omega_{ie}\cos L\cos\Psi_0$, $B = -K\omega_{ie}\cos L\cos\Psi_0$, $\alpha_i = \Omega t_i$ and the effects of error are overlooked then as follows:

$$\omega_{outi}(t_i) = A\cos\alpha_i + B\sin\alpha_i. \quad (5)$$

By using least squares parameter estimation, estimates of A and B can be derived as follows:

$$\begin{cases} \hat{A} = \dfrac{\sum\limits_{i=1}^{n}\sin^2\alpha_i \sum\limits_{i=1}^{n}\omega_{\text{out}i}\cos\alpha_i - \sum\limits_{i=1}^{n}\omega_{\text{out}i}\sin\alpha_i \sum\limits_{i=1}^{n}\sin\alpha_i\cos\alpha_i}{\sum\limits_{i=1}^{n}\sin^2\alpha_i \sum\limits_{i=1}^{n}\cos^2\alpha_i - \left(\sum\limits_{i=1}^{n}\sin\alpha_i\cos\alpha_i\right)^2} \\[4ex] \hat{B} = \dfrac{\sum\limits_{i=1}^{n}\cos^2\alpha_i \sum\limits_{i=1}^{n}\omega_{\text{out}i}\sin\alpha_i - \sum\limits_{i=1}^{n}\omega_{\text{out}i}\cos\alpha_i \sum\limits_{i=1}^{n}\sin\alpha_i\cos\alpha_i}{\sum\limits_{i=1}^{n}\sin^2\alpha_i \sum\limits_{i=1}^{n}\cos^2\alpha_i - \left(\sum\limits_{i=1}^{n}\sin\alpha_i\cos\alpha_i\right)^2} \end{cases} \tag{6}$$

At this point, the estimate of the initial azimuth is as follows:

$$\Psi_0 = \text{ac}\tan\left(-\hat{B}/\hat{A}\right). \tag{7}$$

When the dynamic north finding is carried out by using this method, relevant parameters can be adjusted flexibly according to different requirements of the precision and real-time requirements, which has advantages of simple operation and system stability.

3.2 Modeling and analysis of algorithms

According to the principle of the least squares estimation algorithm used in the dynamic north-seeking of the fiber optic gyroscope, models of the algorithm are built by using the MATLAB Simulink simulation tool. Specific parameters of the models are set as follows: the scale factor of the FOG $K_s = 0.81$, the latitude is $34°16'$, the earth rotation angular rate $\omega_e = 15.04\,108\,°/h$, and the initial included angle between the fiber optic gyroscope sensing axis and true north is $10°$. For the complexity of the specific model of algorithm and space limitations, this paper focuses on the analysis and discussion of simulation results. In the dynamic north-seeking system of the FOG, when hardware resources are determined, the rotary speed and sampling frequency are two important parameters affecting the north-seeking precision. Simulation analysis is carried out aiming at the effect of the variety of both parameters on the north-seeking precision by using the model above.

Firstly, under the condition that the sampling frequency and other parameters remain constant, the speeds of the rotating platform are taken respectively as $1°/s$, $2°/s$, $5°/s$, and $10°/s$. The calculated outputs of dynamic north-finding system of the fiber-optic gyro are shown in Fig. 2.

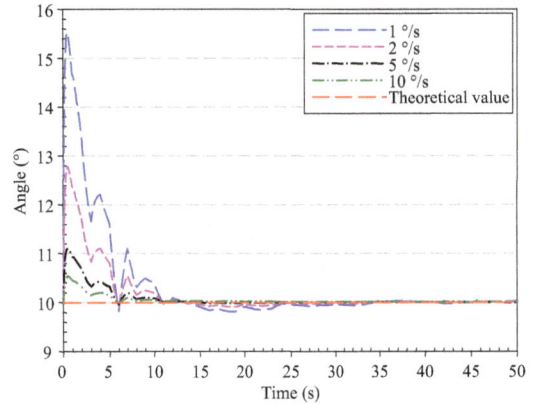

Fig. 2 Speed of the simulation results.

It can be seen from the simulation results in Fig. 2 that in the initial stage of the north-finding system, the deviation between the output value and the theoretical value ($10°$) is large, and $20\,s$ later, the output is close to the theoretical value and tends to be steady. In addition, according to the simulation results, it can be seen that when the platform rotates at different speeds, steady-state errors between the final steady-state output of the north-seeking system and the theoretical value are different. Different steady-state errors are calculated when the rotary speed of the turntable varies from $1°/s$ to $10°/s$, which are shown in Fig. 3.

The results in Fig. 3 show that when other parameters of the FOG remain unchanged, the rotary speed of the platform has an obvious effect on the precision of the north-seeking. It will result in a large error when the speed is too high or too low. According to the parameters of the FOG used in the experiment, when the speed of the rotating platform is approximately between 4.5 °/s and 8.5 °/s, the theoretical precision of the north-seeking can be limited within a range of about ± 10″. Therefore, there is necessity to determine an appropriate speed in the practical application of the FOG dynamic north-seeking system.

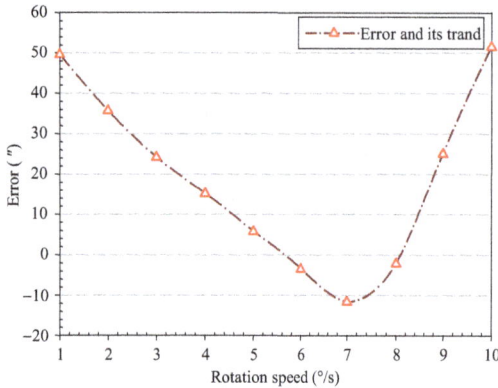

Fig. 3 Speed north errors.

The above results in Fig. 3 show that when other parameters of the FOG remain unchanged, the rotary speed of the platform has an obvious effect on the precision of the north-seeking. It will result in a large error when the speed is too high or too low. According to the parameters of the FOG used in the experiment, when the speed of the rotating platform is approximately between 4.5 °/s and 8.5 °/s, the theoretical precision of the north-seeking can be limited within a range of about ± 10″. Therefore, there is necessity to determine an appropriate speed in the practical application of the FOG dynamic north-seeking system.

In addition, the sampling frequency of the FOG output has a certain effect on the north-seeking precision. When all other parameters are constant, the sampling frequency of the models is set as 10 Hz, 20 Hz, 50 Hz, and 100 Hz for simulation, and the results of different sampling frequencies are shown in Fig. 4.

According to the simulation results, it can be seen that when the sampling frequency is low, the output error of the north-seeking system is large. As the sampling frequency increases step by step, the steady-state error gradually decreases to a steady state. It is shown in Fig. 5 that how error trend between the output of north-seeking system and the theoretical

value varies as the sampling frequency varies.

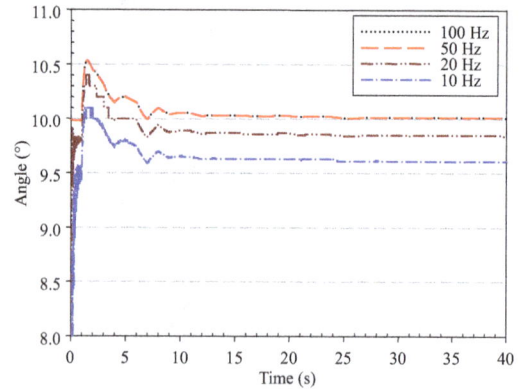

Fig. 4 Sample frequencies of the simulation results.

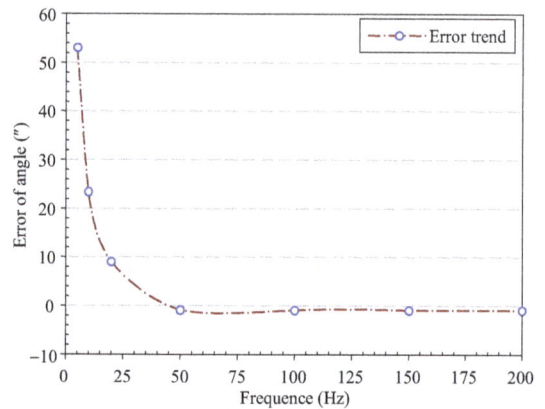

Fig. 5 Error with sampling frequencies of north trend.

According to the above calculations, it is shown in Fig. 5 that when the sampling frequency is lower than 50 Hz, the north-finding error is large, but the steady-state error of the FOG north finding system is stabilized at about −1″ when the sampling frequency is higher than 50 Hz. So the high sampling frequency will not improve the precision, instead it will increase the cost. Therefore, the sampling frequency of the FOG north finding system is usually determined by the maximum sampling frequency of the FOG output.

Therefore, in summary, when the FOG is used in the dynamic north finding, the best turntable rotation speed and output sampling frequency should be determined by the precision and performance parameters of the FOG, in order to obtain the higher precision of the north finding with less time. Under the FOG parameters in this paper, when the angular rate of the rotating platform is approximately between 4.5 °/s and 8.5 °/s and the sampling frequency is 50 Hz, it will get higher north-seeking precision.

As a new type of all solid-state angular rate sensor, the fiber optic gyroscope is very suitable to be the main inertial measurement unit of the north-seeking orientating system. Compared with the traditional static north-finding method, the dynamic north-seeking scheme of the FOG can effectively control the gyro constant drift and random drift, reduce the search time, and improve the precision of the north-finding solution. Based on the analysis of the principle of the dynamic north-seeking with the continuous rotary FOG, this paper focuses on the analysis of the principle of the least squares parameter estimation algorithm, and the model of the dynamic north-seeking system is built by using the Simulink simulation tool. The simulation and optimization analysis of the sampling frequency and rotary speed of the turntable are carried out, which are the two key factors affecting the precision of the north finding. The simulation results show that it will affect the north-seeking precision when the speed of the rotating platform is too low or too high, so there is a speed range of high precision. As the sampling frequency of the system gradually increases, the steady-state error of the north-seeking gradually decreases to a constant. But too high sampling frequency will not improve the precision, instead, it will increase the cost. Therefore, the conclusions of this paper can provide the theoretical reference and simulation basis for achieving high precision and practical application of the dynamic north finding of the FOG.

Acknowledgment

The authors are grateful to all of the subjects who participated in this research.

References

[1] O. Celikel, "Application of the vector modulation method to the north finder capability gyroscope as a directional sensor," *Measurement Science & Technology*, 2011, 22(22): 035203.

[2] Q. X. Jiang, X. B. Chen, X. H. Ma, and H. Wang, "North seeker using single axis FOG," *Journal of Chinese Inertial Technology*, 2010, 18(2): 165–169.

[3] S. W. Lloyd, S. H. Fan, and M. J. F. Digonnet, "Experimental observation of low noise and low drift in a laser-driven fiber optic gyroscope," *Journal of Lightwave Technology*, 2013, 31(13): 2079–2085.

[4] D. B. Liu, J. Y. Liu, and J. Z. Lai, "Study of single cycle fast dynamic north seeking algorithm based on fiber optic gyroscope," *Transducer & Microsystem Technologies*, 2007, 26(11): 61–64.

[5] H. L. Ma, X. H. Yu, and Z. H. Jin, "Reduction of polarization-fluctuation induced drift in resonator fiber optic gyro by a resonator integrating in-line polarizers," *Optics Letters*, 2012, 37(16): 3342–3344.

[6] L. F. Wang, Y. Xie, G. L. Yang, and Y. Y. Wang, "Error analysis of the strap-down north-finder based on the fiber optic gyroscope," *Opto-Electronic Engineering*, 2011, 38(5): 46–51.

[7] X. H. Yu, H. L. Ma, and Z. H. Jin, "Improving thermal stability of a resonator fiber optic gyro employing a polarizing resonator," *Optics Express*, 2013, 21(1): 358–369.

[8] Z. H. Jin, G. H. Zhang, H. Mao, and H. L. Ma, "Resonator micro optic gyro with double phase modulation technique using an FPGA-based digital processor," *Optics Communications*, 2012, 285(5): 645–649.

[9] H. G. Xu and Z. B. Guo, "Research on rotating FOG north-finder," *Piezoelectrics & Acoustooptics*, 2010, 32(2): 38–41.

[10] N. El-Sheimy, H. Hou, and X. Niu, "Analysis and modeling of inertial sensors using Allan variance," *IEEE Transactions on Instrumentation & Measurement*, 2008, 57(1): 140–149.

[11] Q. X. Jiang, X. B. Chen, X. H. Ma, and H. Wang, "North seeker using single axis FOG," *Journal of Chinese Inertial Technology*, 2010, 18(2): 165–169.

[12] X. Q. Gou, L. Huang, and W. Liu, "High precision north determining scheme based on FOG bias stability," *Journal of Chinese Inertial Technology*, 2009, 17(3): 258–260.

[13] Z. Q. Wang, J. Y. Zhao, M. J. Xie, and F. D. Gao, "Design and accuracy analysis for a fast high precision independence north-seeking," *Acta Armamentarii*, 2008, 29(2): 164–168.

[14] X. Q. Gou, J. Wei, G. Wang, W. Liu, and X. Q. Han, "Study on rate fiber optic gyroscope north-finder slop compensating arithmetic," *Acta Photonica Sinica*, 2007, 36(12): 2342–2345.

15

Optimization of Top Coupling Grating for Very Long Wavelength QWIP Based on Surface Plasmon

Guodong WANG[1*], Junling SHEN[2], Xiaolian LIU[1], Lu NI[2], and Saili WANG[2]

[1]*School of Physics and Electronic Information Engineering, Henan Polytechnic University, No.2001, Shiji road, Jiaozuo, 454003, China*

[2]*School of Electrical Engineering and Automation, Henan Polytechnic University, No.2001, Shiji road, Jiaozuo, 454003, China*

*Corresponding author: Guodong WANG   E-mail: wgd@hpu.edu.cn

Abstract: The relative coupling efficiency of two-dimensional (2D) grating based on surface plasmon for very long wavelength quantum well infrared detector is analyzed by using the three-dimensional finite-difference time domain (3D-FDTD) method algorithm. The relative coupling efficiency with respect to the grating parameters, such as grating pitch, duty ratio, and grating thickness, is analyzed. The calculated results show that the relative coupling efficiency would reach the largest value for the 14.5 μm incident infrared light when taking the grating pitch as 4.4 μm, the duty ratio as 0.325, and the grating thickness as 0.07 μm, respectively.

Keywords: Very long wavelength; QWIP; surface plasmon; 2D grating

1. Introduction

The very long wavelength infrared (VLWIR) spectral range from 14 μm to 20 μm is of great importance for space applications [1, 2]. Quantum well infrared photodetectors (QWIPs) based on the GaAs/AlGaAs material system have been studied in detail and have already been used in large format focal plane arrays [3–6]. Because of its mature growth and processing technology, QWIPs have been investigated as an alternative for VLWIR detection [7, 8].

However, VLWIR QWIPs have a lower performance than that of HgCdTe detectors and suffer from low response photocurrent, partly because of the lower coupling efficiency of top metal grating. Therefore, it is quite necessary to improve the performances of VLWIR QWIPs.

Since Ebbesen [9] discovered the interesting extraordinary optical transmission of periodic holes array perforated in the metal film, the surface plasmon (SP) has attracted a lot of research interests. SP has been applied in different kinds of devices, such as semiconductors lasers, quantum dot solar cells, and infrared detectors [10–12]. In the very long wavelength infrared band, the SP propagation length is large enough to effectively reach the entire quantum well region, and the optical loss is relatively small. So if SP is applied in VLWIR QWIP, the coupling efficiency of grating and response photocurrent will be improved evidently.

In this paper, the coupling grating based on SP for the VLWIR QWIPs is simulated by the finite-difference time domain method (FDTD). The relative coupling efficiencies with respect to different grating parameters are compared, and the optimal parameters of grating have been obtained.

2. Simulation model

The three-dimensional finite-difference time domain method (3D-FDTD) is strictly a numerical algorithm for solving Maxwell's equations. It can be applied to simulate the SP waves and electric field component distribution. The simulated structure is shown in Fig. 1. From the bottom to top, the QWIP is composed of GaAs substrate, n-type GaAs bottom contact layer, GaAs/AlGaAs multiple quantum well layer, n-type GaAs top contact layer, and grating layer (periodic hole array). The source is a normally incident plane wave. For convenience, the growth direction of the MQW layer is set to z-coordinate, the bottom of grating layer is set to z=0, the device plane is set to x-y plane, and the central point of x-y plane is set to x = 0, y = 0. P and D respectively represent the grating pitch and the diameter of the hole. Supposing the light incident from the top, the coupling efficiency of x-y at z point can be expressed as [13]

$$\eta(z)=\frac{\iint E_z^2(x,y,z)dxdy}{\iint E_{in}^2(x,y,z)dxdy} \quad (1)$$

where E_{in} represents the electric field component of the incident light, and E_z represents the electric field component along the z axis in the x-y plane.

One attraction is that because of the limitation of computer, all the MQW region is not selected to integration. Accordingly, a reasonable integral area is selected and fixed, so the relative coupling efficiency is obtained. This is only due to the limitations of numerical calculation, but does not affect the results of the physical characteristics of

the performance. In this paper, we fix the integration area as $9.6\,\mu m \times 9.6\sqrt{3}\,\mu m$ in the center of x-y plane.

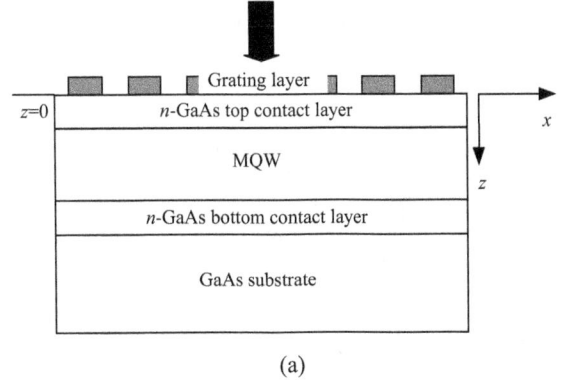

(a)

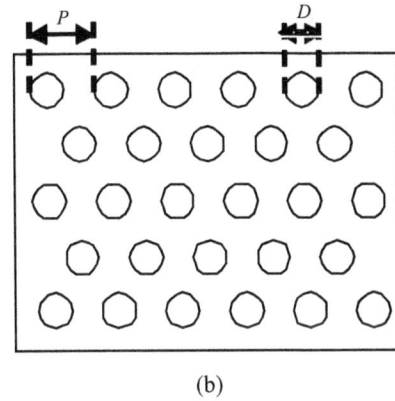

(b)

Fig. 1 Modelling structure of VLWIR QWIP: (a) device structure and coordinate direction and (b) structure of periodic hole array.

3. Calculation and analysis

Firstly, by setting the incident light as TE mode plane wave ($E_z = 0$) and the central wavelength of incident source as 14.5 μm, the grating parameters as grating pitch P = 2.8 μm, the diameter of hole D = 1.6 μm, the thickness of grating layer L = 0.08 μm, and the E_z field in x-y plane are calculated by FDTD. The E_z distribution at z = 0.11 μm is illustrated in Fig. 2. As can be seen from Fig. 2, the direction of infrared light propagation has been changed obviously, and E_z is concentrated on the position in correspondence with the grating holes.

Further calculation shows that the intensity of the light in different x-y planes is decreased with an increase in z, which follows the exponential law, as shown in Fig. 3. In other words, it can be concluded that the farther away from the grating is, the lower intensity of light in x-y plane is. It also fits well with the propagation properties of SP wave. So it can be proved that E_z in the x-y plane is indeed excited by SP.

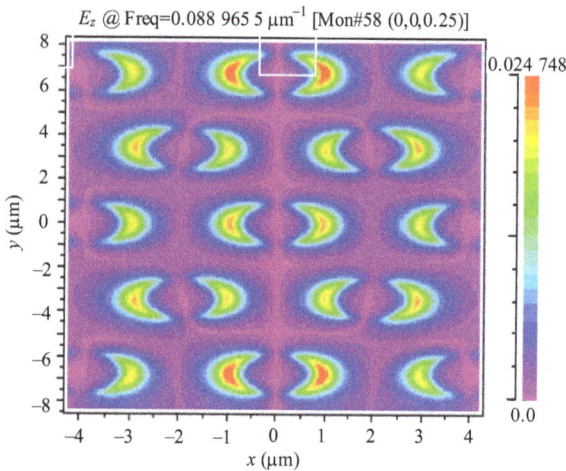

Fig. 2 Distribution of E_z at z = 0.11 μm x-y plane.

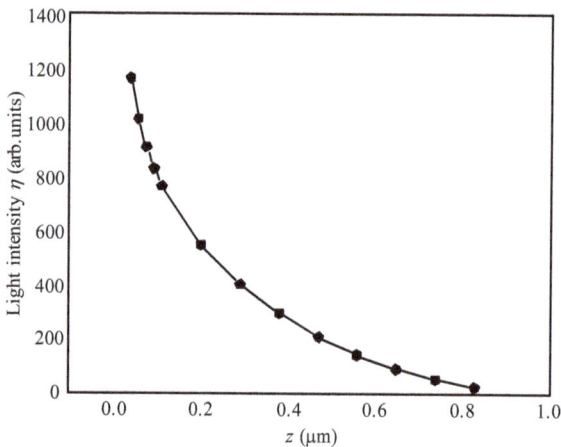

Fig. 3 Distribution of E_z at different x–y planes.

Secondly, in order to obtain the more accurate results, the relative coupling efficiency of grating is calculated by (1) in the fixed integral area which is described above. Figure 4 shows the relative coupling efficiency with respect to the grating pitch. It can be seen that the grating coupling efficiency reaches the maximum when the grating pitch is 4.4 μm. This result is in agreement with the grating equation.

And then, by taking the grating pitch as

$P = 4.4$ μm, the relative coupling efficiency is calculated when the duty ratio D/P is changed. The calculated results are shown in Fig. 5. It shows that the peak value of relative coupling efficiency is achieved when the duty ratio is taken as 0.325.

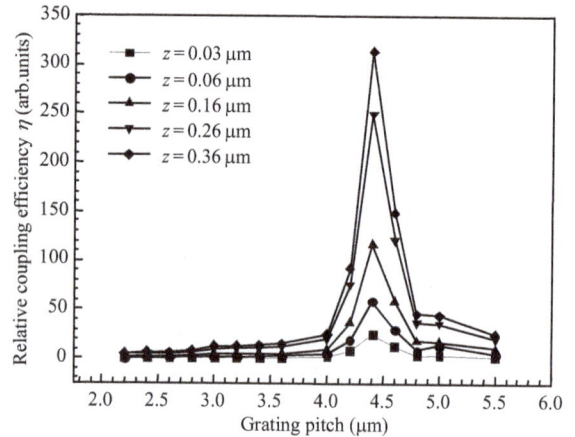

Fig. 4 Relative coupling efficiency with respect to grating pitch.

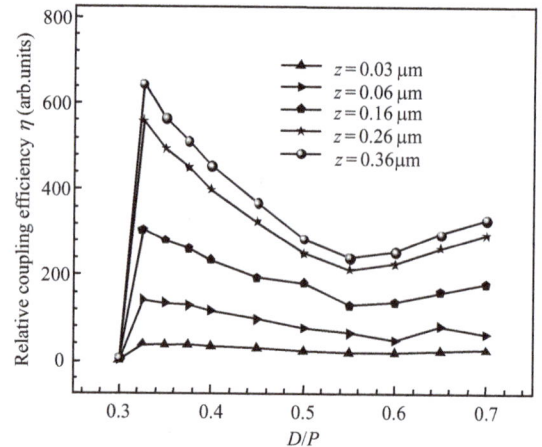

Fig. 5 Relative coupling efficiency with respect to duty ratio D/P.

Then, by taking duty ratio as $D/P = 0.325$, the relative coupling efficiency is calculated with respect to the grating thickness. The calculated results are shown in Fig. 6. It can be concluded that the largest relative coupling efficiency will be obtained when the grating thickness is 0.07 μm.

Finally, the optimized parameters of coupling grating for 15.4 μm input infrared light is obtained, which is illustrated as the grating pitch 4.4 μm, the duty ratio 0.325, and the grating thickness 0.07 μm.

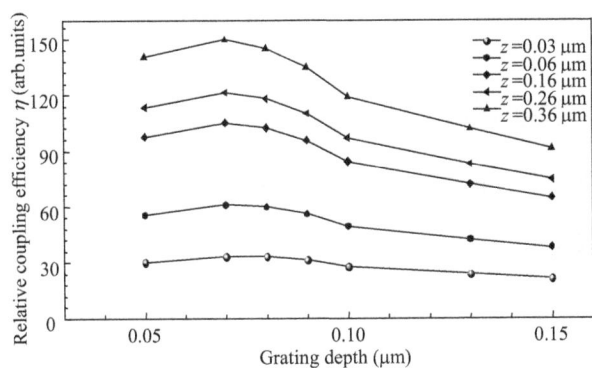

Fig. 6 Relative coupling efficiency with respect to grating thickness.

4. Conclusions

In conclusion, the relative coupling efficiency of top 2D metal coupling grating based on SP for the VLWIR QWIP is calculated by FDTD. The optimized parameters grating is obtained by maximizing the coupling efficiency for different grating parameters. This work is beneficial to enhancing the performance of the very long wavelength QWIP focal plane array devices.

Acknowledgement

This work was supported by the National Natural Science Foundation of China under Grant No. U1304608, the Outstanding Youth Funding of Henan Polytechnic University under Grant No. J2013-05, and Program for Innovative Research Team of Henan Polytechnic University under Grant No. T2015-3.

References

[1] Y. Wei, W. Q. Ma, Y. H. Zhang, J. L. Huang, Y. L. Cao, and K. Cui, "High structural quality of type II InAs/GaSb superlattices for very long wavelength infrared detection by interface control," *IEEE Journal of Quantum Electronics*, 2012, 48(4): 512–515.

[2] A. M. Hoang, G. Chen, R. Chevallier, A. Haddadi, and M. Razeghi, "High performance photodiodes based on InAs/InAsSb type-II superlattices for very long wavelength infrared detection," *Applied Physics Letters*, 2014, 104(25): 251105–1–251105–4.

[3] A. Rogalski, "Recent progress in infrared detector technologies," *Infrared Physics & Technology*, 2011, 54(3): 136–154.

[4] S. D. Gunapala, D. Z. Ting, C. J. Hill, J. Nguyen, A. Soibel, S. B. Rafol, *et al.*, "Large area III-V infrared focal planes," *Infrared Physics & Technology*, 2011, 54(3): 155–163.

[5] G. D. Wang, S. L. Dai, and H. Zhang, "Optimization fo top coupling grating for mid-wave quantum well infrared photodetector," *Chinese Optics Letters*, 2012, 10(B06): 188–189.

[6] W. Lu, L. Li, H. L. Zheng, W. L. Xu, and D. Y. Xiong, "Development of an infrared detector: quantum well infrared photodetector," *Science China Physics, Mechanics & Astronomy*, 2009, 52(7): 969–977.

[7] X. H. Liu, X. H. Zhou, N. Li, L. Wang, Q. L. Sun, K. S. Liao, *et al.*, "Effects of bias and temperature on the intersubband absorption in very long wavelength GaAs/AlGaAs quantum well infrared photodetectors," *Journal of Applied Physics*, 2014, 115(12): 124503.

[8] A. Berurier, A. Nedelcu, V. Gueriaux, T. Bria, A. D. Rossi, X. Marcadet, *et al.*, "Optimization of broadband (11–15 μm) optical coupling in quantum well infrared photodetectors for space applications," *Infrared Physics & Technology*, 2011, 54(3): 182–188.

[9] C. Genet and T. W. Ebbesen, "Light in tiny holes," *Nature*, 445(7123): 39–46.

[10] W. Wu, A. Bonakdar, and H. Mohseni, "Plasmonic enhanced quantum well infrared photodetector with high detectivity," *Applied Physics Letters*, 2010, 96(16): 161107-1–161107-3.

[11] C. C. Chang, Y. D. Sharma, Y. S. Kim, J. A. Bur, R. V. Shenoi, S. Krishna, *et al.*, "A surface plasmon enhanced infrared photodetector based on InAs quantum dots," *Nano Letters*, 2010, 10(5): 1704–1709.

[12] Q. C. Weng, L. Li, J. Chen, J. Wen, and D. Y Xiong, "The metal grating coupling of long wavelength quantum well infrared photodetectors: surface plasmon effect," *Journal of Infrared & Millimeter Waves*, 2011, 30(5): 415–418.

[13] K. Wang, W. H. Zheng, G. Ren, X. Y. Du, M. X. Xing, and L. H. Chen, "Design and optimization color quantum well infrared photodetectors coupled photonic crystal layer," *Acta Physica Sinica*, 2008, 57(3): 1730–1735.

16

Effects of Rubber Shock Absorber on the Flywheel Micro Vibration in the Satellite Imaging System

Changcheng DENG[1,2], Deqiang MU[1,3], Xuezhi JIA[1], and Zongxuan LI[1]

[1]*National & Local United Engineering Research Center of Small Satellite Technology, Changchun Institute of Optics, Fine Mechanics, and Physics ,Chinese Academy of Sciences, Changchun, 130033, China*

[2]*University of Chinese Academy of Sciences, Beijing, 100039, China*

[3]*Changchun University of Technology, Changchun, 130012, China*

[*]Corresponding author: Changcheng DENG   E-mail: changcheng0211@163.com

Abstract: When a satellite is in orbit, its flywheel will generate micro vibration and affect the imaging quality of the camera. In order to reduce this effect, a rubber shock absorber is used, and a numerical model and an experimental setup are developed to investigate its effect on the micro vibration in the study. An integrated model is developed for the system, and a ray tracing method is used in the modeling. The spot coordinates and displacements of the image plane are obtained, and the modulate transfer function (MTF) of the system is calculated. A satellite including a rubber shock absorber is designed, and the experiments are carried out. Both simulation and experiments results show that the MTF increases almost 10%, suggesting the rubber shock absorber is useful to decrease the flywheel vibration.

Keywords: Micro vibration; flywheel; rubber shock absorber; integrated modeling; MTF; isolation experiment

1. Introduction

During photographing of the in-orbit satellite, all kinds of motion of the moving components of the satellite, such as the rotation of the reaction flywheel, the jet of the propulsor, and the adjustment of solar panels, will cause the jitter response of the camera, thus affecting the image quality. The jitter response, also called the micro vibration system, cannot be measured and controlled by the attitude control system [1]. The rotation of the flywheel will have a relatively large impact on the image quality and is a main issue on the satellite imaging system [2].

To reduce the influence of micro vibration on imaging, different types of vibration control

technologies are used [3], namely as passive and active vibration isolations, vibration absorption, vibration resistance, and dynamic designs. Among them, the vibration isolation method is most widely used. The principle of the vibration isolation method is adding a resilient liner (such as spring, rubber mats, and blankets) between the object and the supporting surface to isolate the vibration [4–6].

A rubber shock absorber [7] has the advantages of compact structure, low cost, good craftwork, etc., which is a very common passive vibration isolator and often used in many fields such as ships, buildings, and bridges. It is also easy to fabricate and meet the required stiffness and strength. Its damping ratio is 0.06 – 0.1 and can absorb mechanical energy, especially high frequency mechanical energy. It can be bonded with metal, forming a multilayered structure to bear load, reduce system stiffness, and change its frequency range. A review on the application of the rubber shock absorber in aerospace can be found in [8]. An accurate and efficient finite element modeling [9] and static characteristics analysis of the rubber shock absorber was proposed by Haiting [12]. Sjoberg investigated rubber shock absorber dynamic modeling and dynamic characteristics [13]. Sjoberg used the method of fractional derivative to establish a simulation model [14]. In this study, a rubber shock absorber is added in the flywheel system to reduce the micro vibration of a satellite imaging system.

The principle of the integrated modeling is to integrate the analysis of the overall system, containing realistic structures, disturbance, optics and controls models, and their mutual interactions [15–17]. It is convenient for electronic transfer of data among all types of analysis, and this method can save operating time, heighten analyzing qualities, and achieve a dynamic front-to-end analysis of the disturbance to performance paths within the satellite. Compared with the traditional design and assessment methods that focus on single machine or single discipline or sub system level, integrated

modeling can provide a systematic and comprehensive performance evaluation and error analysis, and guide systemic design. The ray tracing is used to calculate the location and direction of the output ray, and the angle between the ray and the optical axis. The calculation is based on the reflection lens displacement and the incident light, according to the refraction law. In this paper, the ray tracing method is used in the integrated modeling to analyze the influence of micro vibration on the satellite image.

An experiment is designed to use real products to verify the influence rubber shock absorber on the micro vibration of the flywheel in the satellite imaging system [18]. The data of vibration are obtained by the ground image acquisition and processing system, and the corresponding performance values are calculated. The results show that the rubber shock absorber can reduce the vibration in the transfer path and give out better image, which provides a guideline for the satellite design.

2. Integrated modeling of the satellite

The integrated modeling of the satellite consists of the modeling of the flywheel characteristic of vibration, the finite element modeling of the entire satellite platform, and the optical modeling based on ray tracing [19]. The process is shown in Fig. 1.

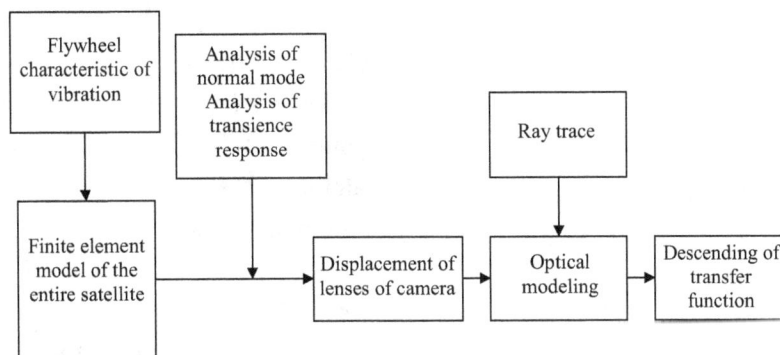

Fig. 1 Schematic of the integrated modeling.

2.1 Modeling of flywheel characteristic of vibration

The micro vibration of the flywheel is caused by static unbalance and dynamic unbalance of the flywheel [20], as shown in Fig. 2. The static unbalance is caused by the deviation of the center of mass of the flywheel rotor from its rotating axis, and the flywheel can be considered as a rotor with two parts, the strict rotational symmetry part and a mass point m_s part at a distance of r_s from the flywheel shaft line. F_r is the rotor inertia force. Dynamic unbalance refers to the uneven distribution of rotor mass which causes the cross product of inertia not zero. The rotor can be considered as two parts, a strict symmetry section part and 2 mass points m_d part at a distance of h in the axial direction. The mass point is at a distance r_d from the rotating axis.

When the rotor rotates, the mass point m_s in static imbalance is subjected to the radial force:

$$F_r = U_s \omega^2 \tag{1}$$

where $U_s = m_s r_s$ is a flywheel mass property, and ω is the angular velocity.

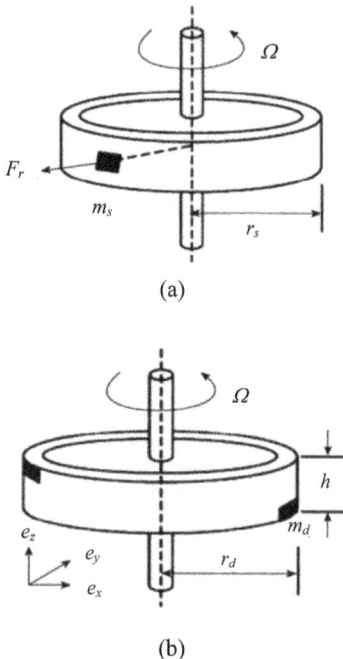

(a)

(b)

Fig. 2 Schematic of two types of the unbalanced flywheel [17]: (a) static unbalanced flywheel and (b) dynamic unbalanced flywheel.

For the dynamic unbalanced flywheel, the force caused by the moment of two mass points m_d is

$$T = U_d \omega^2 \tag{2}$$

where $U_d = m_d r_d h$ is a dynamic unbalance property.

2.2 Finite element model of the entire satellite

The finite element structure model is the basis of dynamic analysis (including the following modal analysis and transient analysis). The dynamics of a multi-degree-of-freedom system (structure) are described in the time domain by the equation [21]:

$$\mathbf{M}\ddot{x} + \mathbf{C}\dot{x} + \mathbf{K}x = \mathbf{F} \tag{3}$$

where $\mathbf{M}$, $\mathbf{C}$, and $\mathbf{K}$ are $N \times N$ dimensional matrices, of which $\mathbf{M}$ is the mass matrix, $\mathbf{C}$ is the damping matrix, and $\mathbf{K}$ is the stiffness matrix; $\mathbf{F}$ is the force matrix; x is the displacement response. Introducing physical modal transformation $x = \mathbf{\Phi}q$ ($\mathbf{\Phi}$ is the $N \times N$ matrix of the modal shape, q is for the modal coordinates). Equation (3) can be decoupled as the modal space equation:

$$\ddot{q} + 2\mathbf{Z}\mathbf{\Omega}\dot{q} + \mathbf{\Omega}^2 q = \mathbf{\Phi}^T \mathbf{F} \tag{4}$$

Supposing retained r modes (three rigid translational modes are not considered), $q \in \mathbf{R}^r$ is modal coordinates; $\mathbf{Z} \in \mathbf{R}^{r \times r}$ is the diagonal damping matrix; $\mathbf{\Phi} \in \mathbf{R}^{n \times r}$ is the mass normalized modal matrix; $\mathbf{\Omega} \in \mathbf{R}^{r \times r}$ is the natural frequency of the diagonal matrix. Equation (5) is the transformed equation to state-space equation of (4):

$$\begin{bmatrix} \dot{x}_p \\ z \\ y \end{bmatrix} = \begin{bmatrix} \mathbf{A}_p & \mathbf{B}_w & \mathbf{B}_u \\ \mathbf{C}_z & \mathbf{D}_{zw} & \mathbf{D}_{zu} \\ \mathbf{C}_y & \mathbf{D}_{yw} & \mathbf{D}_{yu} \end{bmatrix} \begin{bmatrix} x_p \\ w \\ u \end{bmatrix} \tag{5}$$

where $x_p = [\mathbf{q}, \dot{\mathbf{q}}]^T$, $w, u \subset \mathbf{F}$, $w \in \mathbf{R}^{n_w}$, $u \in \mathbf{R}^{n_u}$, w is the disturbance input, and u is the control input; $z \in \mathbf{R}^{n_z}$ is the response output of the satellite; $y \in \mathbf{R}^{n_y}$ is the control measure output. The system matrix is

$$\begin{bmatrix} \mathbf{A}_p & \mathbf{B}_w & \mathbf{B}_u \\ \mathbf{C}_z & \mathbf{D}_{zw} & \mathbf{D}_{zu} \\ \mathbf{C}_y & \mathbf{D}_{yw} & \mathbf{D}_{yu} \end{bmatrix} = \begin{bmatrix} 0 & \mathbf{I} & 0 & 0 \\ -\mathbf{\Omega}^2 & -2\mathbf{Z} & \mathbf{\Phi}^T\mathbf{\beta}_w & \mathbf{\Phi}^T\mathbf{\beta}_u \\ \mathbf{\Phi}\mathbf{\beta}_z & 0 & 0 \\ \mathbf{\Phi}\mathbf{\beta}_y & 0 & 0 \end{bmatrix} \tag{6}$$

where $\boldsymbol{\beta}_w$, $\boldsymbol{\beta}_u$, $\boldsymbol{\beta}_z$, and $\boldsymbol{\beta}_y$ are the modal selection matrices.

Parameters of the camera in the modeling are: focus $f = 8\,mm$; $F\#$ (F-number) = 13.3; wavelength $500\,nm - 800\,nm$. The finite element modeling of the satellite is completed by Patran/Nastran MSC software. The satellite model has 33893 nodes and 21856 elements. The model is used for the following simulation analysis to obtain displacements of all lenses. There is no displacement constrain in the model.

2.3 Ray tracing

The positions of the misaligned lenses are obtained after the vibration, and an optical path transmission is modeled with the ray tracing method in the study. By using kinematic concepts, the position of each lens is described with two coordinate systems, which are the lens coordinate system $o_m x_m y_m z_m$ and the implicated coordinate system $oxyz$. When the optical system is not subject to vibration, the lens coordinate system and implicated coordinate system are identical, and this is in the ideal position. When the vibration occurs, the lens coordinate system diverts from the implicated coordinate system. The lens coordinate system follows the lens, and the implicated coordinate system follows the base. Supposing the light transfers in reflection and the misaligned lens is parabolic, the misaligned lens is explained by an equation of lens in the lens coordinate, which is

$$2pz_m = x_m^2 + y_m^2. \tag{7}$$

The optical path transmission is based on the implicated coordinate. The equation of the misaligned lens relative to the implicated coordinate system is obtained by twice rotation and once translation of coordinates. Two points are taken from the center of up and low surfaces as shown in Fig. 3, i.e., Points 1 and 2, so the connection line of Points 1 and 2 is the optical axis, which can describe the displacement of the lens. Point 1 is the origin of the lens coordinate system, and the connecting direction of the two points is the z_m direction of the coordinate axis. When Point 1 and the connecting direction of the line are determined, the position of the misaligned lens can be obtained. Because a parabolic surface is rotated, there is no need to consider the rotation of the lens.

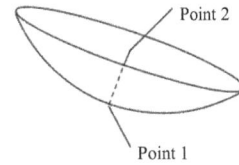

Fig. 3 Key points of lens.

The equation of the misaligned lens can be calculated as in [22]:

$$\begin{aligned}
F(x, y, z) = {} & 2p((x - \Delta x_1)\sin\beta + \cos\beta((z - \Delta z_1)\cos\alpha + \\
& (y - \Delta y_1)\sin\alpha)) - ((x - \Delta x_1)\cos\beta - \sin\beta((z - \Delta z_1)\cos\alpha + \\
& (y - \Delta y_1)\sin\alpha))^2 - ((y - \Delta y_1)\cos\alpha - (z - \Delta z_1)\sin\alpha)^2 \\
= {} & 0
\end{aligned} \tag{8}$$

where α is the angle rotated around axis x of coordinate $oxyz$, to receive coordinate $ox'y'z'$; β is the angle rotated around axis y' of coordinate $ox'y'z'$, to receive coordinate $ox''y''z''$, as shown in Fig. 4; $(\Delta x_1, \Delta y_1, \Delta z_1)$ is the misaligned displacement of Point 1; $(\Delta x_2, \Delta y_2, \Delta z_2)$ is the misaligned displacement of Point 2. Assuming

$$\begin{cases} a = \Delta x_2 - \Delta x_1 \\ b = \Delta y_2 - \Delta x_1 \\ c = \Delta z_2 - \Delta x_1, \end{cases} \tag{9}$$

then

$$\sin\alpha = \frac{b}{\sqrt{b^2 + c^2}}$$

$$\sin\beta = \frac{a}{\sqrt{a^2 + b^2 + c^2}}. \tag{10}$$

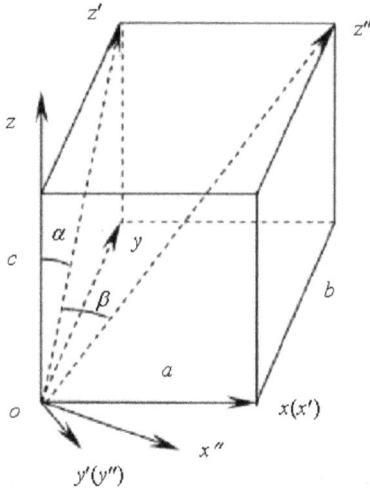

Fig. 4 Rotation of the coordinate system.

The ray tracing method for a single misaligned reflector application is shown in Fig. 5. When the displacements of Points 1 and 2 and the incident ray equation are given, the Matlab solving function is used to obtain the intersection between the misaligned lens and the incident ray. The vector in the direction of the light reflection is calculated by the vector reflection law, thus the angle between the reflection light and the optical axis is obtained.

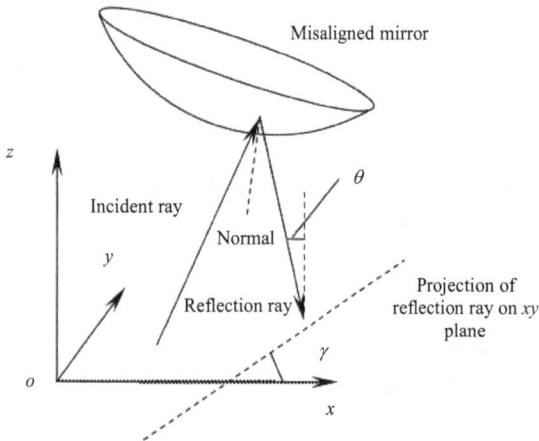

Fig. 5 Ray reflected by the misaligned lens.

The procedures of the ray tracing method for modeling a single misaligned lens optical transmission path are as follows:

(1) Obtain the equation of the misaligned lens by the coordinate transform $F(x,y,z)=0$.

(2) Calculate the intersection between the incident ray and misaligned lens, and the equation of

the incident ray is

$$\frac{x-x_{c-1}}{I_x}=\frac{y-y_{c-1}}{I_y}=\frac{z-z_{c-1}}{I_z} \qquad (11)$$

where $I=\{I_x, I_y, I_z\}$ is the vector of the incident ray.

(3) Obtain the normal vector $N=\{\frac{\partial F}{\partial x},\frac{\partial F}{\partial x},\frac{\partial F}{\partial z}\}$ of the intersection (x_c, y_c, z_c), and N is normalized by the equation $n=\frac{N}{|N|}$.

(4) Calculate the direction vector $R=I-2n(n\cdot I)$ [23] of the reflected ray by the reflection equation in vectors, where $R=\{R_x, R_y, R_z\}$ is the vector of the reflected ray, n is the normalized vector of the lens at the intersection, and the equation of the reflected ray is

$$\frac{x-x_c}{R_x}=\frac{y-y_c}{R_y}=\frac{z-z_c}{R_z}. \qquad (12)$$

(5) Obtain the angle between the reflected ray and optic axis, $\theta=\arccos\frac{R\cdot n_z}{|R||n_z|}$, where $n_z=\{0, 0, -1\}$.

Because the reflected light is downward, the angle between the vertical downward direction and reflected ray can be acquired. The reflected ray is projected to the xy plane, and then the angle γ between the projection line and x axis is obtained. When the ray tracing method is applied to the plural misaligned lens, the above five steps will be applied to each misaligned lens in turn.

3. Design of rubber shock absorber

Only one flywheel is used in the study, which has a total mass of 3.585 kg including four screws on the bracket (the mass of the bracket is 1.11 kg). It is installed on the y load plate, and each screw is mounted with a rubber shock absorber, as shown in Fig. 6.

The limit static stress σ of the rubber shock absorber can be calculated as follows:

$$\sigma=\frac{F'}{S'} \qquad (13)$$

where F' is the static load of the shock absorber; S'

is the minimum bearing area of the shock absorber. For the low damping material, σ is 1.8 MPa, thus the minimum area S' of the shock absorber is 6.39 mm^2 for a static load of 11.5 N. The screw is M5 then [24].

Fig. 6 Rubber shock absorber.

The natural frequency of the shock absorber f_n is

$$f_n = \frac{1}{2\pi}\sqrt{\frac{K'g}{G}} \qquad (14)$$

where K' is the dynamic stiffness of the shock absorber; g is the acceleration of gravity; G is the gravity of the shock absorber. According to technical indicators on the satellite equipment, the vibration frequency ranges are from 10 Hz to 2000 Hz. If the vibration frequency meets $f > \sqrt{2}f_n$ (suppose its natural frequency $f_n = 10$ Hz), rubber shock absorber damping takes effect, and the total dynamic stiffness calculated by (14) is $K' = 18516.33$ N/m.

The complex shape of the shock absorber should be taken into account as a result of the parallel and series connection of some simple shapes. Thus the total rigidity of the connection is written as

$$k_c = k_{c1} + k_{c2} + \cdots \qquad (15)$$

$$\frac{1}{k_b} = \frac{1}{k_{b1}} + \frac{1}{k_{b2}} + \cdots. \qquad (16)$$

For easy calculation, the stress-strain relationship of the rubber damper is linear, then

$$K_S = \frac{SE_c}{h} \qquad (17)$$

where K_S is the static stiffness of the shock absorber, and $\frac{K'}{K_S}$ is generally $1.2 - 2$; S is the effective load area of the shock absorber; E_c is the effective compression modulus of shock absorber; h is the height of the shock absorber.

The effective compression modulus can be expressed mathematically as

$$E_c = E_0(1 + 2K''C^2) \qquad (18)$$

where E_0 is the elastic modulus of rubber materials; C is a shape factor of the rubber shock absorber, which is the ratio between the loaded area and not loaded area, and can be calculated by $C = \frac{D}{4h'}$, where D is the diameter, and h' is the height of the shock absorber. To ensure the stability of the shock absorber, not bending, the condition $\frac{h'}{D} < 1$ is required; K'' is the correction factor of the material properties, which ranges from 0.5 to 1.

Assuming $\frac{K'}{K_S} = 1.5$, then the total stiffness is $K_S = 12344.22$ N/m. Supposing four shock absorbers bear the load uniformly, the static stiffness of each shock absorber is $K'_S = 3086.05$ N/m according to (15); the effective compression modulus of each shock absorber is $E_c = 0.163$ MPa according to (17); the elastic modulus of the damper material is $E_0 = 108.67$ kPa according to (18) and the shock absorber structure size (as shown in Fig. 7). The elastic modulus of the rubber is about 100 kPa, so it can be used [25].

The rubber is made in 703 Institute, and the damping ratio of the rubber shock absorber is 0.1. The dimension of the rubber shock absorber is used in the following simulation and experimental design.

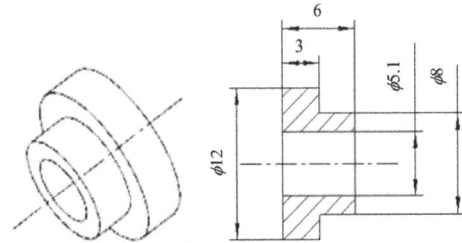

Fig. 7 Structure of the rubber shock absorber.

4. Vibration simulation analysis

4.1 Modal analysis

Modal analysis is to determine the dynamic characteristics of the camera, which provides given order natural frequencies and mode shapes, investigate the dynamic stiffness of the camera, and assess its other dynamic performances [26]. This is the basis of other dynamic response analyses (such as transient analysis). From the modal analysis, the weakness of structural rigidity is obtained which is used for the design optimization. The first few modals and shapes of the entire satellite are shown in Table 1. The modal is low, but it meets the design. The coordinate system is that the optical axis is z axis, which consists of axes x, y, and z, and follows the right-handed coordinate system.

Table 1 Modal and shapes.

Modal order	Frequency (Hz)	Shapes
1	15.93	Swinging about yz plane
2	17.242	Swinging about xz plane
3	30.618	Twisting about z axis
5	35.694	Swinging about xy plane

4.2 Transient analysis

The static and dynamic unbalanced forces and moments are tested by the HR-FP3402 force platform as shown in Fig. 8(a). Figures 8 (b) and 8(c) show an example of the force results (at the speed of 3000 rpm) for the flywheel without and with isolation.

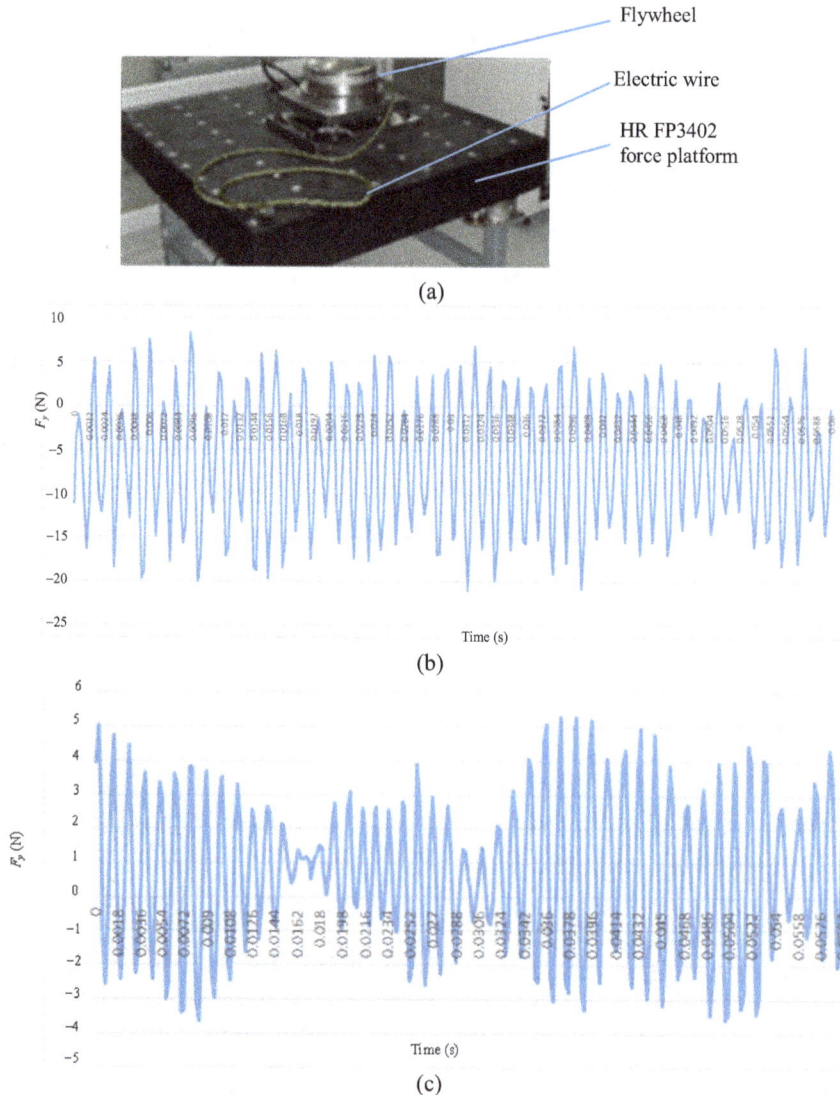

(a)

(b)

(c)

Fig. 8 Test platform and tested force results of the flywheel: (a) test platform for the force and moment of the flywheel, (b) test result F_y of the flywheel without isolation, and (c) test result F_y of the flywheel with isolation.

The forces and moments of the flywheel at various speeds are tested. Only the disturbance data at 3000 rpm are imported into Patran, because the flywheel runs mostly at this speed. The transient displacements of all the lenses and focus plan along the x axis and y axis are obtained by the Nastran calculation, as shown in Figs. 9 and 10. The meaning of the related nodes is shown in Table 2. The displacement data are the inputs of calculation of the modulate transfer function (MTF) in the next section.

Table 2 Meaning of the node.

Node	Lens
11658	Second lens
12479	Primary lens
12480	Focal plane
12481	Focusing lens
12482	Folding lens
12483	Third lens

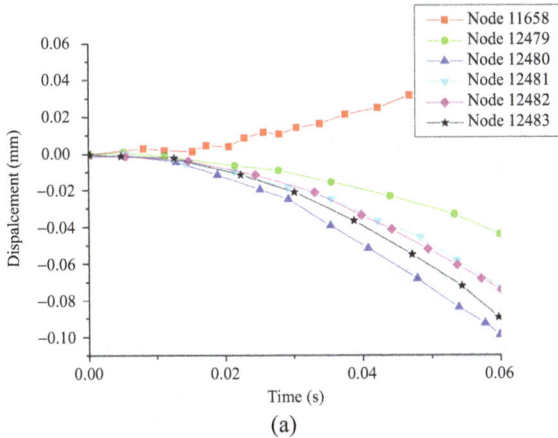

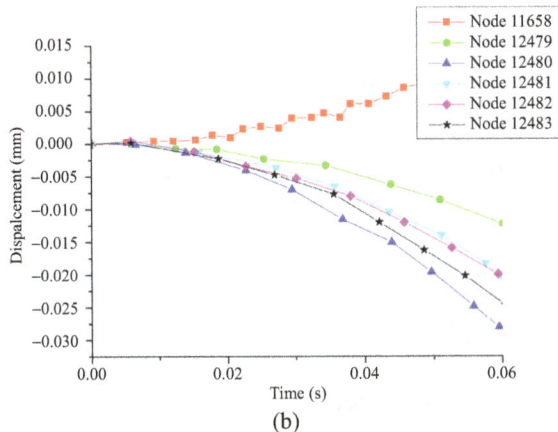

Fig. 9 Displacement in the x axis: (a) without the rubber shock absorber and (b) with the rubber shock absorber.

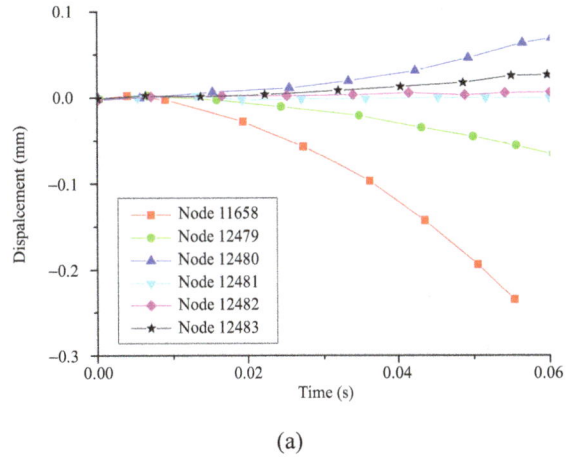

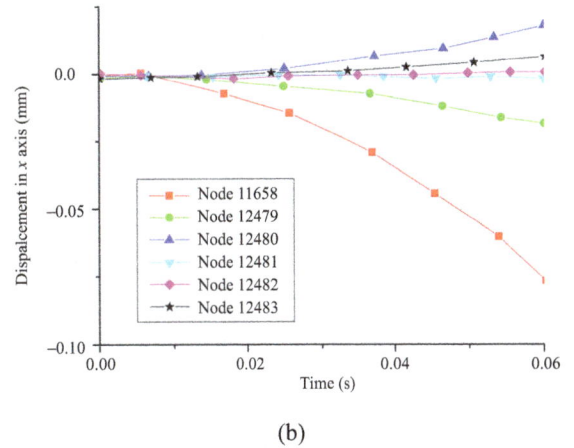

Fig. 10 Displacement in the y axis: (a) without the rubber shock absorber and (b) with the rubber shock absorber.

4.3 Calculation of MTF

In exposure, displacements of the primary lens, second lens, third lens, folding lens, focusing lens, and the focal plane x, y directions are imported into the Matlab program from Patran. Then each set of vibration data is sent to Zemax via the dynamic link. The vibration information about the displacements of the light spot in the image plane is obtained at the sampling time by ray tracing. By repeating each sampling data in integration time, the displacement of the light spot image at the image plane in the exposure can be obtained. The displacement may be interpreted as the image shift probability distribution of the sample generated by vibration. The imaging MTF of vibration is calculated with the optical transfer function (OTF) formula (19) and formula of N-order statistical moments (20) [28], and the flowchart is shown in Fig. 11.

$$\text{OTF}(\omega)=\sum_{n=0}^{\infty}\frac{m_n}{n!}(-\text{j}\omega)^n \qquad (19)$$

$$m_n^x=\frac{1}{S}\sum_{i=1}^{s}x_i^n \qquad (20)$$

where OTF is the optical transfer function; $\{x_i\}\,(i=1,2,3,\cdots,S)$ is the sampling sequence; m_n^x is for the N-order statistical moments of movement.

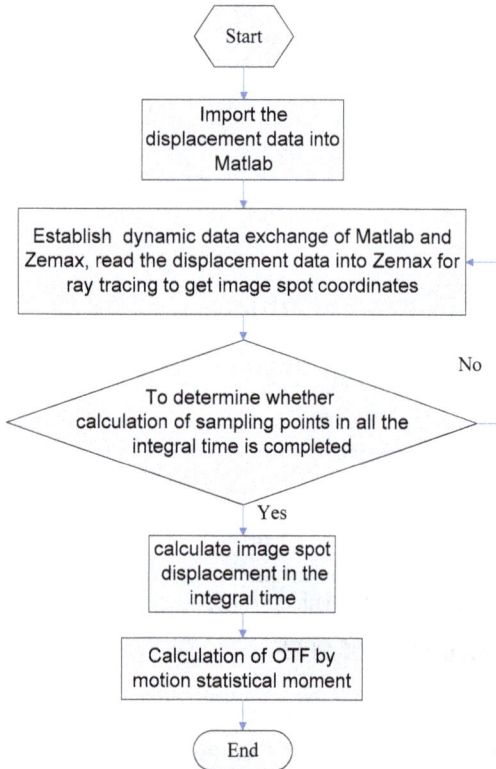

Fig. 11 Flowchart of MTF calculation.

The spot displacements in the focal plane in the x axis and y axis are shown in Fig. 12. The MTF is 1 in the ideal case without vibration. The calculated MTF without the rubber shock absorber at the Nyquist frequency (57 lp/mm) is 0.88 along the x axis and is 0.90 along the y axis as shown in Fig. 13. Generally, the MTF of vibration can be considered as the average of two axes, i.e., 0.89. With the rubber shock absorber added, the MTF at the Nyquist frequency is 0.97 along the x axis and is 0.99 along the y axis, so the whole MTF of the system is 0.98, increased by 10.1% compared with the system without the rubber shock absorber.

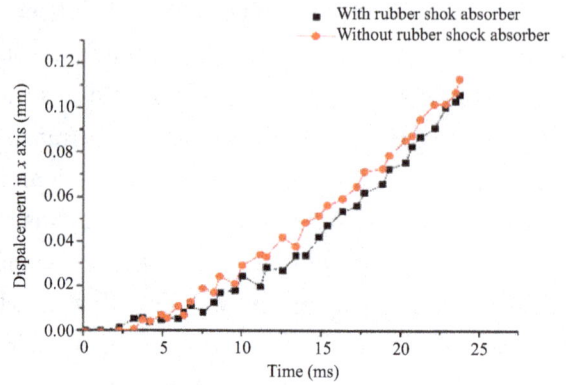

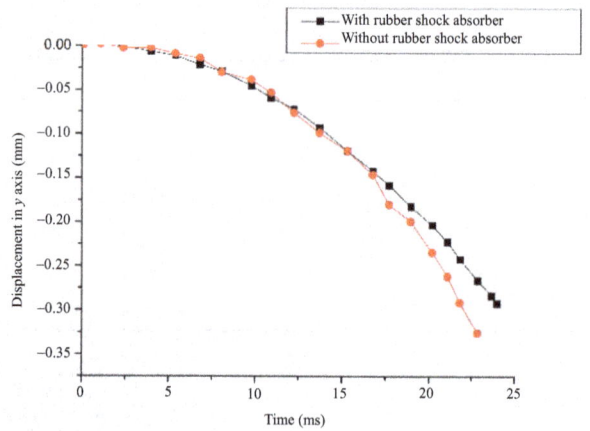

Fig. 12 Spot displacement in the focal plane: (a) in the x axis and (b) in the y axis.

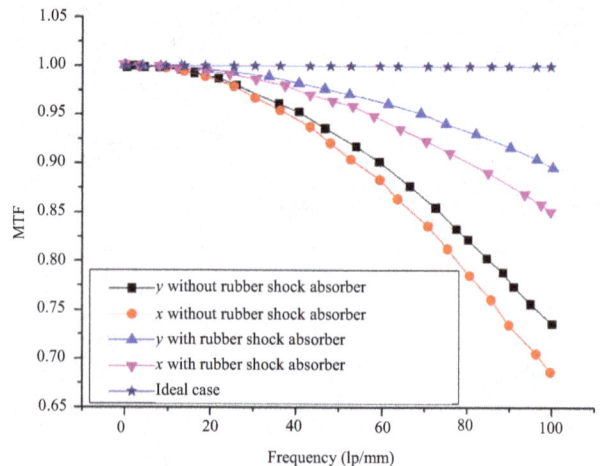

Fig. 13 Effects of vibration on the MTF.

5. Experiment

5.1 Experimental system

The principle of the experiment is plotted in

Fig. 14. The experimental equipment consists of a satellite containing a camera, a rubber shock absorber, a reaction flywheel, a collimator, a crane, a set of charge coupled device (CCD) components, a calibration target, an alignment equipment, a data acquisition system, and a computer.

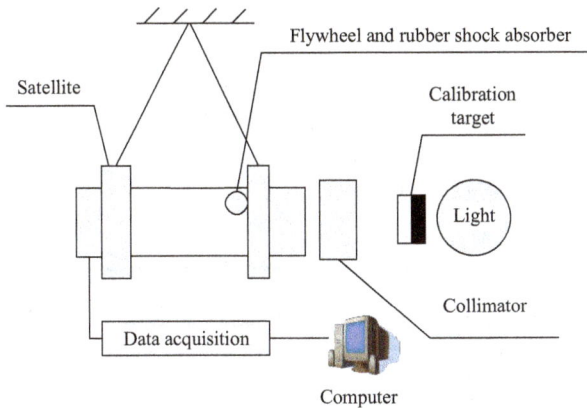

Fig. 14 Schematic of experimental system.

The flywheel rotating axis is vertical to the optic axis. The working conditions with the rubber shock absorber and without the rubber shock absorber were investigated. To intimate the real flying condition of satellite and the gravity environment in space, the satellite is suspended with a soft sling [29, 30].

Before experiment, we need to check that the bolts are screwed down and circuit is expedited. When the experiment starts, the flywheel is powered on, the speed of flywheel increases gradually from 0 to 3000 rpm, and the speed of the flywheel is controlled by the software platform. When the speed reaches 300 rpm, the CCD starts to sample, and then the speed increases by 200 rpm by step.

A static target is illuminated by a light source, and the light passes through collimator and lenses of the camera. The light is focused on the focal plane, and the CCD sensor is used for imaging. When the flywheel speed is stable, the images are sampled at a corresponding speed. The data are acquired by the ground image acquisition and processing system, and the corresponding MTF is calculated. The data

are analyzed and calculated in the time domain and frequency domain, outputing graphs in 2 dimensions, to describe the changes in related parameters of the vibration. In the process of camera imaging, the effect of the reaction flywheel on the imaging quality of the camera is observed by varying the reaction wheel speed.

Two coordinate systems are used in the experiments: the satellite and the flywheel coordinate systems. In the satellite coordinate system, the z axis is the optic axis, the x axis is the direction of flight, and the y axis is decided following the right-handed coordinate system. In the flywheel coordinate system, the z axis is the rotation axis, and x and y axes are decided by the right-handed coordinate system.

5.2 Experimental result

It is shown in Fig. 15 that the amplitude limit in the time domain is 0.4 pixel to −0.4 pixel (1 pixel = 0.2″) at the speed of 3000 rpm without the rubber shock absorber, while the amplitude limit with the rubber shock absorber is 0.2 pixel to −0.2 pixel. The spectrum can be obtained by the Fourier transform of the results in Fig. 15 and is shown in Fig. 16. With the rubber shocker absorber added, the amplitude decreases in all the ranges of frequency, and the summit of amplitude decreases 87.5%, from 0.2386 at 537.1 Hz to 0.02978 at 85.45 Hz. Figure 17 shows the MTF at the Nyquist frequency as a function of the started phase. The MTF is varied at different starting phases, which is because the time of exposure is very short and the time of vibration is relatively long. It is also shown that the MTF with the rubber shock absorber is generally higher than that without the rubber shock absorber. At 0.5 (starting phase), the MTF increases from 0.8897 to 0.97 by adding the rubber shock absorber, increasing by 9%. It can be compared with Fig. 14, providing the MTF is taken at the Nyquist frequency. The simulated and experimental results of the MTF with and without the rubber shock absorber are listed in Table 3, and both the simulation and experiment show the rubber shock absorber increases the MTF almost 10%, and the experimental results matches well with the simulation values.

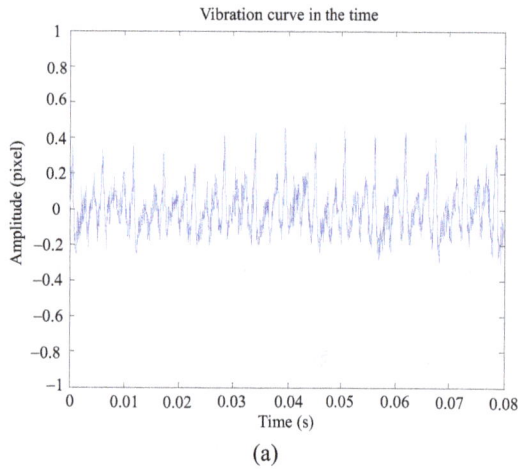
(a)

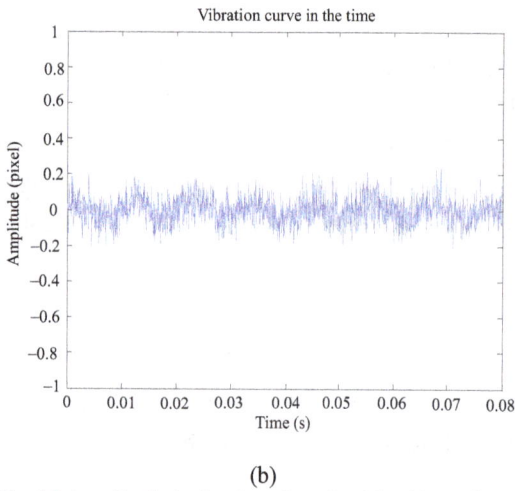
(b)

Fig. 15 Amplitude in the time domain: (a) without the rubber shock absorber and (b) with the rubber shock absorber.

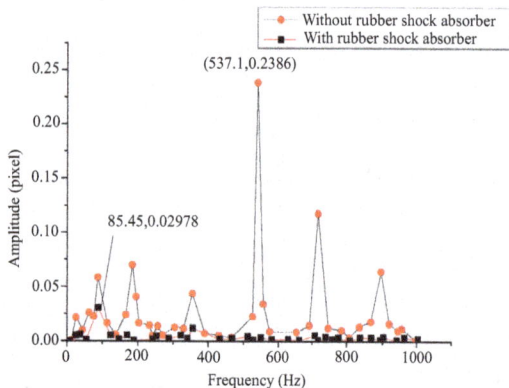

Fig. 16 Amplitude in frequency.

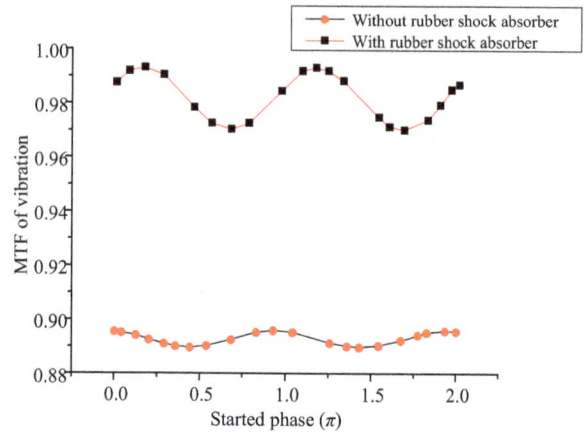

Fig. 17 MTF of vibration.

Table 3 MTF contrast.

Case	MTF of simulation at Nyquist frequency	MTF of experiment at Nyquist frequency
Without rubber shock absorber	0.89	0.8897
With rubber shock absorber	0.98	0.97

6. Conclusions

To reduce the effect of micro vibration of the flywheel on the camera imaging system, a rubber shock absorber is used in the satellite system, and its effects are analyzed by the simulation and experimental method. In the simulation, an integrated modeling analysis based on ray tracing is developed. A real product is designed and tested. The experimental results match well with the simulated ones, and both show that the MTF increases almost 10 %, which confirms the rubber shock absorber is helping to decrease the effect of the flywheel vibration on the camera images.

Acknowledgment

The author would like to thank the great help from Changchun Iinstitute of Optics, Fine Mechanics, and Physics, Chinese Academy of Sciences during the experiment.

References

[1] Z. Wei, D. Li, Q. Luo, and J. Jiang, "Modeling and analysis of a flywheel microvibration isolation system for spacecrafts," *Advances in Space Research*, 2015, 55(2): 761-777.

[2] D. O. Lee, G. Park, and J. H. Han, "Experimental study on on-orbit and launch environment vibration isolation performance of a vibration isolator using bellows and viscous fluid," *Aerospace Science and Technology*, 2015, 45: 1-9.

[3] C. Liu, X. Jing, S. Daley, and F. Li, "Recent advances in micro-vibration isolation," *Mechanical System and Signal Processing*, 2015, 56(1): 55-80.

[4] V. G. Geethamma, R. Asaletha, N. Kalarikkal, and S. Thomas, "Vibration and sound damping in polymers," *Resonance*, 2014, 19(9): 821-833.

[5] M. Abdulhadi, "Stiffness and damping cofficients of rubber," *Archive of Applied Mechanics*, 1985, 55(6): 421-427.

[6] S. E. Klenke and T. J. Baca, "Structural dynamics test simulation and optimization for aerospace components," *Expert Systems with Applications*, 1996, 11(4): 82-89.

[7] J. C. Dixon, *The shock absorber handbook*. New York: SAE International, 2007.

[8] J. Njuguna and K. Pielichowski, "The role of advanced polymer materials in aerospace," *Research Gate*, 2013: 1-48.

[9] A. Dall'Asta and L. Ragni, "Nonlinear behavior of dynamic systems with high damping rubber devices," *Engineering Structure*, 2008, 30(12): 3610-3618.

[10] D. W. Nelson and N. W. Nelson, "Finite element analysis in design with rubber," *Chemistry and Technology*, 1990, 63(3): 368-406.

[11] T. J. R. Hughes, *The finite element method: linear static and dynamic finite element analysis*. New Jersey: Prentice Hall, 2000.

[12] L. Chen, "Numerical methods for analysing static characteristics of rubber isolator," *Journal of Vibration and Shock*, 2005, 25(123-124): 56-61.

[13] M. Sjoberg, "On dynamic properties of rubber isolators," Ph.D. dissertation, Kungliga Tekniska högskolan, 2002.

[14] M. Sjoberg, "Rubber isolators measurements and modelling using fractional derivatives and friction," *SAE Technical Paper*, 2000, 1(3518): 133-144.

[15] M. D. Lieber, "Space-based optical system performance evaluation with integrated modeling tools," *SPIE*, 2004, 5420: 85-96.

[16] D. M. LoBosco, C. Blaurock, S. J. Chung, and D. W. Miller, "Integrated modeling of optical performance for the Terrestrial Planet Finder structurally connected interferometer," *SPIE*, 2004, 5497: 278-289.

[17] O. L. D. Weck, D. W. Miller, G. J. Mallory, and G. E. Mosier, "Integrated modeling and dynamics simulation for the next generation space telescope (NGST)," *SPIE*, 2000, 4013: 920-934.

[18] W. Zhou and D. Li, "Experimental research on a vibration isolation platform for momentum wheel assembly," *Journal of Sound and Vibration*, 2013, 332(5): 1157-1171.

[19] D. W. Miller, O. L. D. Weck, and G. E. Mosier, "Framework for multidisciplinary integrated modeling and analysis of space telescope," *Integrated Modeling of Telescopes*, 2002, 4757: 1-18.

[20] L. M. Elias, F. G. Dekens, I. Basdogan, and L. A. Sievers, "Methodology for modeling the mechanical interaction between a reaction wheel and a flexible structure," *SPIE*, 2003, 4852: 541-555.

[21] D. O. Lee, J. S. Yoon, and J. H. Han, "Development of integrated simulation tool for jitter analysis," *International Journal of Aeronautical and Space Sciences*, 2012, 13(1): 64-73.

[22] A. S. Glassner, "An introduction to ray tracing," *Morgan Kaufmann Publishers*, 1989, 34(2): 417-417.

[23] M. Katz, *Introduction to geometrical optics*. New Jersey: World Scientific, 2002.

[24] H. T. Yang, J. Z. Cao, Z. Y. Fan, and W. N. Chen, "The research of the high precision universal stable reconnaissance platform in near space," *International Symposium on Photoelectronic Detection and Imaging*, 2011, 8196(3): 111-116.

[25] S. Hadden, T. Davis, P. Buchele, J. Boyd, and T. L. Hintz, "Heavy load vibration isolation system for airborne payloads," *SPIE*, 2001, 4332: 171-182.

[26] B. Zhang, X. Wang, and Y. Hu, "Integrated modeling and optical jitter analysis of a high resolution space camera," *SPIE*, 2012, 8415: 841508-1-841508-7.

[27] O. Hadar and N. S. Kopeika, "Numerical calculation of MTF for image motion: exprerimental verification," *SPIE*, 1992, 1697: 183-197.

[28] O. Hadar, I. Dror, and N. S. Kopeika, "Real-time numerical calculation of optical transfer function for

image motion and vibration. Part 1: experimental verification," *Optical Engineering*, 1997, 33(2): 566−578.

[29] W. Zhou, L. Dongxu, Q. Luo, and K. Liu, "Analysis and testing of microvibrations produced by momentum wheel assemblies," *Chinese Journal of Aeronautics*, 2012, 25(4): 640−649.

[30] W. Y. Zhou, G. S. Aglietti, and Z. Zhang, "Modelling and testing of a soft suspension design for a reaction/momentum wheel assembly," *Journal of Sound and Vibration*, 2011, 330(18): 4596−4610.

17

Microfiber Bragg Grating for Temperature and Strain Sensing Applications

Jie TIAN[1], Shuhui LIU[1], Wenbing YU[2*], and Peigang DENG[1]

[1]*Laboratory of Optical Information Technology, Wuhan Institution of Technology, Wuhan, 430205, China*
[2]*School of Electronic Information, Shanghai DianJi University, Shanghai, 201306, China*
*Corresponding author: Wenbing YU   E-mail: yuwb@sdju.edu.cn

Abstract: Fiber Bragg grating is inscribed on microfiber with femtosecond laser pulses irradiation. The microfiber is fabricated by stretching a section of single mode fiber over a flame. Periodic grooves are carved on the microfiber by the laser as have been observed experimentally. The microfiber Bragg grating is demonstrated for temperature and strain sensing, and the strain sensitivity is improved with decreased diameters of the microfibers.

Keywords: Fiber Bragg gratings; fiber optics sensors; femtosecond laser micromachining

1. Introduction

There has been increasing interest in the research of microfiber these years due to its many unique properties such as large evanescent field, configurability, and strong confinement of the conducted light [1]. These distinctive features have been exploited in a wealth of applications ranging from telecommunication devices to sensors [2, 3], and from optical manipulation to high Q resonators. Fiber sensors, based on the various fiber components, have been developed in many forms [4–9], and are exhibiting outstanding performances in the field of physical and bio-chemical sensing. Fiber Bragg grating (FBG) is among the most important optical components which has a wide variation of applications [10–13]. Here we fabricate an FBG in the microfibers with diameter down to several micrometers, and the temperature and strain responses of the microfiber FBG are investigated by the experiment.

2. Inscription of FBG in microfiber

The microfiber is produced by the use of a flame torch and a translation stage with a flame-brush method. The flame is placed under the single mode fiber (SMF) to heat the fiber while the fiber is stretched by two translation stages. By controlling the position of the flame, microfibers with diameters from 5 μm to 50 μm are fabricated. Figure 1(a) shows the graphic of a microfiber with a diameter of 10 μm.

The FBG is fabricated with the phase mask method. The femtosecond laser used for the fabrication process is a Ti-sapphire laser system (Spectrum Physics) with pulse duration of 50 fs and repetition rate of 1 kHz at 800 nm. The laser beam is focused into the microfiber through a phase mask (Stocker Yale) by a cylindrical lens. The fiber is located in a distance of 2 mm from the phase mask, and the position of the fiber is adjusted by a high precision 3-axis translation stage. The pulse energy for the fabricating process is about 350 μJ. The exposure time varies from several ms to tens of ms for microfibers with different diameters. We use an amplified spontaneous emission (ASE) light source and an optical spectrum analyzer to collect the spectra of FBG.

Figure 1(b) shows the periodic grating pattern on the microfiber that we observed from a microscope. As can be seen from the figure, the femtosecond exposure is likely to have caused surface ablation on the microfiber, which leads to the many grooves on the microfiber.

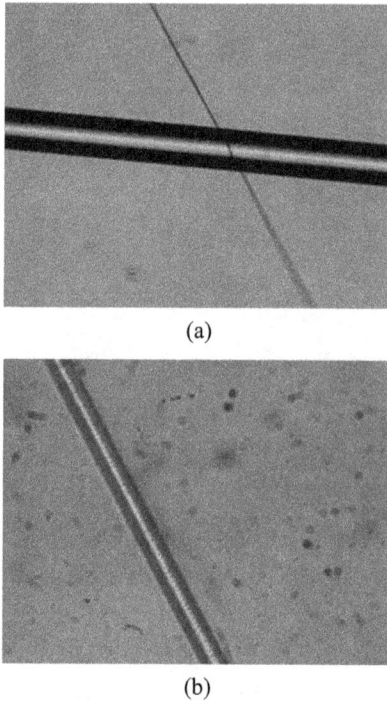

(a)

(b)

Fig. 1 Femtosecond laser fabrication of microfiber: (a) a section of microfiber with diameter of about 10 μm and (b) periodic grating structure on a microfiber.

The reflection and transmission spectrum of FBG in microfiber with a diameter of 35 μm is shown in Fig. 2(a). Very strong reflection is obtained by such a grating. From Fig. 2(b) we can see that the resonant wavelength of FBG in a 35 μm diameter fiber has a blue shift versus the FBG in an SMF fabricated with the same laser and phase mask. This is because a thinner fiber leads to a drop in the effective refractive index of the modes propagating in the fiber, thus a decrease in λ_B, which can be derived from the Bragg condition [10]:

$$\lambda_B = 2n_{\text{eff}} \cdot \Lambda \qquad (1)$$

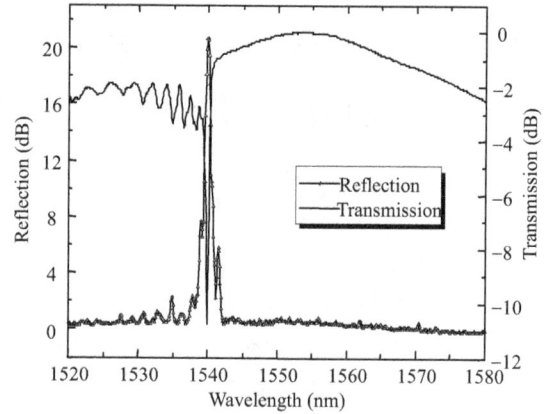

(a)

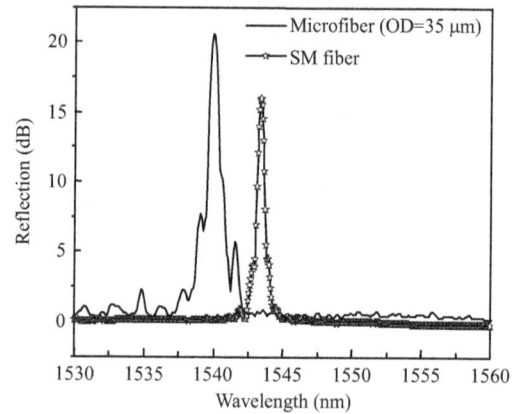

(b)

Fig. 2 Spectrum of microfiber FBG: (a) transmission and reflection spectrum of FBG in a microfiber and (b) reflection spectrum of single mode fiber FBG and microfiber FBG.

3. Temperature and strain sensing

A group of FBGs in microfibers with diameters of 9 μm, 35 μm, and 50 μm are fabricated. These samples are heated from room temperature to 400℃ by use of a tube furnace, and the reflection spectra are recorded by an increment of 50℃. Figure 3(a) shows the wavelength shift of FBG in microfiber of 50 μm diameter, from which we can see that there is an obvious red-shift of the Bragg wavelength, which can be explained by the thermal-optic effect and thermal expansion of the grating. The temperature induced shift of the Bragg wavelength λ_B can be written as [13]:

$$\Delta\lambda_B = \lambda_B \left(\frac{1}{\Lambda} \frac{\partial \Lambda}{\partial T} + \frac{1}{n_{\text{eff}}} \frac{\partial n_{\text{eff}}}{\partial T} \right) \Delta T = \lambda_B (\alpha_\Lambda + \alpha_n) \Delta T \quad (2)$$

where ΔT is the temperature change, α_n is the thermo-optic coefficient, and α_A is the thermo-expansion coefficient. Temperature changes influence two factors: the temperature-induced refractive index variation and the temperature-induced grating pitch variation. The former is the dominating factor of the wavelength variation.

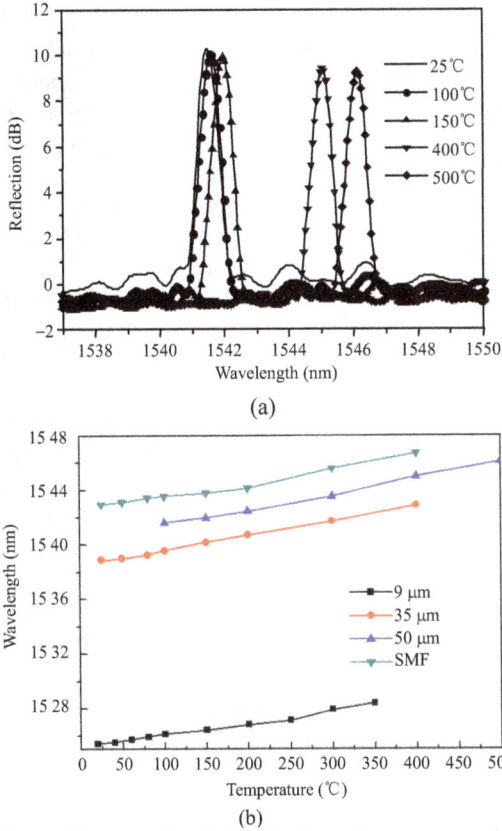

(a)

(b)

Fig. 3 Wavelength shift and thermal response of FBG: (a) wavelength shift of FBG (50 μm) during heating up and (b) thermal response of FBG with fiber diameters of 9 μm, 35 μm, 50 μm, and an SMF.

Thermal responses of FBGs of different fiber diameters are shown in Fig. 3(b). From Fig. 3(b) we can see that there are no dramatic differences among the temperature sensitivities (around 11 pm/℃) of these FBGs. The results are close to the temperature sensitivity of FBG in a single mode fiber, since the shrinking of the fiber diameter hardly causes changes to the thermo-optic coefficient of the fiber material.

The strain sensitivity of the microfiber Bragg grating is investigated by a setup shown in Fig. 4. The microfiber with FBG in its center is fixed between two precision translation stages where the

microfiber is stretched by driving the screw of the stage along the fiber length, and the tensile elongation could be read directly from the scale on the screw.

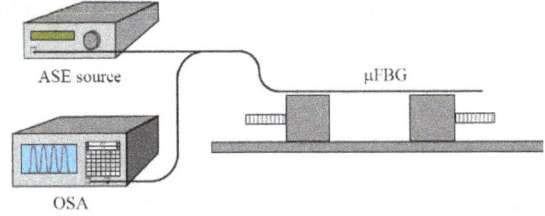

Fig. 4 Setup for strain sensitivity measurement.

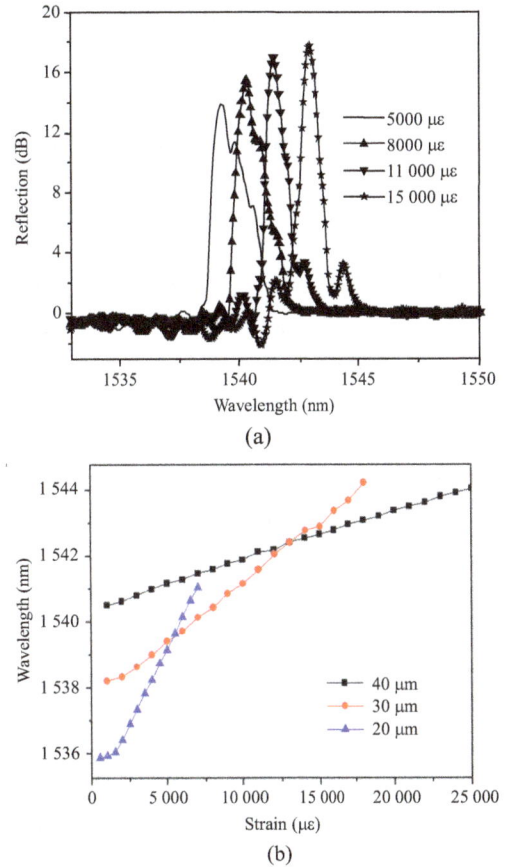

(a)

(b)

Fig. 5 Strain test of FBGs: (a) wavelength shift of FBG with fiber diameter of 30 μm under strain test and (b) a comparison of tensile strain responses of FBGs in different diameter fibers.

Here we also employ a group of experiments on microfibers with the diameters of 20 μm, 30 μm, and 40 μm for comparison. Figure 5(a) shows the wavelength shift of FBG with fiber diameter of 30 μm when the FBG is stretched. We can see that there is a large change in the resonant wavelength from 1539 nm to 1543 nm when the strain employed on the microfiber increases from 5000 με to 15000 με, which indicates a strain sensitivity of

0.35 pm/με. Later, another two FBG samples with diameter of 20 μm and 40 μm are tested to made comparison, which is shown in Fig. 5(b). It can be seen that a decrease in fiber diameter has a significant contribution to improving tensile strain sensitivity.

The strain sensitivities of these FBGs are 0.228 pm/με, 0.35 pm/με, and 0.8 pm/με for the fiber diameter of 40 μm, 30 μm, and 20 μm, respectively. It should be mentioned that the large value of strain induced in our experiment (tens of thousands of με) is due to a tremendous decrease in the fiber cross-section area, which means that a little tensile elongation would induce a massive strain in the microfiber. This unique character indicates a potential application for the sensing of tiny force.

4. Conclusions

We present the fabrication of Bragg grating on microfibers with diameter down to 9 μm. The microfiber is made by stretching single mode fibers over a flame. Femtosecond laser pulse is used for the grating inscription, and periodic grooves are carved on the microfiber as we have observed experimentally. The microfiber Bragg grating is demonstrated for temperature and strain sensing, and the strain sensitivity can be improved by decreasing the diameter of the microfiber.

Acknowledgment

This work was supported by the Wuhan Science and Technology Bureau under Grant No. 2015010101010002, the Natural Science Foundation of Hubei Province under Grant No. 2014CFB770, the Science Foundation of Wuhan Institute of Technology under Grant No. k201616, and the Headmaster Foundation of Wuhan Institute of Technology under Grant No. 2016066.

References

[1] J. Lou, L. Tong, and Z. Ye, "Modeling of silica nanowires for optical sensing," Optics Express, 2005, 13(6): 2135–2140.

[2] H. Xuan, W. Jin, and S. Liu, "Long-period gratings in wavelength-scale microfibers," Optics Letters, 2010, 35(1): 85–87.

[3] S. Liu, Z. Wang, M. Hou, J. Tian, and J. Xia. "Asymmetrically infiltrated twin core photonic crystal fiber for dual-parameter sensing," Optics & Laser Technology, 2016, 82: 53–56.

[4] S. Liu, Y. Wang, M. Hou, J. Guo, Z. Li, and P. Lu, "Anti-resonant reflecting guidance in alcohol-filled hollow core photonic crystal fiber for sensing applications," Optics Express, 2013, 21(25): 31690–31697.

[5] M. Hou, Y. Wang, S. Liu, J. Guo, Z. Li, and P. Lu, "Sensitivity-enhanced pressure sensor with hollow-core photonic crystal fiber," Journal of Lightwave Technology, 2014, 32(23): 4035–4039.

[6] M. Hou, Y. Wang, S. Liu, Z. Li, and P. Lu, "Multi-components interferometer based on partially-filled dual-core photonic crystal fiber for temperature and strain sensing," IEEE Sensors Journal, 2016, 16(16): 6192–6196.

[7] S. Liu, N. Liu, M. Hou, J. Guo, Z. Li, and P. Lu, "Direction-independent fiber inclinometer based on simplified hollow core photonic crystal fiber," Optics Letters, 2013, 38(4): 449–451.

[8] S. Liu, N. Liu, Y. Wang, J. Guo, Z. Li, and P. Lu, "Simple in-line M-Z interferometer based on dual-core photonic crystal fiber," IEEE-Photonics Technology Letters, 2012, 24(19): 1768–1770.

[9] S. Liu, J. Tian, N. Liu, J. Xia, and P. Lu, "Temperature insensitive liquid level sensor based on anti-resonant reflecting guidance in silica tube," Journal of Lightwave Technology, 2016, 34(22): 5239–5243.

[10] X. Fang, C. R. Liao, and D. N. Wang, "Femtosecond laser fabricated fiber Bragg grating in microfiber for refractive index sensing," Optics Letters, 2010, 35(7): 1007–1009.

[11] X. Shu, K. Chisholm, I. Felmeri, K. Sugden, A. Gillooly, L. Zhang, et al., "Highly sensitive transverse load sensing with reversible sampled fiber Bragg gratings," Applied Physics Letters, 2003, 83(15): 3003–3005.

[12] Y. Li, W. Chen, H. Wang, N. Liu, and P. Lu, "Bragg gratings in all-solid Bragg photonic crystal fiber written with femtosecond pulses," Journal of Lightwave Technology, 2011, 29(22): 3367–3371.

[13] N. Liu, Y. Li, Y. Wang, H. Wang, W. Liang, and P. Lu, "Bending insensitive sensors for strain and temperature measurements with Bragg gratings in Bragg fibers," Optics Express, 2011, 19(15): 13880–13891.

Theoretical and Experimental Investigation into the Influence Factors for Range Gated Reconstruction

Sing Yee CHUA[1], Xin WANG[1*], Ningqun GUO[1], and Ching Seong TAN[2]

[1]School of Engineering, Monash University Malaysia, Jalan Lagoon Selantan, Bandar Sunway, 47500 Selangor, Malaysia
[2]Faculty of Engineering, Multimedia University, Jalan Multimedia, 63000 Cyberjaya, Selangor, Malaysia
*Corresponding author: Xin WANG   E-mail: wang.xin@monash.edu

Abstract: Range gated is a laser ranging technique that has been applied in various fields due to its good application prospects. In order to improve the effectiveness of this method, influence factors contributing to the system performance should be well understood. Thus this paper performs theoretical and experimental investigation to comprehend the effects caused by multiple factors on range gated reconstruction. Our study focuses on the distance, target reflection, and acquisition time step parameter where their impacts on the quality of range reconstruction are analyzed. The presented experimental results show the expected trends of range error to support the validity of our theoretical model and discussion which can be used in future improvement works.

Keywords: Laser ranging; reflection; sensor

1. Introduction

Over the past decades, laser ranging has been a popular approach in optical metrology because of its unique characteristics of non-contact and non-destructive nature [1]. As of today, laser ranging has been applied in various fields such as oceanic and environmental research, surveillance, industry, and day-to-day applications [2].

Range gated is a laser ranging technique operates based on time-of-flight (TOF) principle to measure the travel time between the emitted laser pulse and the pulse reflected from the target. Pulsed laser and sensor's gate are controlled simultaneously to capture the reflected pulse where range r is determined from the round trip time t and the speed of light c.

$$r = \frac{ct}{2}. \qquad (1)$$

Range gated has been a promising method in applications such as target detection and recognition [3], night vision [4, 5], underwater [6, 7], and three-dimensional (3D) imaging [8, 9]. Besides, continuous development in laser, sensor, signal processing, and computer technology further improves the cost effectiveness of this approach. The good application prospects motivate the study into the influence factors for range gated reconstruction which can contribute to improve the system performance.

In a range gated system, laser pulse interacts with the target surface to generate a backscatter signal which contains the key information for range reconstruction. Hence, the quality of range reconstruction strongly relies on the reflected laser pulse from the target which undergoes changes along the propagation. Essentially, the detected laser pulse is affected by the laser source, sensor, target, and atmospheric effect [11]. These parameters could change the characteristics and cause variations in the reflected laser pulse which directly impact the accuracy of range determination. The importance of laser intensity profile [11, 12], distance interference [10], sensor [13, 14], and scattering effects [16] were discussed in various literatures.

In this paper, range gated reconstruction is analyzed theoretically and experimentally to obtain a comprehensive understanding and relationship between the influence factors and ranging performance. In Section 2, a brief of range gated technique is given, and theoretical derivation and analysis of 3D range gated reconstruction model are presented. On the other hand, the experimental setup for our investigation is described in Section 3. The impact of multiple influence factors is analyzed and discussed in Section 4. Finally, a conclusion is given in Section 5.

2. Theoretical derivation of range gated reconstruction model

Range gated approach operates based on TOF concept by measuring the round trip time between the emitted laser pulse and the pulse echo resulting from its reflectance off the target. The working principle of a range gated imaging system using time slicing technique is illustrated in Fig. 1. Pulsed laser is used as the illumination source, and gated camera is time delayed to open only for a very short duration normally in nanoseconds or picoseconds to capture the reflected image slice from the target over a distance. Synchronization between the laser and gated camera is particularly important. Camera gate remains closed when the laser pulse is emitted towards target. Camera gate is configured to open at the designated delayed time to capture the visible time slice reflected in the form of intensity image.

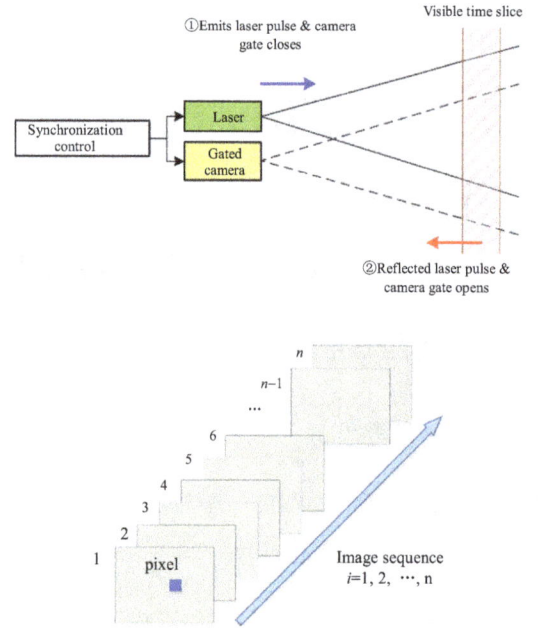

Fig. 1 Operating principle of range gated imaging system.

Based on time slicing technique [8, 9, 12], the camera gate $G(t)$ is delayed by time t_i with a time step t_{step} to acquire an image sequence $i=1, 2, 3, \cdots, n$. Intensity captured in an image pixel $I_i(x, y)$ is the incident energy of reflected laser pulse $P_r(t)$ integrated when the camera gate is opened for a t_{gate} time which can be expressed as

$$I_i(x,y) = \int P_r\left(t - \frac{2r}{c}\right)G(t-t_i)dt. \qquad (2)$$

Typically, the camera gate $G(t)$ is assumed as constant; hence the pixel intensity relies on the reflected laser energy $P_r(t)$. From laser detection and ranging (LADAR) range equation, the received signal P_r is defined as [17]

$$P_r = \frac{\eta_{sys}\eta_{atm}D^2\rho AP_t}{r^2\theta_R(\theta_t r)^2} \qquad (3)$$

where P_r and P_t are the received and transmitted signals across range r, and η_{sys} and η_{atm} represent system efficiency factor and atmospheric transmission loss caused by absorption and scattering. D is the diameter of receiver aperture, and ρ is the target surface reflectivity. θ_t represents the laser transmitter beam diameter and angular divergence, and θ_R is the solid angle over which radiation is dispersed upon reflection.

Assume the target surface area A is equal to the projected area of laser beam [17]

$$A = \frac{\pi \theta_t^2 r^2}{4}.$$ (4)

Equation (3) can be simplified as

$$P_r = \frac{\pi \eta_{sys} \eta_{atm} D^2}{4r^2} \frac{\rho}{\theta_R} P_t$$ (5)

where ρ/θ_R corresponds to the target reflection characteristics which we can represent with a bidirectional reflection distribution function (BRDF) model [18] where K_S and K_D are the specular and diffuse reflection constants, θ is the angle of incidence and reflection, s is the surface slope, and m is the diffusivity coefficient.

$$BRDF = \frac{K_S}{\cos^6 \theta} \exp\left(\frac{-\tan^2 \theta}{s^2}\right) + K_D \cos^m \theta.$$ (6)

Gaussian form is commonly assumed for temporal function of the transmitted laser pulse $P_t(t)$ where P_o represents the transmitted power and σ_p denotes the standard deviation of laser pulse [9, 13]:

$$P_t(t) = \frac{P_o}{\sqrt{2\pi}\sigma_p} \exp\left(\frac{-t^2}{2\sigma_p^2}\right).$$ (7)

Accordingly, (5) can be written as

$$P_r(t) = \frac{\pi \eta_{sys} \eta_{atm} D^2}{4r^2} \frac{P_o}{\sqrt{2\pi}\sigma_p} \exp\left(\frac{-t^2}{2\sigma_p^2}\right).$$
$$\left[\frac{K_S}{\cos^6 \theta} \exp\left(\frac{-\tan^2 \theta}{s^2}\right) + K_D \cos^m \theta\right].$$ (8)

Using time slicing technique, the summation of radiant energy in the image pixel can be seen as the integration over time slices $\int dt_i / t_{step}$ as the time step for image acquisition is much smaller than the laser pulse width and camera gate [9]:

$$I(x,y) = \sum_i I_i(x,y) = \frac{\int I_i(x,y)dt_i}{t_{step}}.$$ (9)

Based on (2), we can further simplify $I(x,y)$ as

$$I(x,y) = \frac{\int P_r(t)dt \int G(\tau)d\tau}{t_{step}}.$$ (10)

By substituting $P_r(t)$ from (8) into (10) and assume

$G(\tau)=1$ when $0 \le \tau \le t_{gate}$, $I(x,y)$ becomes

$$I(x,y) = \frac{\pi \eta_{sys} \eta_{atm} D^2}{4r^2 t_{step}} \cdot$$
$$\left[\frac{K_S}{\cos^6 \theta} \exp\left(\frac{-\tan^2 \theta}{s^2}\right) + K_D \cos^m \theta\right] \cdot$$
$$\frac{P_o}{\sqrt{2\pi}\sigma_p} \int_{-\infty}^{\infty} \exp\left(\frac{-t^2}{2\sigma_p^2}\right) dt \int_0^{t_{gate}} d\tau.$$ (11)

This equation eventually resolved into

$$I(x,y) = \frac{\pi \eta_{sys} \eta_{atm} D^2}{4r^2} P_o \frac{t_{gate}}{t_{step}} \cdot$$
$$\left[\frac{K_S}{\cos^6 \theta} \exp\left(\frac{-\tan^2 \theta}{s^2}\right) + K_D \cos^m \theta\right].$$ (12)

Signal to noise ratio (SNR) is an important parameter in analyzing the system performance [19]. SNR is defined as the ratio between the reflected intensity and the associated noises. For our range gated reconstruction, SNR can be expressed as follows after substitute $\sum_i I_i$ from (9) [10]:

$$SNR = \frac{\sum_i I_i}{\sqrt{\sum_i (\delta I_i)^2}} = \frac{\int I_i dt_i}{t_{step}\sqrt{\sum_i (\delta I_i)^2}}.$$ (13)

Uncertainty in the two-way travel time is given by the acquisition time step t_{step}; hence the expected range error can be written as

$$\delta r = \frac{c}{2} t_{step} = \frac{c}{2} \frac{\int I_i dt_i}{SNR \sqrt{\sum_i (\delta I_i)^2}}.$$ (14)

Average range $<r>$ and two-way travel time $<t>$ of an image pixel (x, y) can be determined from the captured intensity over an image sequence $i=1, 2, 3, \cdots, n$ using weighted average method:

$$<r>(x,y) = \frac{c<t>}{2} = \frac{c}{2} \frac{\sum_{i=1}^n I_i(x,y)t_i}{\sum_{i=1}^n I_i(x,y)}.$$ (15)

The calculated range $<r>$ strongly relies on the reflected intensity which is influenced by various factors as shown in (12). In addition, range accuracy is impacted by SNR which is proportional to the reflected intensity when the system noise level remains unchanged. Generally, the decreased SNR results in higher range errors. Based on the range gated reconstruction model derived, the relationship between the reflected intensity I and various influence factors in the system is shown as well as their impact to the SNR and range error. Experimental study involves a few factors including the distance, target reflection, and acquisition time step to validate our theoretical discussion.

3. Experimental setup

In order to investigate the effects induced by various influence factors, an experimental setup as illustrated in Fig. 2 is used. A pulsed diode pumped solid state Q-switched Nd:YAG laser that operates at wavelength 532 nm with output energy up to 1mJ is used. Silicon high speed biased non-amplified photodetector with active diameter of 400 μm and <300 ps rise/fall time is used to detect the laser pulses in the emitting or reflecting direction. Photodetector transforms the optical pulse into the usable signal for analysis via oscilloscope.

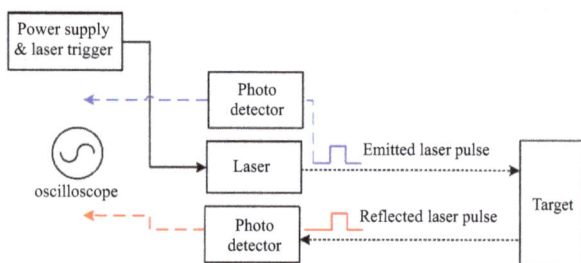

Fig. 2 Schematic diagram of experimental setup to capture the emitted and reflected laser pulse for investigation.

A backscatter signal is produced after the emitted laser pulse interacts with the target surface and is received by the detector in the form of time function. Two-way travel time across the distance between target and the detector is determined from the time difference between the emitted and reflected laser pulse. Correspondingly, the distance or range r can be obtained based on the TOF principle. For our study, the experiment is designed

to focus on three factors: distance, target reflection, and acquisition time step where the reflected intensity and range error are analyzed.

4. Analysis of influencing factors

4.1 Distance

The reflected intensity model derived as (12) shows that the reflected laser energy underlies an inverse range-squared dependency. Using the experimental setup described in Section 3, variation of the reflected intensity across distance is studied. The analyzed results are summarized in Fig. 3 where it clearly shows an inverse range-squared relationship of the reflected intensity [10].

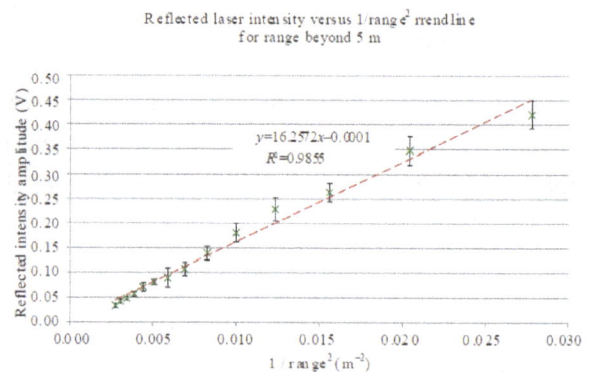

Fig. 3 Measured reflected laser intensity versus $1/range^2$ trendline [10].

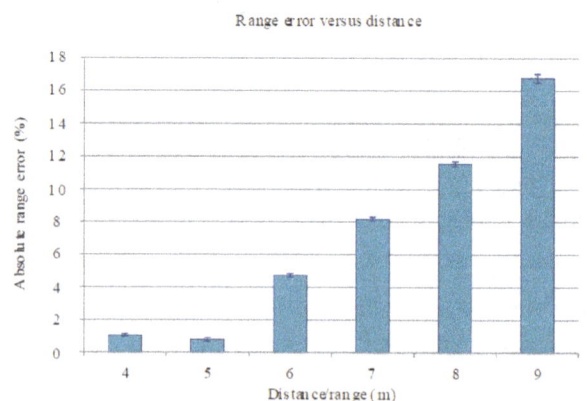

Fig. 4 Comparison of range error versus distance/range under the same constant setup condition.

Because of the reduced intensity over distance, SNR decreases, and we expect higher range error as deduced from (14). Figure 4 shows the range error calculated using weighted average method based on

30 measurements captured at different distances. The data sets are acquired under the same setup condition to ensure the range error is not influenced by other parameters in the system. The results show that an increase in distance causes a proportional decrease in the reflected intensity and leads to increasing range error as observed which agrees well with our theoretical discussion.

4.2 Target reflection

Reflected intensity strongly depends on the characteristics of the target surface [20]. Although Lambertian target (ideal diffuse surface) is commonly assumed due to its simplicity, target reflection is in fact far more complicated, and BRDF concept is normally used to describe that. Our theoretical model has adopted a BRDF model given by (6) which consists of specular and diffuse reflection to analyze the characteristics of reflected intensity in this study. Reflection off a rough surface returned in many directions leads to diffuse reflection while reflection from a smooth surface remains concentrated with the angle of reflection which causes specular reflection. Any target surface practically exhibits mixture of specular and diffuse behavior per surface properties such as roughness and absorption level.

Simulation based on the BRDF model is shown in Fig. 5 where four examples of target surface model are compared. These include two extreme cases of pure specular and pure diffuse surface models, and two examples of mixed components surface with different ratios of surface glint to diffuse behaviour given by specular and diffuse reflection constants, i.e. K_S/K_D. The amplitude of the reflection is maximum when angle of incidence $\theta=0$ degree and decreases when θ increases, adheres to the BRDF model. As a result, the decreased intensity causes the reduced SNR which gives rise to range error.

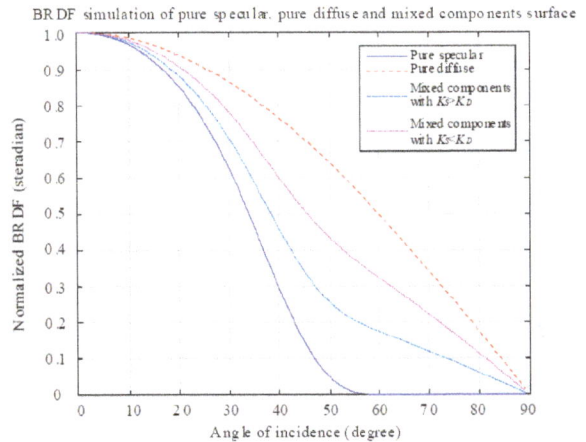

Fig. 5 BRDF simulation as a comparison of different target surface models.

For our experimental study, various target surface materials and roughnesses are tested. Figure 6 compares the range error for target surfaces captured at 5 m, and the results are analyzed based on average of 30 measurements. From the results, we observe that the range error is higher for rough and weak reflective surfaces as compared to smooth and strong reflective surfaces [21] where these surfaces can be modeled using BRDF described in our theoretical model. In addition, the effect of angle variation is evaluated for various target surfaces where the corresponding range error is shown in Fig. 7. It can be clearly seen that the range error is minimum at zero angle of incidence $\theta=0$ degree and increases with the angle of incidence in general. This has demonstrated the angular dependency which agrees with the theoretical model discussed.

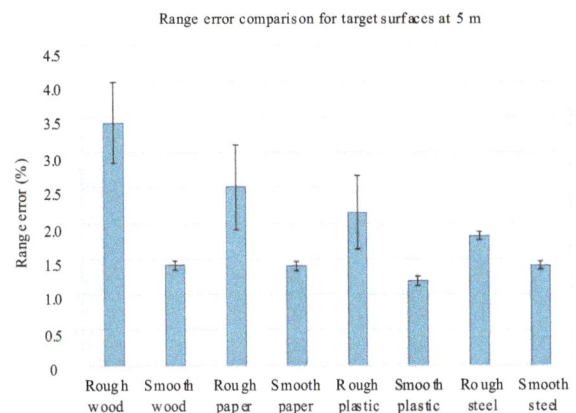

Fig. 6 Comparison of range error for target surfaces with different reflectivity and roughness.

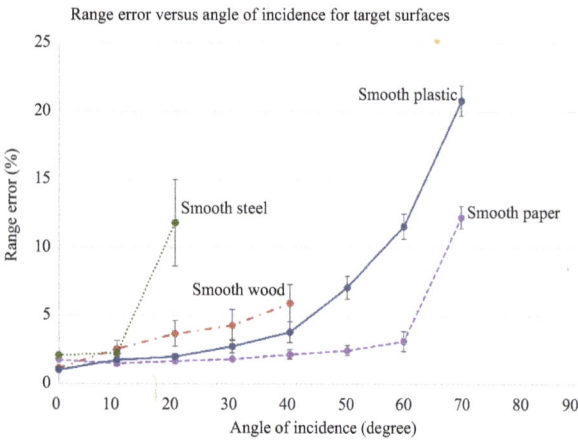

Fig. 7 Comparison of range error versus angle of incidence for target surfaces with different reflectivity.

4.3 Acquisition time step

From (12) and (14), it can be seen that the reflected intensity is inversely proportional to the time step used to acquire a series of image slices and error in the calculated range δ_r shows direct dependency on the time step parameter. Under the same setup condition where all parameters are regarded as constants, range error is expected to increase with time step value in theory. Figure 8 shows the range error trend analyzed based on average of 30 measurements. This set of experimental result clearly points out that a smaller time step should be selected to gain higher accuracy. However, the choice of the time step used is a trade-off between range accuracy and processing cost in terms of time and effort which should be taken into consideration.

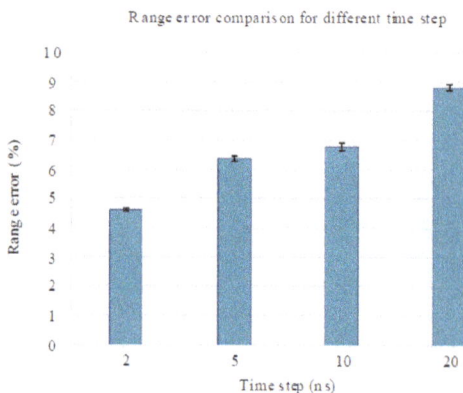

Fig. 8 Comparison of range error for different time steps.

5. Conclusions

In summary, this paper has demonstrated the influence of multiple factors on range gated reconstruction through theoretical and experimental investigation. Based on the operating principle of time slicing technique, LADAR, and BRDF, theoretical derivation of range gated reconstruction model is presented. Range accuracy shows dependency on the SNR which is proportional to the reflected laser intensity when the system noise level remains unchanged.

Impact on the accuracy of range reconstruction is studied from the perspective of distance, target reflection, and acquisition time step. Each influence factor is analyzed theoretically, and experimental investigation is performed to validate the theoretical discussion. It is concluded that our experimental results agree well with the theoretical analysis where the expected range error trends are shown.

The presented findings establish a comprehensive understanding of multiple influence factors which may benefit various applications and serve as references to perform correction or compensation. In future, follow-up improvement of range reconstruction can be proposed and additional effects caused by illumination, sensor, and noise can be included.

Acknowledgment

The authors gratefully acknowledge the support of funding from Ministry of Higher Education, Malaysia under the Grant No: FRGS/1/2016/STG02/MUSM/02/1.

References

[1] P. Huke, R. Klattenhoff, and C. V. Kopylow, "Novel trends in optical non-destructive testing methods," *Journal of the European Optical Society Rapid Publications*, 2013, 8(13): 1–7.

[2] M. C. Amann, T. Bosch, M. Lescure, R. Myllylä, and M. Rioux, "Laser ranging: a critical review of usual

techniques for distance measurement," *Optical Engineering*, 2001, 40(1): 10–19.

[3] Y. Liu, W. Zhang, T. Xu, J. He, F. Zhang, and F. Li, "Fiber laser sensing system and its applications," *Photonic Sensors*, 2011, 1(1): 43–53.

[4] A. Velten, T. Willwacher, O. Gupta, A. Veeraraghavan, M. G. Bawendi, and R. Raskar, "Recovering three-dimensional shape around a corner using ultrafast time-of-flight imaging," *Nature Communications*, 2012, 3(745): 1–8.

[5] D. Monnin, A. L. Schneider, F. Christnacher, and Y. Lutz, "A 3D outdoor scene scanner based on a night-vision range-gated active imaging system," in *Third International Symposium on 3D Data Processing, Visualization, and Transmission*, Chapel Hill, NC, pp. 938–945, 2006.

[6] X. W. Wang, Y. Zhou, S. T. Fan, J. He, and Y. L. Liu, "Range-gated laser stroboscopic imaging for night remote surveillance," *Chinese Physics Letters*, 2010, 27(9): 94–97.

[7] J. Busck, "Underwater 3-D optical imaging with a gated viewing laser radar," *Optical Engineering*, 2005, 44(11): 6456–6468.

[8] C. Tan, G. Seet, A. Sluzek, and D. He, "A novel application of range-gated underwater laser imaging system (ULIS) in near-target turbid medium," *Optical and Lasers Engineering*, 2005, 43(9): 995–1009.

[9] J. Busck and H. Heiselberg, "Gated viewing and high-accuracy three-dimensional laser radar," *Applied Optics*, 2004, 43(24): 4705–4710.

[10] S. Y. Chua, X. Wang, N. Guo, and C. S. Tan, "Range compensation for accurate 3D imaging system," *Applied Optics*, 2016, 55(1): 153–158.

[11] B. Höfle and N. Pfeifer, "Correction of laser scanning intensity data: Data and model-driven approaches," *ISPRS Journal of Photogrammetry and Remote Sensing*, 2007, 62(6): 415–433.

[12] X. W. Wang, Y. F. Liu, and Y. Zhou, "Triangular-range-intensity profile spatial-correlation method for 3D super-resolution range-gated imaging," *Applied Optics*, 2013, 52(30): 7399–7406.

[13] S. Y. Chua, X. Wang, N. Guo, C. S. Tan, T. Y. Chai, and G. G. L. Seet, "Improving three-dimensional (3D) range gated reconstruction through time-of-flight (TOF) imaging analysis," *Journal of European Optical Society Rapid Publications*, 2016, 11: 16015-1–16015-6.

[14] B. Fu, K. Yang, J. Rao, and M. Xia, "Analysis of MCP gain selection for underwater range-gated imaging applications based on ICCD," *Journal of Modern Optics*, 2010, 57(5): 408–417.

[15] X. Wang, L. Hu, Q. Zhi, Z. Chen, and W. Jin, "Influence of range-gated intensifiers on underwater imaging system SNR," *Proc. SPIE*, 2013, 8912: 89120E.

[16] M. Laurenzis, F. Christnacher, D. Monnin, and T. Scholz, "Investigation of range-gated imaging in scattering environments," *Optical Engineering*, 2012, 51(6): 061303.

[17] R. D. Richmond and S. C. Cain, Direct-detection LADAR systems. U. S. A.: SPIE Press, 2010: 1–26.

[18] O. Steinvall, "Effects of target shape and reflection on laser radar cross sections," *Applied Optics*, 2000, 39(24): 4381–4391.

[19] D. Kong, J. Chang, P. Gong, Y. Liu, B. Sun, X. Liu, et al., "Analysis and improvement of SNR in FBG sensing system," *Photonic Sensors*, 2012, 2(2): 148–157.

[20] S. S. Patil and A. D. Shaligram, "On-line defect detection of aluminum coating using fiber optic sensor," *Photonic Sensors*, 2015, 5(1): 72–78.

[21] S. Y. Chua, X. Wang, N. Guo, C. S. Tan, and T. Y. Chai, "Effects of target reflectivity on the reflected laser pulse for range estimation," in *Progress In Electromagnetics Research Symposium Proceedings*, Prague, Czech Republic, pp. 2695–2699, 2015.

19

Acoustic Emission Localization Based on FBG Sensing Network and SVR Algorithm

Yaozhang SAI[1*], Xiuxia ZHAO[2], Dianli HOU[1], and Mingshun JIANG[3]

[1]*School of Information and Electrical Engineering, Ludong University, Yantai, 264025, China*

[2]*TIANRUN CRANKSHAFT CO., LTD, Wendeng, 264400, China*

[3]*School of Control Science and Engineering, Shandong University, Jinan, 250061, China*

*Correspond author: Yaozhang SAI　　　E-mail: saiyaozhang@163.com

Abstract: In practical application, carbon fiber reinforced plastics (CFRP) structures are easy to appear all sorts of invisible damages. So the damages should be timely located and detected for the safety of CFPR structures. In this paper, an acoustic emission (AE) localization system based on fiber Bragg grating (FBG) sensing network and support vector regression (SVR) is proposed for damage localization. AE signals, which are caused by damage, are acquired by high speed FBG interrogation. According to the Shannon wavelet transform, time differences between AE signals are extracted for localization algorithm based on SVR. According to the SVR model, the coordinate of AE source can be accurately predicted without wave velocity. The FBG system and localization algorithm are verified on a 500 mm×500 mm×2 mm CFRP plate. The experimental results show that the average error of localization system is 2.8 mm and the training time is 0.07 s.

Keywords: Fiber Bragg grating; acoustic emission localization; support vector regression; carbon fiber reinforced plastics

1. Introduction

Carbon fiber reinforced plastics (CFRP) have been widely applied in aircraft industry [1, 2]. But due to the impact and material fatigue, various barely visible damages can appear [3, 4]. The damages seriously degrade the reliability of CFRP structure and affect structure safety [5, 6]. Due to above reasons, damages should be timely located. The acoustic emission (AE) technology is an important damage detection means with great potential. Various damages of CFRP can cause AE phenomenon. According to the AE source localization, we can determine the location of damages.

Due to weight and size, the traditional AE piezoelectric sensor is difficult for damages detection of aircraft. In recent years, the fiber Bragg grating (FBG) sensor is researched for detecting AE signals [7, 8]. Because of light weight and small size of the FBG, it is appropriate for the AE detection in aircraft. Many methods are researched for the AE localization. Lu *et al.* [9] used the particle swarm optimization (PSO) algorithm to achieve the AE localization in an aluminum alloy plate. The localization error is less than 10mm. But the algorithm is not appropriate for the AE localization of anisotropic materials. Mostafapour *et al.* [10] developed a wavelet transform and cross-time frequency spectrum technology for the AE source localization, and high localization accuracy was obtained. Wave velocity is necessary for this technology. Therefore, it is only appropriate for the AE localization of an isotropic plate on which wave velocity is easily acquired. Fu *et al.* [11] used the back propagation neural network to realize the AE localization. But the algorithm uses wave velocity which is difficult to be obtained in the CFRP structure. Xiao *et al.* [12] applied the beamforming method to locate AE source in the CFRP plate. However, the wave velocity is still necessary. Jiang *et al.* [13] applied least squares support vector machine for the AE localization without wave velocity. The average localization error was 6.78 mm. But the frequency dispersion, which seriously affects the localization accuracy, is not been considered. Kim *et al.* [14] used the least squares support vector machine (LSSVM) to locate AE source, and better localization accuracy was obtained. However, the method is only used in an aluminum alloy plate.

In this paper, an FBG sensing network and support vector regression (SVR) algorithm are developed for the AE source localization on the CFRP plate. A high-speed FBG interrogation system is designed to obtain AE signals. The Shannon wavelet transform is applied to extract narrow-band signals of AE signals for AE localization, which reduces the influence of frequency dispersion on the AE localization. According to time differences of narrow-band signals, the SVR algorithm, which does not use wave velocity, is applied to locate AE source with high accuracy. The FBG sensing system and localization algorithm are verified on the CFRP plate. The experimental results show that the localization system is practical and efficient.

2. Localization algorithm

In many AE localization methods of CFRP, wave velocity is necessary. However, the CFRP material is anisotropic. Wave velocities of different directions are difference. If wave velocity is used for localization, it must be accurately measured in different directions. However, the work is very

difficult in practical applications. In this paper, we only use time differences of AE signals. The SVR algorithm is applied to forecast the coordinate of AE source.

2.1 Shannon wavelet transform

The frequency dispersion of AE wave and noise can seriously affect the calculation of time difference. Therefore, the Shannon wavelet transform is applied to extract narrow-band signals of AE signals for time differences.

Shannon wavelet transform can be expressed as

$$WT(x,a,b) = \frac{1}{\sqrt{a}} \int_{-\infty}^{+\infty} u(x,t)\psi^*\left(\frac{t-b}{a}\right)dt \quad (1)$$

where a and b are scale factor and time factor, respectively. $\psi(t)$ is the Shannon wavelet function which is given by

$$\psi(t) = \sqrt{f_b}\, \mathrm{sin}\, c(f_b t)e^{2\pi i f_c t} \quad (2)$$

where $f_b = \omega_b/2\pi$ and $f_c = \omega_c/2\pi$. The Fourier transform of the Shannon wavelet function is expressed as

$$\Psi(\omega) = \begin{cases} \sqrt{\dfrac{2\pi}{\omega_b}}, & \omega_c - \dfrac{\omega_b}{2} < \omega \le \omega_c + \dfrac{\omega_b}{2} \\ 0, & \text{others} \end{cases} \quad (3)$$

where $\Psi(\omega)$ is the rectangular window function of which the bandwidth and central frequency are ω_b and ω_c, respectively [15]. The narrow-band signals can be obtained by the Shannon wavelet transform. The AE signal $u(x,t)$ can be defined as

$$u(x,t) = e^{-j(k_1 x - \omega_1 t)} + e^{-j(k_2 x - \omega_2 t)} \quad (4)$$

where k_1 and k_2 are wave numbers. Introduction:

$$\Psi^*(a\omega) = \Psi^*(a\omega_1) = \Psi^*(a\omega_2)$$
$$\frac{k_2 - k_1}{2} = \Delta k, \quad \frac{\omega_2 - \omega_1}{2} = \Delta\omega \quad (5)$$
$$\frac{k_1 + k_2}{2} = k_0, \quad \frac{\omega_1 + \omega_2}{2} = \omega_0.$$

After the Shannon wavelet transform, the module value of AE signals can be obtained as

$$|WT(x,a,b)| = \sqrt{2a}|\Psi(a\omega_0)|\sqrt{1+\cos(\Delta\omega b - \Delta k x)}. \quad (6)$$

When $b = \Delta k / \Delta \omega$, the module value is the maximum. The time differences of AE signals are obtained by the peak times of module values of narrow-band signals.

2.2 SVR model

The basic idea of the SVR model is to map low-dimensional data into high-dimensional feature space by using a nonlinear mapping function, then to perform linear regression in this space. Given a training set of data $G=\{(x_i, y_i)\}(i=1, 2, \cdots, N)$, where x_i is an input vector, and y_i is an expected value. In the AE localization, x_i is a time differences vector of AE signals, and y_i is the ordinate or the abscissa of AE source. The regression function is expressed as

$$f(x) = w\phi(x_i) + b \qquad (7)$$

where $\phi(x_i)$ is a mapping function, w is a weight vector, and b is an error value. w and b are confirmed by minimizing the regularized risk function which is as follows:

$$\min \frac{1}{2}\|w\|^2 + C\frac{1}{N}\sum_{i=1}^{N} L[f(x_i) - y_i]$$

$$s.t. \quad L[f(x_i) - y_i] \qquad (8)$$

$$= \begin{cases} |f(x_i) - y_i| - \varepsilon, & |f(x_i) - y_i| \geq \varepsilon \\ 0, & |f(x_i) - y_i| < \varepsilon \end{cases}$$

where C is a penalty factor, $L(\cdot)$ is an ε-non-sensitive loss function, and ε is an ε-intensive loss parameter. Two relaxation factors ξ_i and ξ_i^*, which control a linearly inseparable boundary, are introduced [16]. So (8) can be rewritten as

$$\min_{w,b,\xi_i,\xi_i^*} \frac{1}{2}\|w\|^2 + C\frac{1}{N}\sum_{i=1}^{N}(\xi_i + \xi_i^*)$$

$$s.t. \quad L[f(x_i) - y_i] \qquad (9)$$

$$= \begin{cases} y_i - w\phi(x_i) - b \leq \varepsilon + \xi_i, & \xi_i \geq 0 \\ w\phi(x_i) + b - y_i \leq \varepsilon + \xi_i^*, & \xi_i^* \geq 0. \end{cases}$$

By introducing Lagrangian multiplier a_i and a_i^*, the above formula can be rewritten as

$$\min_a \frac{1}{2}\sum_{i,j=1}^{N}(a_N^* - a_N)(a_j^* - a_j)K(x_i, x_j) +$$

$$\varepsilon\sum_{i=1}^{N}(a_i^* + a_i) - \sum_{i=1}^{N}(a_i^* - a_i)$$

$$s.t.\begin{cases} w = \sum_{i=1}^{N}(a_i^* - a_i)x_i \\ \sum_{i=1}^{N}(a_i^* - a_i) = 0 \\ 0 \leq a_i^*, a_i \leq C \end{cases} \qquad (10)$$

where $K(x_i, x_j)$ is a kernel function [17]. We use the radial basis function (RBF). The resulting regression function of SVR model can be expressed as

$$f(x) = \sum_{i=1}^{N}(a_i^* - a_i)K(x_i, x) + b. \qquad (11)$$

The coordinate of AE source includes the ordinate and the abscissa, but the above SVR model can only output one valve. Two SVR models are built to respectively predict the ordinate and the abscissa of AE source.

2.3 AE localization process

The specific localization process, which is realized by the Shannon wavelet transform and SVR, is as follows:

(1) The coordinate, on which AE experiments are implemented and training data is obtained, is determined.

(2) AE experiments are implemented on training data points. AE signals are acquired. According to the Shannon wavelet transform, time differences of narrow-band signals of AE signals are extracted. Time differences and coordinates are used as training data.

(3) According to training data, SVR localization models, which output the ordinate and the abscissa of AE source, are built. According to cross validation, models are tested to confirm that models can be used for AE localization.

3. AE source localization

3.1 Experimental setup

AE localization experiments are implemented on a 300 mm×300 mm monitoring area of CFRP plate, whose dimension is 500 mm×500 mm×2 mm. Four FBG sensors are stuck on four corners of the CFRP plate, respectively. The FBG interrogation system includes FBG sensors, tunable narrow-band laser, fiber-optic coupler, optical circulator, photoelectric converter, amplifier, and data acquisition system, as shown in Fig. 1. The sampling frequency of FBG interrogation system is 5 MHz. The reflectance wavelengths of all FBG sensors are close to 1565.213 nm and errors are less than 0.01 nm. When FBGs are stuck, a definite pretension force is applied to them. The central wavelength of tunable narrow-band laser is 1565.238 nm, and wavelength variation is less than ±1 pm. The output power is 1 mW. The wavelength of narrow-band light source lies in the linear edge of FBG reflectance spectrum. The narrow-band laser demodulation technology is applied to obtain AE signals [7]. AE signals are generated by steel ball impact. The AE localization system is shown in Fig. 2.

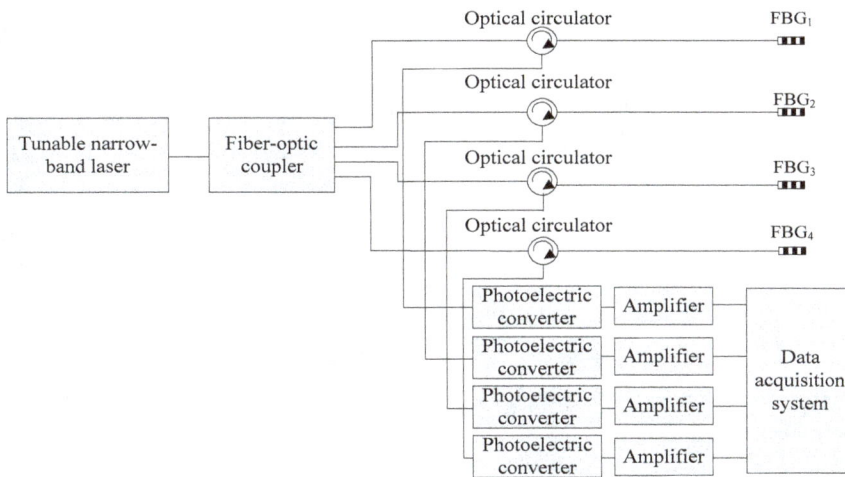

Fig. 1 FBG interrogation system.

3.2 AE localization experiment

The coordinates of four FBGs respectively are (0, 300), (300, 300), (300, 0) and (0, 0) in the monitoring area, as shown in Fig. 3. The monitoring area is divided into 36 small areas. Every small area is 50 mm×50 mm. The central point of a small area is chosen to carry out AE experiments and obtain AE experiments data for the training SVR model. The central point of big area, which is composed of four adjacent small areas, is chosen for testing data of the SVR model.

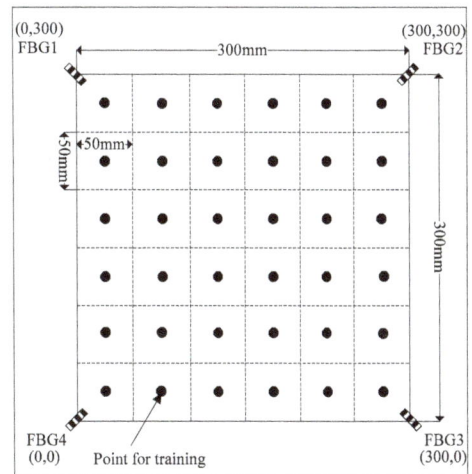

Fig. 2 AE localization experiment system.

Fig. 3 AE experimental layout.

On the above points, AE experiments are implemented. AE signals are obtained by FBG sensors. AE signals, which are acquired by carrying out AE experiments on the closest point of FBG_1, are shown in Fig. 4. Figure 5 is the frequency spectrum of AE signals of FBG_1. According to the frequency spectrum, the AE signal is a wideband signal. The frequency range is 20 kHz–200 kHz. According to frequency dispersion effect and signal attenuation, the accurate time differences between AE wide-band signals are difficult to be extracted. So AE narrowband signal, whose central frequency is 40 kHz, is extracted by the Shannon wavelet transform. At the same time, the module value of narrowband signal is calculated. According to the peaks of module values, time differences between AE signals can be obtained, as shown in Fig. 6. For the training SVR model, time differences between FBG₁ signals and other sensors signals are used for input, and the coordinates of points for training are used for output. According to the above process, SVR models for AE localization are built.

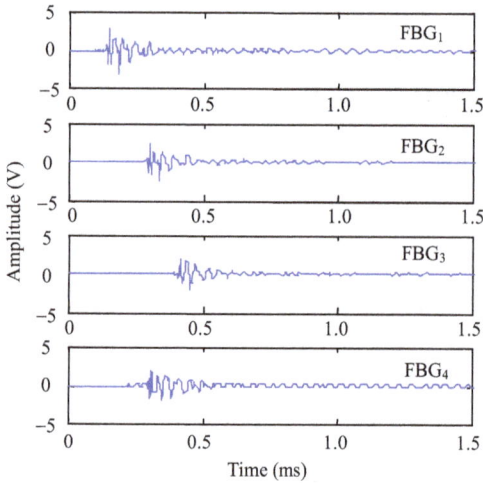

Fig. 4 AE signals.

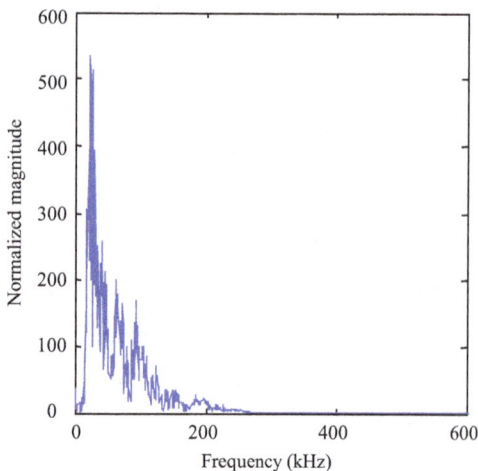

Fig. 5 Frequency spectrum of AE signals.

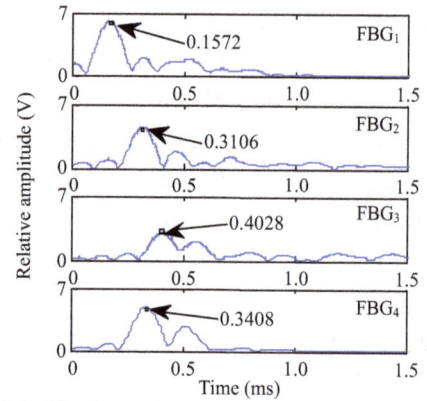

Fig. 6 Model values of narrowband signals.

For testing the SVR model for the AE localization, AE experiments are carried out on arbitrary ten testing points. Time differences are inputted, and the prediction coordinates of testing point are outputted. For making the contrast analysis of SVR and neural network, RBF neural network is applied for the AE localization. The testing results of SVR and RBF neural network are shown in Fig. 7. According to the radial error principle, the testing errors are calculated and shown in Fig. 8. The maximum errors of SVR and RBF are 7.2 mm and 6 mm, and the minimum errors are 0.4 mm and 0.8 mm, respectively. The average errors of SVR and RBF neural network are 2.8 mm and 3.7 mm, respectively. At the same time, the training times of SVR and RBF neural network are 0.07 s and 0.84 s, respectively. Compared with the RBF neural network, the localization accuracy of SVR is higher, and the training time is less. The above calculation results are given by MATLAB 7.1 and computer with AMD 3.0 GHz CPU.

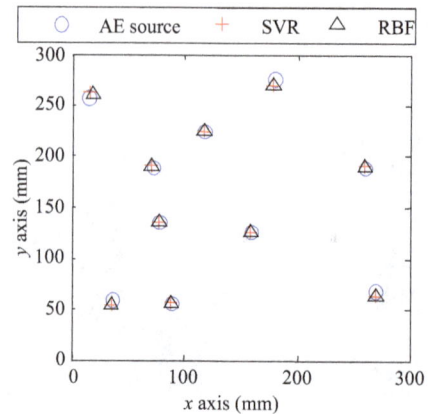

Fig. 7 Result of testing experiments.

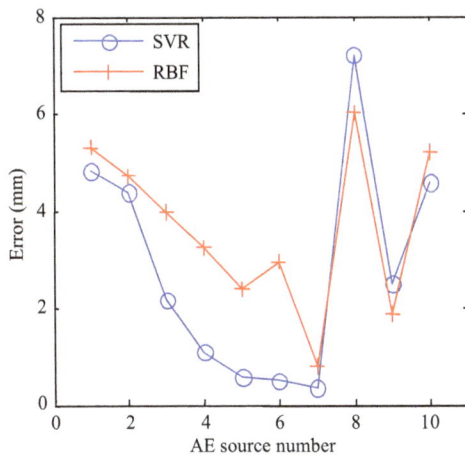

Fig. 8 Error of testing experiment.

4. Conclusions

In this paper, an AE localization system based on FBG sensing network and SVR algorithm is proposed. According to the narrowband laser demodulation technology, AE signals can be obtained by FBG sensors. The Shannon wavelet transform is applied for time differences between AE signals. According to time difference and SVR, the AE source is accurately located. The FBG system and localization algorithm are verified on the 500 mm×500 mm×2 mm CFRP plate. The average error of localization system is 2.8 mm, and training takes 0.07 s. Therefore, this paper proposes a practical AE localization system of CFRP plate for damage detection

Acknowledgment

This research is supported by the National Natural Science Foundation of China (Grant Nos. 61503218 and 41472260), the Fundamental research funds of Shandong University, China under Grant Nos. 2014YQ009, 2016JC012, and the Young Scholars Program of Shandong University 2016WLJH30.

References

[1] M. Grujicic, B. Pandurangan, W. C. Bell, C. F. Yen, and B. A. Cheeseman, "Application of a dynamic-mixture shock-wave model to the metal–matrix composite materials," *Materials Science and Engineering: A*, 2011, 528(28): 8187–8197.
[2] A. Katunin, K. Dragan, and M. Dziendzikowski, "Damage identification in aircraft composite structures: a case study using various non-destructive testing techniques," *Composite Structures*, 2015, 127(9): 1–9.
[3] F. Otero, S. Oller, S. X. Martinez, and O. Salomon, "Numerical homogenization for composite materials analysis. Comparison with other micro mechanical formulations," *Composite Structures*, 2015, 122(4): 405–416.
[4] H. Singh, K. K. Namala, and P. Mahajan, "A damage evolution study of E-glass/epoxy composite under low velocity impact," *Composites Part B: Engineering*, 2015, 76(8): 235–248.
[5] J. Zhang and X. Zhang, "An efficient approach for predicting low-velocity impact force and damage in composite laminates," *Composite Structures*, 2015, 130(12): 85–94.
[6] P. Yang, S. S. Shams, A. Slay, B. Brokate, and R. Elhajjar, "Evaluation of temperature effects on low velocity impact damage in composite sandwich panels with polymeric foam cores," *Composite Structures*, 2015, 129(11): 213–223.
[7] D. Pang and Q. Sui, "Response analysis of ultrasonic sensing system based on fiber Bragg gratings of different lengths," *Photonic Sensors*, 2014, 4(3): 281–288.
[8] Z. Jin, M. Jiang, Q. Sui, F. Zhang, and L. Jia, "Acoustic emission source linear localization based on an ultra-short FBGs sensing system," *Photonic Sensors*, 2014, 4(2):152–155.
[9] S. Lu, M. Jiang, Q. Sui, H. Dong, Y. Sai, and L. Jia, "Acoustic emission location on aluminum alloy structure by using FBG sensors and PSO method," *Journal of Modern Optics*, 2015, 63(8):1–8.
[10] A. Mostafapour, S. Davoodi, and M. Ghareaghaji, "Acoustic emission source location in plates using wavelet analysis and cross time frequency spectrum," *Ultrasonics*, 2014, 54(8): 2055–2062.
[11] T. Fu, Z. Zhang, Y. Liu, and J. Leng, "Development of an artificial neural network for source localization using a fiber optic acoustic emission sensor array," *Structural Health Monitoring*, 2015, 14(2): 168–177.
[12] D. Xiao, T. He, Q. Pan, X. Liu, J. Wang, and Y. Shan, "A novel acoustic emission beamforming method with two uniform linear arrays on plate-like structures," *Ultrasonics*, 2014, 54(2): 737–745.
[13] M. Jiang, S. Lu, Y. Sai, Q. Sui, and L. Jia, "Acoustic emission source localization technique based on least squares support vector machine by using FBG

sensors," *Journal of Modern Optics*, 2014, 61(20): 1634–1640.

[14] K. Kim and Y. Lee, "Acoustic emission source localization in plate-like structures using least-squares support vector machines with delta t feature," *Journal of Mechanical Science and Technology*, 2014, 28(8): 3013–3020.

[15] Y. Sai, M. Jiang, Q. Sui, S. Lu, and L. Jia, "Multi-source acoustic emission localization technology research based on FBG sensing network

and time reversal focusing imaging," *Optik*, 2016, 127(1): 493–498.

[16] Y. Liu and R. Wang, "Study on network traffic forecast model of SVR optimized by GAFSA," *Chaos, Solitons and Fractals*, 2016, 89(3): 153–159.

[17] W. Zhang, W. Hong, Y. Dong, G. Tsai, J. Sung, and G. Fan, "Application of SVR with chaotic GASA algorithm in cyclic electric load forecasting," *Energy*, 2012, 45(1): 850–858.

20

Optical Fiber Grating Vibration Sensor for Vibration Monitoring of Hydraulic Pump

Zhengyi ZHANG[*], Chuntong LIU, Hongcai LI, Zhenxin HE, and Xiaofeng ZHAO

Department Two, Rocket Force University of Engineering, Xi'an, 710025, China

[*]Corresponding author: Zhengyi ZHANG   E-mail: 18809231139@163.com

Abstract: In view of the existing electrical vibration monitoring traditional hydraulic pump vibration sensor, the high false alarm rate is susceptible to electromagnetic interference and is not easy to achieve long-term reliable monitoring, based on the design of a beam of the uniform strength structure of the fiber Bragg grating (FBG) vibration sensor. In this paper, based on the analysis of the vibration theory of the equal strength beam, the principle of FBG vibration tuning based on the equal intensity beam is derived. According to the practical application of the project, the structural dimensions of the equal strength beam are determined, and the optimization design of the vibrator is carried out. The finite element analysis of the sensor is carried out by ANSYS, and the first order resonant frequency is 94.739 Hz. The vibration test of the sensor is carried out by using the vibration frequency of 35 Hz and the vibration source of 50 Hz. The time domain and frequency domain analysis results of test data show that the sensor has good dynamic response characteristics, which can realize the accurate monitoring of the vibration frequency and meet the special requirements of vibration monitoring of hydraulic pump under specific environment.

Keywords: Fiber Bragg grating; uniform strength beam; vibrating monitoring; vibration sensor

1. Introduction

The mechanical vibration of the hydraulic system is one of the key factors that affect the reliability of the hydraulic system. Analyzing and monitoring the mechanical vibration of hydraulic pump has a positive theoretical and practical significance in the vibration and fault diagnosis of the hydraulic system [1, 2]. At present, the traditional hydraulic pump vibration monitoring technology mainly uses electrical sensors as the signal acquisition unit to turn the vibration signal into electrical signal, by monitoring the electrical signal to indirectly reflect the size of the vibration, which is now the most widely used vibration monitoring method [3, 4]. The advantages of this method are that the technology is mature, and the sensitivity is higher than the mechanical measurement. But it is difficult to achieve insulation measurement and very vulnerable to electromagnetic interference, which runs a high risk of causing accidents.

Based on the optical fiber sensing technology, the vibration monitoring system of the hydraulic pump is based on the optical fiber, and the light wave is the information carrier. The structure of the monitoring system is greatly simplified [5]. The optical fiber itself has the advantages of electrical insulation, anti electromagnetic interference, and good environmental adaptability, and can be highly integrated with the modern communication equipments, which provides a new technical approach for the vibration monitoring of the hydraulic system [6, 7]. The vibration signal sensing and detection technology of the hydraulic pump is the core of the whole monitoring system, and its performance directly affects the stability, sensitivity, and reliability of the whole monitoring system [8–10]. However, the vibration sensor which is widely used at present is mostly based on the switch quantity detection sensor unit integration, and the sensor network has not been formed, thus the price is high.

The fiber Bragg grating (FBG) is a kind of optical passive devices using spatially periodic refractive formation in the fiber core distribution to change the wave propagation behavior in the fiber. In addition to the advantages of the general optical fiber sensor, one can also use its unique advantages of wavelength encoding, through the vibration sensor with a specially designed structure [11, 12]. The FBG vibration sensor has a higher accuracy and lower price than the long period grating (LPG) vibration sensor and interferometric vibration sensor. In view of the above background, a kind of the FBG vibration sensor based on the equal intensity beam is studied and designed in this paper, which has an important application value in the vibration monitoring and fault diagnosis of the hydraulic pump.

2. FBG vibration sensing principle and analysis

2.1 Basic principle of FBG strain sensing

According to the coupled mode theory, the refractive index perturbation of the FBG cycle is only affected by a narrow range of wavelengths, i.e., the Bragg condition is only [13]

$$\lambda_B = 2n_{eff}\Lambda . \tag{1}$$

The light wave can be reflected by the grating, the rest of the transmission spectrum is not affected, and the FBG acts as a mirror or a filter. Equation (1) shows that the central wavelength λ_B changes with the grating period Λ and the effective index of refraction of the core n_{eff}. Strain is one of the physical parameters which is directly sensitive to the FBG, which affects the central reflection wavelength λ_B by the changes of the photoelastic effect and the grating period:

$$\Delta\lambda_B = (1 - P_e)\varepsilon\lambda_B \tag{2}$$

where P_e is the effective elastic optic coefficient of the fiber, and ε is the axial strain. When the FBG is affected by the external stress field, by means of the variation of the wavelength demodulation device to measure the central wavelength, we can accurately obtain the corresponding parameters of the external role information [14], which is the basic principle of the FBG strain sensor. By using the special structure design, we can use the FBG to realize the sensing and measurement of vibration and other physical parameters [15].

2.2 FBG vibration sensing principle of cantilever beam structure

Equation (2) shows that if the strain of the FBG ε changes periodically with time, the model can be used to measure the vibration periodically. The vibration element based on the cantilever beam structure has the advantages of simple structure, easy manufacture, and good bending property, so it is often used as the basic structure of the FBG vibration sensor [16, 17]. Because the gate area of the FBG with a certain length (8 mm – 10 mm) and the surface strain of the equal strength beam for pure bending can ensure that the tensile stress and compressive stress of each part of grating are the same, the chirp effect will not happen because of the local stress of FBG.

Based on the above analysis, this paper uses the equal intensity beam as the elastic element of the FBG vibration sensor. In order to analyze the mechanical properties of the strength beam and further deduce the sensitivity and natural frequency of the vibration sensor, let L, b, and h be the length, width, and thickness of the equal strength beam, respectively, E be the elastic modulus of the beam, and F be the end force of the beam. The structure and principle of the FBG vibration sensor are shown in Fig. 1.

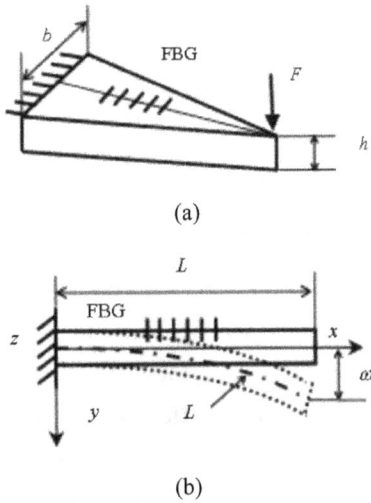

(a)

(b)

Fig. 1 Schematic of the FBG vibration sensor structure and beam bending with force: (a) sensor structure and (b) schematic of beam bending with force.

According to the structure shown in Fig. 1, the FBG is attached to the central axis of the equal strength beam along the axis, and the vibrator (mass) is fixed at the free end of the beam. When the external vibration causes the acceleration a of the vibrator to change, the oscillator will produce a periodic force F, which will be transformed into a periodic dynamic strain. The vibration information can be obtained by detecting the variation of the FBG central wavelength on the surface of the beam. Since the vibrational strain at the maximum amplitude of the grating reaches the maximum, and the central wavelength of the grating returns to the initial value when the vibrator returns to the equilibrium position, the frequency reflected by the change in the FBG wavelength is the vibration frequency to be measured.

According to the mechanical principle of the equal strength beam, the strain [18] of each point on the beam can be obtained by

$$\varepsilon = \frac{6L}{Ebh^2}F = KF \tag{3}$$

where K is the strain stress sensitivity of the equal strength beam. The quality of the cantilever beam is m, the vibration acceleration is a, and the force of the free end of the beam is $F=ma$. By means of (2) and (3), the sensitivity of the FBG vibration sensor S is

$$S = \frac{\Delta\lambda_B}{a} = \frac{6(1-P_e)mL}{Ebh^2}\lambda_B . \tag{4}$$

Sensitivity S is used to describe the relationship between the change in the FBG wavelength and the measured vibration acceleration a. The length of the mass block is L_m, and the length of the beam is about L. According to the knowledge of mechanics, the natural frequency of the FBG vibration sensor ω_0 is

$$\omega_0 = \sqrt{\frac{Ebh^3}{L\left(2L^2 + 6LL_m + 3L_m^2\right)m}} . \tag{5}$$

The sensitivity and natural frequency in (4) and (5) are two important parameters in determining the performance of the FBG vibration sensor, and the correct choice of the two parameters is essential for the design and test of the effect of the vibration sensor.

3. Structure design and finite element analysis of FBG vibration sensor

3.1 Structure design

According to (4) and (5), the sensitivity of the sensor S and the natural frequency ω_0 are related to the size, material, and the quality of the cantilever beam. Because the natural frequency of the sensor should be higher than the upper limit of the sensor system, in order to improve the upper limit of the frequency of the vibration sensor, the natural frequency of the sensor should be increased as much as ω_0. However, increasing the natural frequency means reducing the sensitivity of the sensor.

Therefore, the design requirements of the FBG vibration sensor is to meet the needs of the frequency measurement range, and the sensitivity should be as high as possible, so it is necessary to combine the actual needs of the above parameters for the rational design and selection.

In the hydraulic system, the hydraulic pump vibration frequency is generally located in the range of 20 Hz – 80 Hz. The vibration sensor is a typical two-order system, and the upper limit of the measured vibration frequency is about 80% of the natural frequency of the sensor. Therefore, the design of the natural frequency of the vibration sensor at 100 Hz or so is to meet the use requirements. In the concrete design, the material of the cantilever beam is chosen as 304 stainless steel, and its young's modulus E is about 193 GPa. Taking into account the length of the beam to reduce L can improve the natural frequency of the sensor and the impact on the sensitivity of small and convenient processing, to meet the premise of FBG paste and determine the length of the beam L being 50 mm and the width of b being 10 mm. The beam thickness h and the oscillator mass m are two important parameters in the design. In order to facilitate the processing, the cantilever beam can be directly used with the thickness of 0.5 mm of the plate, so the design is focused on the quality of the oscillator m optimization and selection. Figure 2 shows the change trend of the vibrator mass m with the natural frequency ω_0 of the cantilever beam when the above design parameters are used for the equal strength beam.

According to the relationship between the vibrator mass and the natural frequency in Fig. 2, we can see when the other parameters are determined, with an increase in the quality m of the oscillator, the natural frequency of the sensor negatively exponentially decreases. According to the calculation, when the natural frequency of the sensor is 100 Hz, the corresponding oscillator mass m is about 7.2 g. At this point, the sensitivity of the FBG

vibration sensor is about $0.05 \, \text{nm}/(\text{m·s}^{-2})$ when the central wavelength λ_B is 1550 nm.

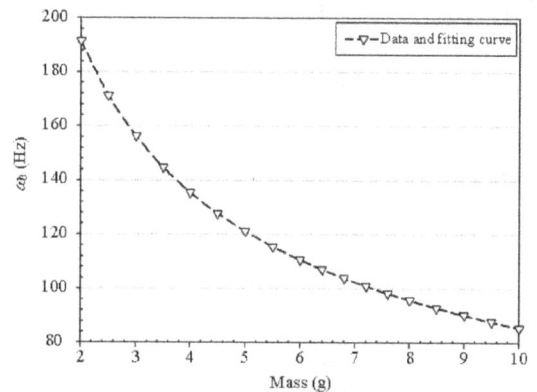

Fig. 2 Relation curve of the sensor natural frequency with the changes in vibrator mass.

3.2 Finite element analysis

In order to further analyze the theoretical model of the vibration sensor, based on the above theoretical design, the finite element and modal analysis of the FBG vibration sensor are carried out by using ANSYS software. According to the design parameters of the equal strength beam and the fixed requirements of the experimental platform, the structural model of the sensor is established. The vibration quality of the sensor is chosen to be a cylindrical stainless steel profile with a diameter of 10 mm, and the material is the same as that of the equal strength beam with a density of 7.93 g/cm³. In order to facilitate the processing, the design height of the vibrator is 12 mm, and the corresponding mass is about 7.4 g. According to the design parameters, a three-dimensional model of the vibration sensor by using Pro/E software is imported into the ANASYS software to mesh for the finite element analysis, and the results are shown in Figs. 3(a) and 3(b).

According to the results of the finite element analysis, the natural frequency of the vibration sensor is 94.739 Hz, and the maximum deflection deformation of the first order resonance is 4.16 mm. The finite element analysis results agree well with the theoretical analysis, which can meet the actual needs.

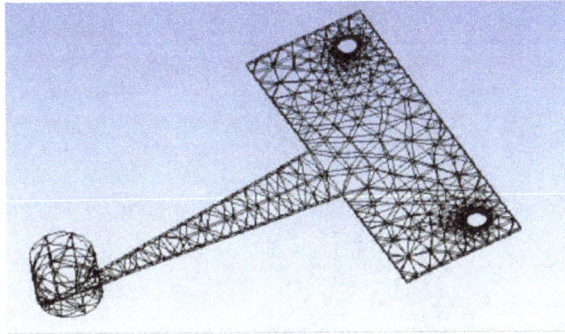

(a)

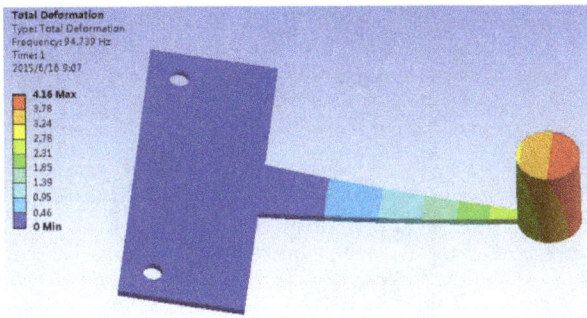

(b)

Fig. 3 Meshing the vibration sensor and finite element analysis: (a) mesh generation and (b) strain distribution of the first-order resonance.

4. Experimental measurement and analysis of the FBG vibration sensor

According to the above theory and the results of the finite element analysis, the structural dimension of the equal intensity beam used in the FBG vibration sensor is determined as: the length of L is 50 mm, the width of b is 10 mm, and the thickness of h is 0.5 mm. The vibrator is made up of 10 mm in diameter, with a height of about 12 mm, which is the same as that of the cantilever beam. The corresponding mass is about 7.4 g. The FBG central wavelength is 1550 nm, the gate length is 10 mm, the reflectivity is higher than 85%, and the fiber type is SMF-28, using epoxy glue FBG along the central axis of the cantilever beam, which is pasted on the surface. In the experiment, first, the FBG vibration sensor is fixed on the test bench, and the vibration frequency is 35 Hz and 50 Hz, respectively. The wavelength range of the FBG demodulator using wave capture series of high-speed demodulation module is 1520 nm – 1570 nm with a resolution of

1 pm and wavelength measurement repeatability of 5 pm. The schematic diagram of the FBG vibration sensing experiment is given in Fig. 4.

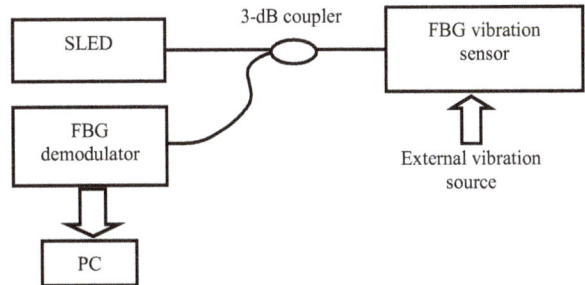

Fig. 4 Principle schematic of the FBG vibration test.

In Fig. 5, the actual measured FBG wavelength variation data are given when the excitation source is excited by 35 Hz, and the spectral curve of the experimental data are processed by fast Fourier transform algorithm (FFT). The sampling frequency of the FBG demodulator is 100 Hz.

According to the experimental data and analysis results, it can be seen that under the excitation frequency of 35 Hz, the change in FBG central wavelength is in the range of 1.2 nm – 1.6 nm. And the central wavelength of the FBG shows significant periodicity. The FFT spectrum analysis shows that it has an obvious peak value near 33 Hz, which is close to the frequency of the vibration source, and the FBG vibration sensor can measure the vibration frequency of the vibration source.

Similarly, in Fig. 6, the actual measured FBG wavelength variation data and the spectral curve after FFT treatment are given by using the 50-Hz excitation source.

According to the experimental data and analysis results, it can be seen that the central wavelength of the FBG also shows a periodic variation rule under the excitation frequency, and the peak frequency obtained by the FFT spectrum analysis is about 48 Hz. Compared with the experiment in Fig. 5, the variation range of the FBG central wavelength in Fig. 6 is only about 0.1 nm, which is mainly due to the different powers of the vibration motor in the experiment. In the above two experiments, the

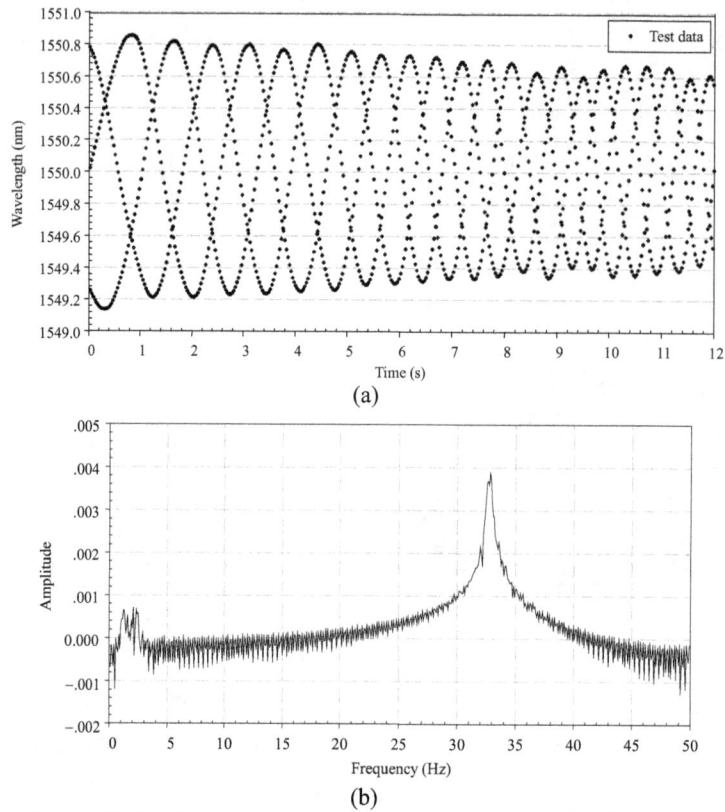

Fig. 5 Test data and FFT analysis with the vibration source at 35 Hz: (a) test data and (b) FFT spectral curve after treatment.

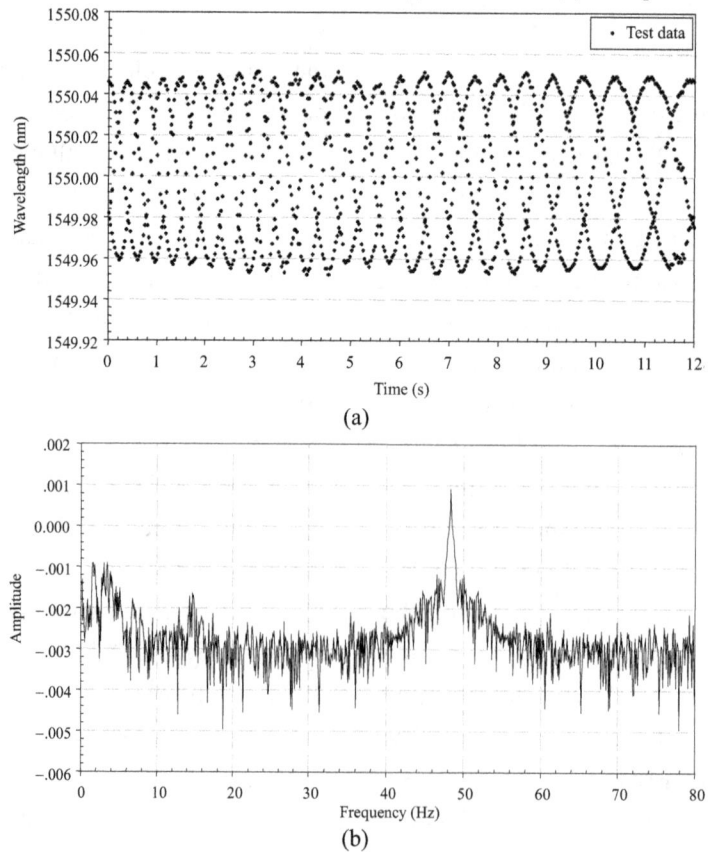

Fig. 6 Test data and FFT analysis with the vibration source at 50 Hz: (a) test data and (b) FFT spectral curve after treatment.

vibration frequency measured by the FBG vibration sensor is slightly smaller than the actual frequency of the vibrating motor, which may be due to the fact that there is no connection between the motor and the test bench. However, the measurement error is small, which does not affect its engineering application.

5. Conclusions

A fiber grating vibration sensor based on the intensity of the beam structure is designed in this paper, so as to solve the problem of hydraulic pump vibration electrical measurement of the vibration sensor, such as the high false alarm rate, vulnerablity to electromagnetic interference, and not easiness to achieve long-term and reliable monitoring and other issues. In this paper, based on the analysis of the vibration theory of the equal strength beam, the principle of FBG vibration tuning based on the equal intensity beam is derived. Combined with the practical application of the project, the structural dimensions and the oscillator mass of the equivalent strength beam are optimized, and the finite element analysis is carried out by using ANSYS software. Through the establishment of the test platform, two different frequencies of the vibration source are used to test the FBG vibration sensor design. Test data analysis results show that the sensor has good response characteristics and can realize the accurate test of vibration frequency. The advantages of the optical fiber sensing technology and the unique advantages of FBG wavelength coding, especially for the vibration measurement and fault diagnosis of the hydraulic pump, have important application prospects.

Acknowledgment

This work was supported by the National Natural Science Foundation of China (No. 41404022) and the Shanxi National Science Foundation (No. 2015JM4128).

References

[1] Q. F. He, C. J. Yao, G. M. Chen, X. H. Chen, and Q. Yang, "Feature extraction method of hydraulic pump vibration signal based on singular value decomposition and wavelet packets analysis," *Journal of Data Acquisition & Processing*, 2012, 27(2): 241–247.

[2] Z. Jing and L. Guo, "Application of adaptive stochastic resonance morphology in hydraulic pump vibration signal feature extraction," *Instrument Technique and Sensor*, 2015, 8(1): 92–95.

[3] J. Sun, H. R. Li, W. G. Wang, and B. H. Xu, "Preprocessing algorithm for vibration signals of a hydraulic pump based upon WMUWD," *Journal of Vibration and Shock*, 2015, 34(21): 93–99.

[4] Y. K. Wang, H. R. Li, and P. Ye, "Preprocessing method of hydraulic pump vibration signals based on FastPW and CNC de-noising," *Journal of Vibration and Shock*, 2014, 33(24): 144–149.

[5] X. F. Zhou, "Study of fiber optical Bragg grating sensing technology," Ph.D. dissertation, Wuhan University of Technology, Wuhan, 2003.

[6] X. Y. Dong and C. L. Zhao, "Chirp rate tunable fiber Bragg grating fibers based on a cantilever beam," *Journal of Optoelectronics Laser*, 2010, 21(10): 1455–1458.

[7] Q. Zhang, T. Zhu, and J. D. Zhang, "Micro-fiber-based FBG sensor for simultaneous measurement of vibration and temperature," *IEEE Photonics Technology Letters*, 2013, 25(18): 1751–1753.

[8] F. X. Zhang, X. L. Zhang, and L. J. Wang, "Study on FBG micro-seismic geophone with high sensitivity and broad bandwidth," *Journal of Optoelectronics Laser*, 2014, 25(6): 1086–1091.

[9] G. Rajan, M. Ramakrishnan, and Y. Semenova, "Analysis of vibration measurements in a composite material using an embedded PM-PCF polar metric sensor and an FBG sensor," *IEEE Sensors Journal*, 2012, 12(5): 1365–1371.

[10] Y. Du, T. G. Liu, and K. Liu, "Research of hybrid fiber sensing network based on FBG and optical frequency domain reflectometry," *Journal of Optoelectronics Laser*, 2013, 24(10): 1900–1905.

[11] Y. Y. Weng, X. G. Qiao, and T. Guo, "A robust and compact fiber Bragg grating vibration sensor for seismic measurement," *IEEE Sensors Journal*, 2011, 12(4): 800–804.

[12] Y. X. Guo, D. S Zhang, and Z. D. Zhou, "Cantilever based FBG vibration transducer with sensitization structure," *Optoelectronics Letters*, 2012, 8(3): 220–223.

[13] Y. Q. Li, Y. Wang, and G. Z. Yao, "Research on a vibration sensor system with temperature compensation using double-matched FBGs," *Journal of Optoelectronics Laser*, 2015, 26(2): 217–223.

[14] G. Xu, Y. T. Dai, and X. L. Jin, "A high-frequency dual-FBG accelerometer and its demodulation method," *Journal of Optoelectronics Laser*, 2011, 22(4): 515–519.

[15] H. L. Wang, H. Q. Zhou, and H. Gao, "Fiber grating acceleration vibration sensor with double uniform strength cantilever beams," *Journal of Optoelectronics Laser*, 2013, 24(4): 635–641.

[16] S. L. Wang, G. H. Xiang, and M. L. Hu, "Design of a novel FBG vibration sensor," *Journal of Optoelectronics Laser*, 2011, 22(4): 515–519.

[17] L. Sun, D. Z. Liang, and H. N. Li, "Analysis and modication of demarcate error of FBG sensor by equal," *Journal of Optoelectronics Laser*, 2007, 18(7): 776–779.

[18] H. Sun, B. Liu, and H. B. Zhou, "A novel FBG high frequency vibration sensor based on equi-intensity cantilever beam," *Chinese Journal of Sensor and Actuators*, 2009, 22(9): 1270–1275.

Ultra-Compact Photonic Crystal Based Water Temperature Sensor

Mahmoud NIKOUFARD[*], Masoud KAZEMI ALAMOUTI, and Alireza ADEL

Department of Electronics, Faculty of Electrical and Computer Engineering, University of Kashan, Kashan, Iran

[*]Corresponding author: Mahmoud NIKOUFARD   E-mail: mnik@kashanu.ac.ir

Abstract: We design an ultra-compact water temperature sensor by using the photonic crystal technology on the InP substrate at the 1.55-μm wavelength window. The photonic crystal consists of rods in a hexagonal lattice and a polymethyl methacrylate (PMMA) background. By using the plane wave expansion (PWE) method, the lattice constant and radius of rods are obtained, 520 nm and 80.6 nm, respectively. With a nanocavity placed in the waveguide, a resonance peak is observed at the 1.55-μm wavelength window. Any change of the water temperature inside the nanocavity results in the shift of the resonance wavelength. Our simulations show a shift of about 11 nm for a temperature change of 22.5 °C. The resonance wavelength has a linear relation with the water temperature.

Keywords: Sensor; water temperature; InGaAsP material; photonic crystal

1. Introduction

Photonic crystals are based on reflection and transmission of light in a periodic structure. This structure is used to avoid the propagation of electromagnetic wave in a certain frequency band, which is called photonic bandgap [1]. This properties and other capabilities like high flexibility design in the nanometer scale and integration with photodetector and laser, have turned photonic crystals to one of the important fields in the nano-scale researches [2–5]. One of the impressive factors for the photonic crystal lattice is the refractive index [6]. Meanwhile, the refractive index is one of the important parameters to identify gases and liquids. Many factors influence the refractive index such as temperature and pressure [7, 8]. Considering the fact that temperature variation

influences the refractive index [9] and the refractive index influences the output wavelength in a photonic crystal lattice, the temperature can easily be determined. In this paper, a water temperature sensor is designed based on photonic crystals with InGaAsP materials and a hexagonal lattice. The background is filled with polymethyl methacrylate (PMMA), and then by making a linear defect in photonic crystals, a path for the propagation of light has been created at the 1.55-μm wavelength. Existence of a large cavity defect in the light path makes a space for water to cross the sensor. One of the advantages of this sensor is that it can be monolithically integrated with a laser source and a photodetector having an ultra-compact water temperature sensor. In addition, the material safety has to be considered to prevent water pollution.

2. Structure of photonic crystal

The schematic of photonic crystal with a hexagonal lattice and a pillar-based structure is depicted in Fig. 1. The layer stack consists of a film layer of InGaAsP with an energy bandgap of E_g=0.992 eV (λ_g=1.25 μm) and a thickness of 500 nm, denoted as Q(1.25), and an upper cladding layer of InP with a thickness of 1.5 μm on the InP substrate. The refractive indices of the InP cladding and InGaAsP [Q(1.25)] film layer are 3.17 and 3.3640, respectively, at the 1.55-μm wavelength. The refractive index of the Q(1.25) film layer is higher than that of the lower cladding and substrate layers [10, 11]. Specifications of layers including the diameter (r), lattice constant (a), and refractive index (n) are depicted in Fig. 1.

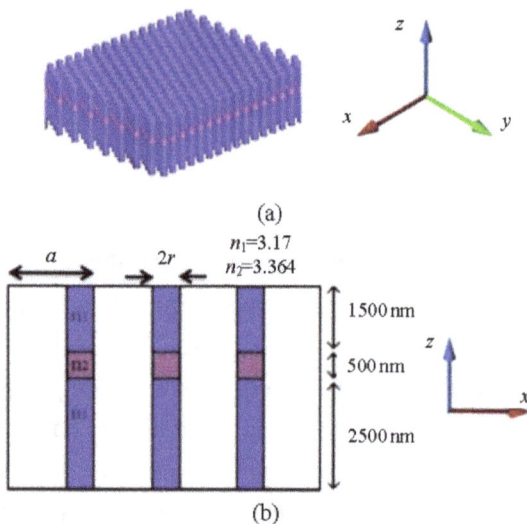

(a)

(b)

Fig. 1 Specifications of layers: (a) 3D-geometry of photonic crystal and (b) x-z view with layer specifications.

Due to the memory and speed limitations of computational computer, the three-dimensional (3D) simulation of the sensor is reduced to a two-dimensional (2D) photonic crystal structure. At first, the effective refractive index of layers is determined to be 3.2634 using COMSOL software. The two-dimensional plane wave expansion method is used to determine the photonic bandgap of the dielectric rods periodically arranged in the

hexagonal lattice for the polarization of the E-field parallel and perpendicular to the dielectric rods (TM and TE modes, respectively). The photonic bandgap diagram for the different radii is plotted in Fig. 2 for two polarization states of TE and TM modes when the radius varies within a 0 nm–250 nm range with a lattice constant of 520 nm. It can be seen that for a rod with a radius of 86 nm, the band gap includes the normalized frequency of 0.3354 equivalent to a wavelength of 1.55 μm. In Fig. 3, the band structure is shown in the K direction for both TE and TM modes.

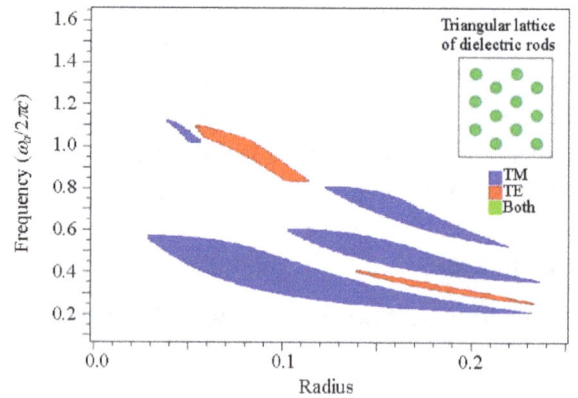

Fig. 2 Normalized frequency bandgap as a function of radius (μm) for TE and TM polarization modes.

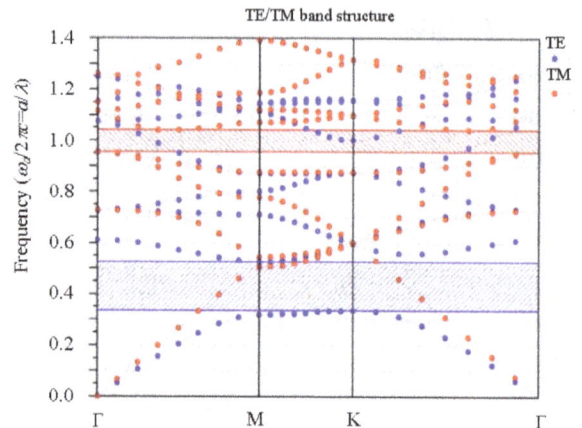

Fig. 3 Band structure in the K direction as a function of the normalized frequency for TE and TM modes.

The main advantage of this layer stack is that it can be integrated with active devices such as the laser and photodetector, as can be seen in Fig. 4. The on-chip integration of the nanosensors, laser source, and photodetector is attractive because of the higher operating temperature of InP-based devices, the possibility of making very compact sensor through the on-chip integration of optical transmitter, and receiver components with the photonic crystal-based sensor.

The laser produces a light wave at the 1.55-μm wavelength launching into the photonic crystal sensor. The light is then detected by the photodetector. The only difference between the water temperature sensor (passive structure) and the laser source and photodetector (active structures) is a layer of InGaAsP with a bandgap wavelength of 1.55 μm [denoted as Q(1.55)]. This layer has a direct bandgap at the 1.55-μm wavelength and a bandgap energy of lower than that of adjacent layers. A direct bias voltage applied to the laser source injects an electron-hole pair to the active layer of Q(1.55), generating the light beam. Whereas, a reversed bias voltage applied to the photodetector converts the absorbed light to the electron-hole pair in the active layer (shown in Fig. 4). This scheme requires a butt-joint technology to fabricate the active and passive structures on a single chip [10, 11].

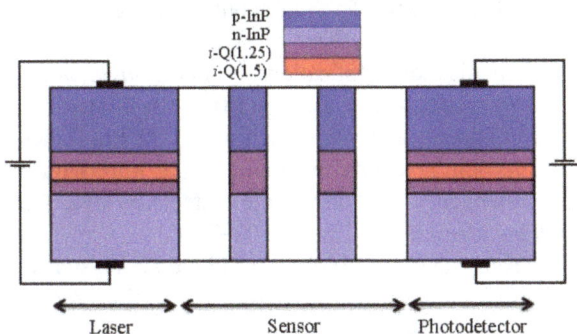

Fig. 4 Schematic view of the water temperature sensor chip including the monolithic integrated laser, photonic-crystal-based sensor, and photodetector.

3. Photonic crystal based water temperature sensor

The photonic crystal-based temperature sensor has an area of 17×17 lattice constants in a hexagonal lattice dielectric-pillar. The background is filled with the PMMA material having a refractive index of 1.4914. By creating a line defect and a great hole

with a radius of 1.05 μm, we have developed a sensor that works at the wavelength window of 1.55 μm. The great hole acts as a resonator allowing water to enter the sensor, which shifts the resonance wavelength of resonator. In addition, the resonance wavelength depends on the water temperature.

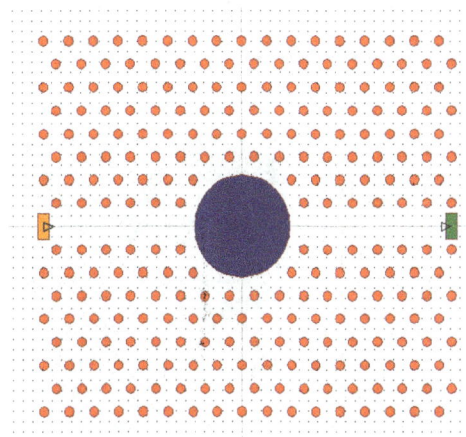

Fig. 5 Photonic crystal structure after creating the line defect (the large cavity in the center of the sensor acts as a resonator).

4. Simulation results

The water temperature sensor is investigated using the finite difference time domain (FDTD) method. Figure 6 shows the field distribution in the photonic crystal sensor at a wavelength of 1.55 μm. A pulsed laser beam is launched toward the temperature sensor. The optical signal is then propagated through the defects and absorbed in the photodetector after the photonic crystal-based sensor. The change in the water temperature causes a change in the refractive index of the large hole, thus resulting in a shift in the resonant wavelength peak. The relationship between the water temperature and refractive index at a pressure of 1 bar is shown in Fig. 7 [6]. Multiple simulations are performed at several temperatures to verify the resonant wavelength peak. The simulation results are presented in Table 1 (see also Fig. 8). It can be observed that the relationship between the water temperature and the resonant wavelength peak is linear in the wide range of 24 ℃ to 55 ℃.

Table 1 Relationship among the resonant wavelength peak, refractive index, and temperature of water in the water temperature sensor.

Temperature (℃)	Refractive index	Wavelength (nm)
24.8	1.33293	1578.888
34.5	1.33177	1578.303
40	1.3302	1572.427
47	1.3290	1569.268
54.34	1.32866	1567.132

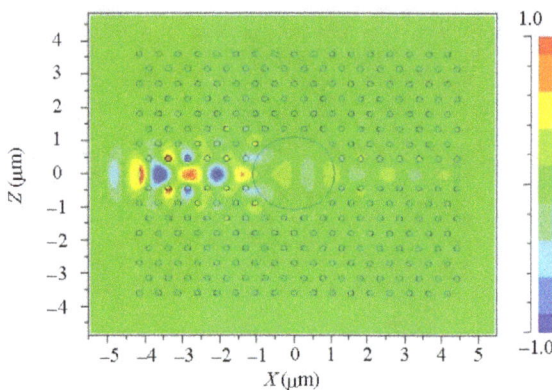

Fig. 6 Electric field distribution in the photonic crystal-based water temperature sensor at the 1.55-μm wavelength.

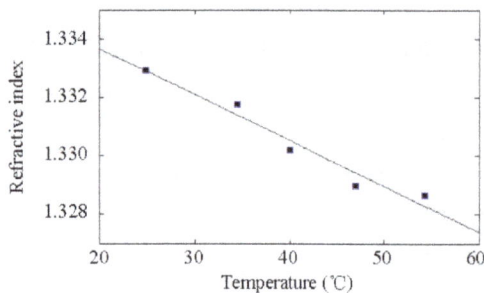

Fig. 7 Refractive index of water as a function of the water temperature with a linear curve fit of $n = -0.00015T + 1.3367$.

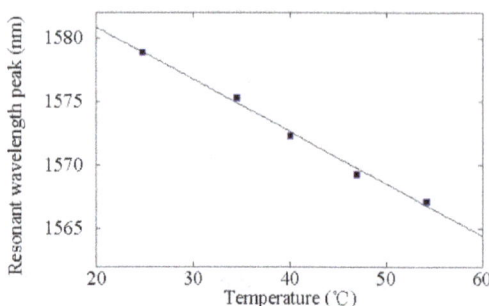

Fig. 8 Resonant wavelength peak as a function of temperature with a linear curve fit of $\lambda = -0.41083T + 1589.1$.

5. Conclusions

In this article, we design an ultra-compact water temperature sensor by using the photonic crystal technology. The main advantage of the presented structure is the capability of monolithic integration with the laser and photodetector by using the butt-joint technology which reduces the size of the sensor to nano-scale. Simulations show that there is a linear relationship between the water temperature and the resonant peak wavelength of the sensor. A water temperature change in the range of 24.8 ℃ to 54.34 ℃ causes a wavelength shift of about 10 nm.

References

[1] J. D. Joannopoulos, S. G. Johnson, J. N. Winn, and R. D. Meade, "Photonic crystals: molding the flow of fight," *Computing in Science and Engineering*, 1995, 3(6): 38–47.

[2] S. Olyaee and F. Taghipour, "Design of new square-lattice photonic crystal fibers for optical communication applications," *International Journal of Physical Sciences*, 2011, 6(18): 4405–4411.

[3] S. Olyaee and F. Taghipour, "Ultra-flattened dispersion hexagonal photonic crystal fibre with low confinement loss and large effective area," *IET Optoelectronics*, 2012, 6(2): 82–87.

[4] C. Kang and S. M. Weiss, "Photonic crystal with multiple-hole defect for sensor applications," *Optics Express*, 2008, 16(22): 18188–18193.

[5] S. Hosseini, M. Haas, D. Plettemeier, and K. Jamshidi, *Integrated optical devices for 3D photonic transceivers*. Germany: Springer International Publishing, 2016.

[6] T. Stomeo, M. Grande, A. Qualtieri, A. Passaseo, A. Salhi, M. De Vittorio, *et al.*, "Fabrication of force sensors based on two-dimensional photonic crystal technology," *Microelectronic Engineering*, 2007, 84(5–8): 1450–1453.

[7] R. M. Waxler and C. E. Weir, "Effect of pressure and temperature on the refractive indices of benzene, carbon tetrachloride and water," *Biochemical & Biophysical Research Communications*, 1973, 51(2):

163–171.

[8] S. Tao, D. Chen, J. Wang, J. Qiao, and Y. Duan, "A high sensitivity pressure sensor based on two-dimensional photonic crystal," *Photonic Sensors*, 2016, 6(2): 137–142.

[9] H. Wang, H. Meng, R. Xiong, Q. Wang, B. Huang, X. Zhang, *et al.*, "Simultaneous measurement of refractive index and temperature based on asymmetric structures modal interference," *Optics Communications*, 2016, 364: 191–194.

[10] L. Xu, M. Nikoufard, X. J. Leijtens, T. De Vries, E. Smalbrugge, R. Nötzel, *et al.*, "High-performance InP-based photodetector in an amplifier layer stack on semi-insulating substrate," *IEEE Photonics Technology Letters*, 20(23): 1941–1943.

[11] K. A. Williams, E. A. J. M. Bente, D. Heiss, Y. Jiao, K. Ławniczuk, X. J. M. Leijtens, *et al.*, "InP photonic circuits using generic integration [Invited]," *Photonics Research*, 2015, 3(5): B60–B68.

22

A Precision Fiber Bragg Grating Interrogation System Using Long-Wavelength Vertical-Cavity Surface-Emitting Laser

Binxin HU[1*], Guangxian JIN[2], Tongyu LIU[2], and Jinyu WANG[1]

[1]*Key Laboratory of Optical Fiber Sensing Technology of Shandong Province, Laser Institute of Shandong Academy of Sciences, Jinan, 250014, China*

[2]*Shandong Micro-sensor Photonics Co. Ltd, Jinan, 250014, China*

*Corresponding author: Binxin HU   E-mail: binxin.hu@sdlaser.cn

Abstract: This paper presents the development of a cost-effective precision fiber Bragg grating (FBG) interrogation system using long-wavelength vertical-cavity surface-emitting laser (VCSEL). Tuning properties of a long-wavelength VCSEL have been studied experimentally. An approximately quadratic dependence of its wavelength on the injection current has been observed. The overall design and key operations of this system including intensity normalization, peak detection, and quadratic curve fitting are introduced in detail. The results show that the system achieves an accuracy of 1.2 pm with a tuning range of 3 nm and a tuning rate of 1 kHz. It is demonstrated that this system is practical and effective by applied in the FBG transformer temperature monitoring.

Keywords: FBG; VCSEL; wavelength interrogation; intensity normalization; peak detection

1. Introduction

For some physical quantities (e.g., temperature, strain, and pressure), fiber Bragg grating (FBG) sensors yield accurate and stable measurements and offer the added benefits of electrically passive operation, immunity to electromagnetic interference, and multiplexing capabilities [1]. FBG sensors therefore have been widely used in mine safety monitoring [2], industrial automation [3], structural health monitoring [4], etc. So far, the external-cavity tunable lasers are commonly adopted in some commercial FBG interrogators (e.g., MOI SM125, NI PXIe-4844, and OSICS ECL 1560). They can deliver an accuracy of about 1 pm and larger capacity over longer fiber lengths owing to its

high-optical power. However, due to their fabrication complexity, these interrogators are generally complicated and expensive [5]. This makes their usage inefficient and impractical in many field applications, such as power transformer temperature monitoring, reservoir water level monitoring, and coal face ground pressure monitoring.

In recent years, vertical-cavity surface-emitting laser (VCSEL) emitting in the 1.3-μm to 1.6-μm wavelength regime, also known as long-wavelength VCSEL, has been highly desirable for FBG sensing applications. It has some unique properties such as wide tuning range, narrow linewidth, low cost, and low power consumption [6]. Therefore, a variety of FBG interrogation solutions using VCSEL have been reported. Huang *et al.* implemented a fast FBG interrogation system using VCSEL [7]. Van Hoe *et al.* designed a portable integrated FBG sensing system based on a single VCSEL [8]. Mizunami *et al.* designed a power-stabilized tunable source using a VCSEL and an erbium-doped fiber amplifier (EDFA) for FBG interrogation [9]. To our knowledge, these solutions usually rely on a linear function of the injection current for the VCSEL wavelength and may be susceptible to laser noise and ultimately lead to a worse throughput in many field applications.

In this paper, the tuning properties of a long-wavelength VCSEL are studied experimentally. The wavelength versus injection current tuning curves is found approximated with a second-order polynomial. Then, the overall design and key techniques of this system including intensity normalization, peak detection, and quadratic curve fitting, are introduced in detail. In particular, the combination of a four-point peak detection method and the real-time calibration using the absorption lines of C_2H_2 realizes a uniform and predicable wavelength calculation. The results show that the system achieves an accuracy of 1.2 pm with a tuning range of 3 nm and a tuning rate of 1 kHz. Finally, it is demonstrated that this system is practical and effective when applied in the FBG transformer temperature monitoring.

2. Methodology

2.1 Tuning properties

We studied the output parameters of a long-wavelength VCSEL module (VERTILAS GmbH, Germany) operating near 1530 nm integrated with a thermoelectric cooler (TEC) and a thermistor. The experimental setup included a Keithley 2420 current source, a Keithley 2510 TEC controller, and an Agilent 86120c multi-wavelength meter. All experiments were performed at room temperature. Figure 1 shows the test results.

(a)

(b)

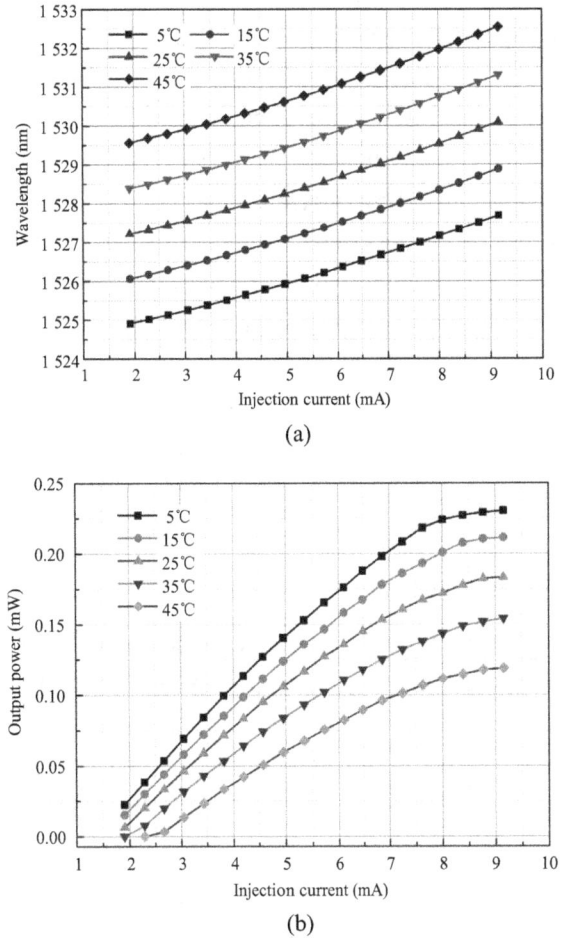

Fig. 1 Tuning curves of VCSEL module: (a) output wavelength and (b) output power vs. injection current at different heat sink temperatures.

It is evident from Fig. 1(a) that there is an approximation of wavelength tuning curves of the VCSEL by a second-order polynomial with respect to an intercept wavelength λ_0 to which a laser is tuned at the injection current and heat sink temperature T:

$$\lambda(I,T) = a(I - I_0)^2 + b(I - I_0) + kT + \lambda_0 \quad (1)$$

where a, b, and k are unknown coefficients. Given that $T = 25\ °C$, $I_0 = 1.907\ mA$, coefficients a, b, and k are calculated as 0.01462 nm/mA^2, 0.29146 nm/mA, and 0.11629 nm/°C, respectively. Since the coefficient k does not depend on the injection current and heat sink temperature, the wavelength of VCSELs can be tuned by the injection current and heat sink temperature independently.

For FBG sensing purposes, an appropriate package with a temperature-controlled heat sink is used to fix the wavelength range by stabilizing the laser temperature. Wavelength tuning across the reflection spectrum of the FBG sensor is then usually realized by current tuning. For VCSELs, this mechanism works up to tuning rates even in the ~100 kHz regime [10]. Moreover, as one can see from Fig. 1(b), the output power of the VCSEL is related to both heat sink temperature and injection current. As the current decreases, so does its output power. To get precise wavelength information, variations in the light intensity of the VCSEL should be corrected by intensity normalization.

2.2 Overall design

Based on the above ideas, we designed a precise FBG sensor interrogation system, as Fig. 2 clearly shows.

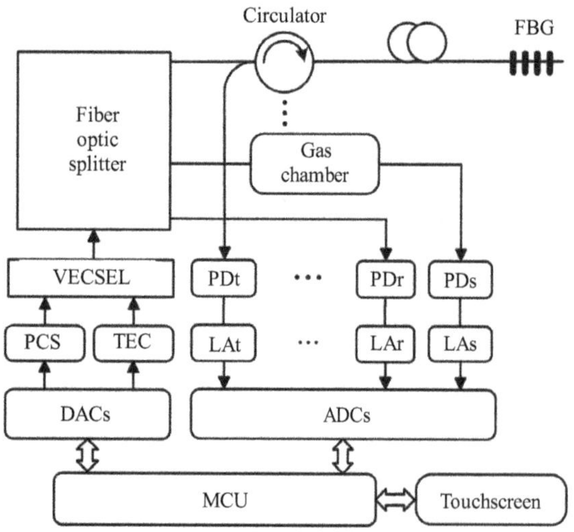

Fig. 2 Block diagram of the FBG sensor interrogation system.

The heat sink temperature of the VCSEL is maintained at a constant value (e.g., 25 ℃) by a TEC controller. A programmable current source (PCS) provides a saw-tooth modulation current (2 mA to 10 mA) with a tuning rate up to 1 kHz for the laser. The tuning signal enters a 1×16 fiber optic splitter splitting the optical signal power evenly into all the output ports. The first light is converted to a

reference current by photodiode r (PDr). The second light passes through a gas chamber full of C_2H_2 and is converted to a standard current by PDs. The third light enters a circulator and reaches the sensor. The reflected light is converted to a test current by PDt. These currents are converted to the corresponding voltages by respective logarithmic amplifiers (LAs) and enter analog-to-digital converters (ADCs). The acquired data are processed by a microcontroller (MCU) to perform such operations as smoothing, normalization, peak detection, fitting, wavelength calculation, and user interface. The wavelength of the sensor is eventually displayed on a touchscreen, or sent to personal computer (PC) for further processing via an Ethernet port.

The MAX1978 serves as the TEC controller. On chip thermal control-loop circuitry can maintain a temperature stability of 0.001 ℃. The PCS is based on the AD8276 difference amplifier and the AD8603 op amp. It can drive output currents of 15 mA without the need for the external transistor or metal oxide semiconductor field effect transistor. The AD5663R, dual 16-bit digital to analog converters (DACs) are used to generate programmable reference voltages for the PCS and the TEC controller. The STM32F429 with ARM Cortex-M4 core is chosen as the MCU. The 180-MHz central processing unit (CPU), the digital signal processor (DSP) instructions, and the floating point unit make it ideal for the measurement and control applications. This device also incorporates rich peripherals such as serial peripheral interface (SPI), universal serial bus (USB), and Ethernet. The LTC1867 are used for data acquisition (DAQ), which are 16-bit, 8-channel, 200 kSPS ADCs. Two LTC1867s together form 16 channels. The MCU is linked with ADCs and DACs via the high-speed SPI interface.

2.3 Key operations

In practice, the reflected light from the FBG sensor often experiences a large amount of losses due to connectors, splices, and fiber attenuations. The PD current is only a few nA. Meanwhile, the PD current can be hundreds of μA when local measurements are taken. The conventional gain control solutions using digital potentiometers require regulation for many times and sometimes fail. Alternatively, the LA (i.e., AD8305) can convert the PD current to a voltage with a precise logarithmic relationship and provides a large dynamic range up to 100 dB, thus more adaptive in field applications [11].

Figure 3 shows the standard signal V_s, the reference signal V_r, and the normalized signal V_n. It can be seen that there is an overestimation of time stamp for peak power of V_s due to the effect of variable light intensity.

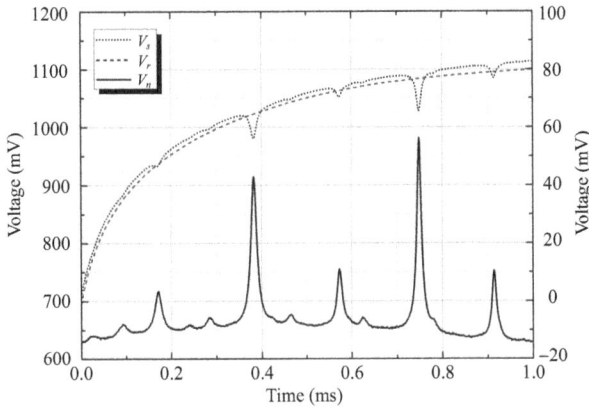

Fig. 3 Spectrums of standard, reference, and normalized signals.

Using the logarithmic transfer function of the LA, the normalized signal V_n can be expressed as

$$V_n = V_s - V_r = 0.2 \log_{10} \frac{I_s}{I_r} \quad (2)$$

where I_s and I_r denote the PD currents from standard and reference channels, respectively. This normalization scheme corrects the unknown variations in the signal amplitude that accounts for the majority of random fluctuations in the average signal due to laser noise resulting in a robust sensor applicable in harsh environments.

Peak detection is another challenge for FBG sensor interrogation. The algorithm of direct peak-detection is relatively simple. But it will lead to large errors when the noise is high. Some curve

fitting techniques (e.g., Gauss curve fitting, Gauss formula nonlinear curve fitting, and artificial neural network) can effectively improve the detection accuracy [12, 13]. But they are generally complicated and are not suitable for micro-controller systems. Therefore, we propose a simple and effective four-point peak detection method that can be described as:

(1) Calculate the half maximum value (V_{hm}) of the FBG reflection spectra in the specified bounds.

(2) Find two points whose ordinate is closest to V_{hm} at the left [(X_1, V_1), (X_2, V_2)] and right slopes [(X_1', V_1'), (X_2', V_2')] of the spectra.

(3) Use the line extrapolation method to get $X_1(V_{hm})$ and $X_2(V_{hm})$:

$$X(V_{hm}) = X_{k-1} + \frac{V_{hm} - V_{k-1}}{V_k - V_{k-1}}, \ V_{k-1} < V_{hm} < V_k. \quad (3)$$

(4) The detection result is thus calculated as

$$X^* = \frac{X_1(V_{hm}) + X_2(V_{hm})}{2}. \quad (4)$$

Using this method, we can get five peak values of the standard signal (see Fig. 3), which are 0.16637 ms, 0.37788 ms, 0.56833 ms, 0.74442 ms, and 0.90994 ms. By referring to HITRAN database, we also get the corresponding wavelengths of the absorption lines (C$_2$H$_2$), which are 1527.44098 nm, 1528.01418 nm, 1528.59374 nm, 1529.17972 nm, and 1529.7721 nm. Using the least squares method for quadratic curve fitting, the relationship between the wavelength and tuning time is shown in Fig. 4. Eventually, we measure the peak value of the reflected signal from the FBG sensor and calculate its wavelength using the relational expression (see Fig. 4).

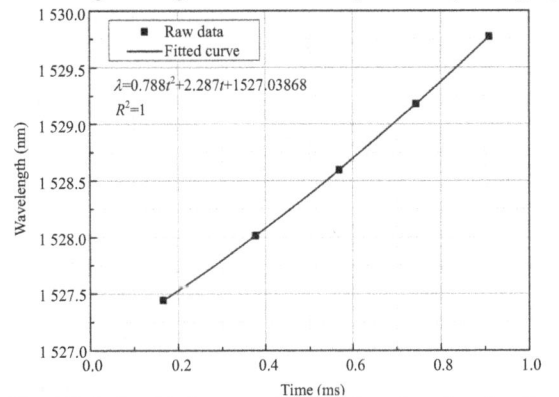

Fig. 4 Relationship between the wavelength and tuning time.

3. Results

3.1 Measurement capabilities

A photograph of the finished FBG interrogation system is shown in Fig. 5. The system is housed in a rack-mountable enclosure with dimensions of 480 mm × 87 mm × 360 mm. VCSEL, PCS, and TEC controllers are mounted on a printed circuit board (PCB), called VCSEL module. The MCU board is mounted lower right of the enclosure along with PCBs providing a touchscreen and the Ethernet interface.

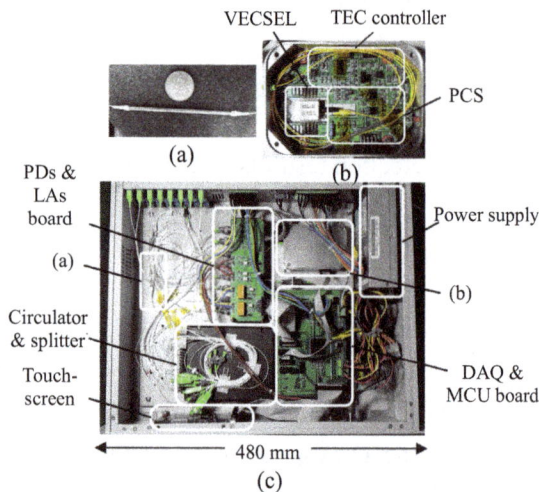

Fig. 5 Interrogation system: (a) gas chamber, (b) VCSEL module, and (c) enclosure and control electronics.

A gas chamber (C_2H_2) was used for wavelength calibration. The test conditions included the heat sink temperature of 25 ℃, current tuning rate of 1 kHz, and current running range of 2 mA to 10 mA. We repeated each measurement 100 times and used the average as the measured value. Table 1 shows the results of calibration. The accuracy is expressed as the absolute error and is found well within 1.2 pm at five standard wavelengths. In particular, the measurement uncertainties for 1528.01418 nm standard wavelength is calculated as about 1.1 pm ($k = 2$).

Another testing was to verify the spectral tuning range. The heat sink temperatures were set to 15 ℃, 25 ℃, and 35 ℃, respectively. The APEX AP204XB optical spectrum analyzer with a 0.04-pm resolution was used to measure the emitted spectrums. The test results are shown in Fig. 6.

Table 1 Results of calibration.

Standard wavelength (nm)	Measured value (nm)	Accuracy (pm)
1527.44098	1527.44039	−0.59
1528.01418	1528.01531	1.13
1528.59374	1528.59357	−0.17
1529.17972	1529.17863	−1.09
1529.7721	1529.77317	1.07

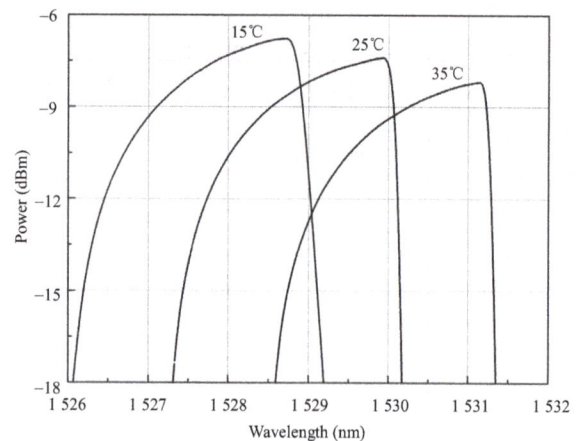

Fig. 6 Spectrum of the VCSEL output signal using current tuning (2 mA to 10 mA) at different heat sink temperatures.

It can be seen that the spectral tuning range is about 3 nm at three different temperatures and can be extended to above 5 nm when the temperature tuning is also applied. To do so, temperature tuning can be used to adjust the laser output near the central wavelength of the FBG. Moreover, this system had 14 separate optical channels. Each channel was typically connected with one FBG sensor.

3.2 Practical application

The power transformer is one of the most important devices in the power transmission and transformation system. The temperature is the main factor affecting the insulation capability [14]. The FBG sensing technology is suitable for high temperature and pressure in the oil and gas environments, thus providing a good way to the real-time monitoring of transformer temperature.

On basis of this idea, an FBG temperature sensor

(~1526.5 nm @ 25 ℃) was taken for example. And we used our interrogation system to monitor its temperature changes. The test setup included a Fluke 1502A thermometer readout together with a 5628 standard platinum resistance thermometer (0 ℃ to 300 ℃, ±0.05 ℃), a BILON CXW-501 thermostatic bath (0 ℃ to 95 ℃), and a BILON HH-05BS thermostatic bath (30 ℃ to 300 ℃).

As shown in Fig. 7, the measurement error is evaluated over a range from 0 ℃ to 230 ℃ and found well within ±0.4 ℃. This can meet high accuracy measurement requirements of transformer temperature monitoring.

The transformer had three phases (A, B, and C). Each phase included a high-voltage (HV) winding and a low-voltage (LV) winding. Besides the iron core and the camp, FBG temperature sensors were

installed inside all of these windings. Our FBG interrogation system served as the temperature monitoring. The monitoring time was from 21:30 to 8:00 one day. The monitoring results are shown in Fig. 8.

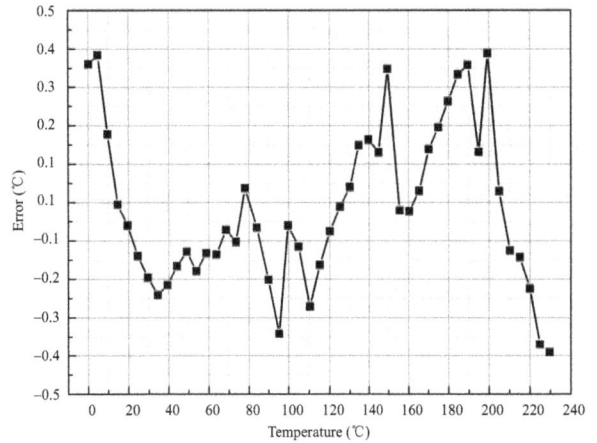

Fig. 7 FBG temperature measurements evaluation.

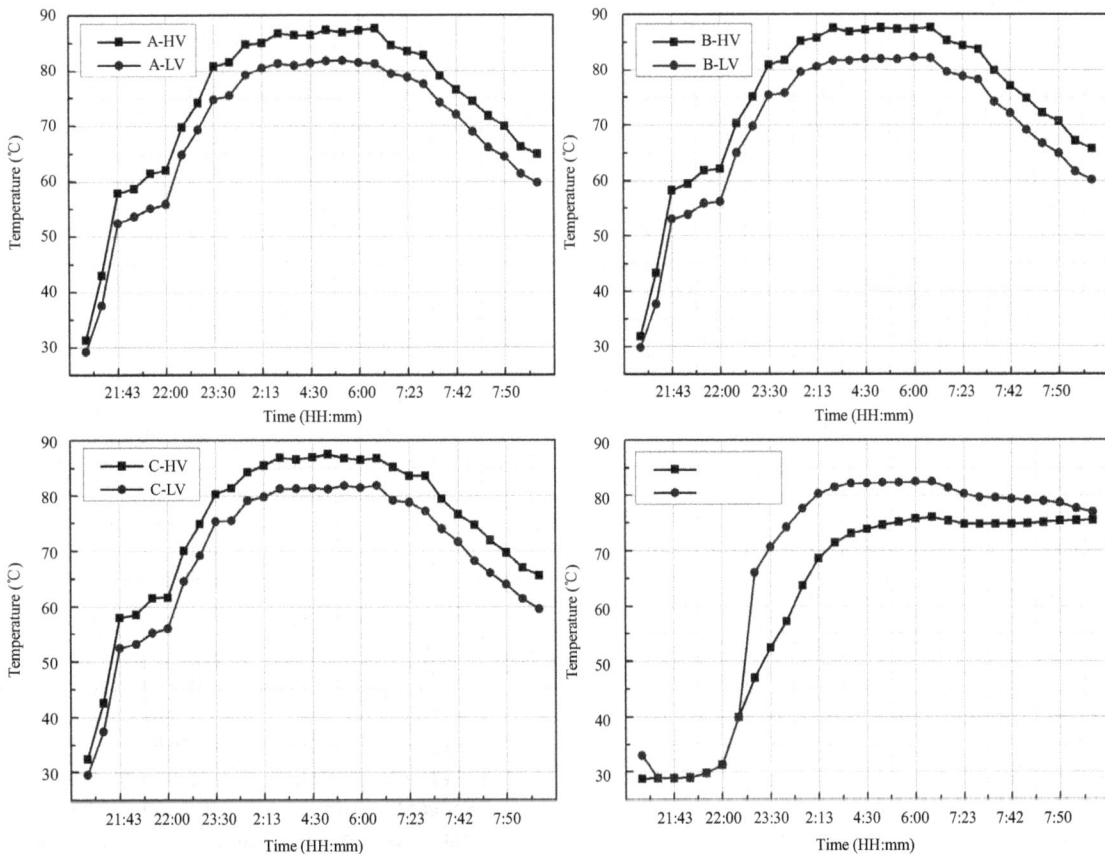

Fig. 8 FBG transformer temperature monitoring results.

It can be seen that the maximum temperature is about 90 ℃ for the HV winding of B phase. All temperatures rise along with the transformer running. As the power cut time (6:30) is approached, the temperatures of all windings gradually decrease to about 60 ℃. Meanwhile, the temperatures of the iron core and clamp are kept about 75 ℃. For an oil-immersed transformer with class-A insulation, the working temperature does not exceed to 105 ℃ [15]. Higher temperature will reduce the quality of mineral oil, thus reducing the transformer life and premature failure.

4. Conclusions

Those FBG interrogators based on external-cavity tunable laser are generally complicated and expensive, thus not suitable for many field applications. Therefore, a cost-effective precision FBG interrogation system using the long-wavelength VCSEL has been developed. The experimental study of tuning properties of a long-wavelength VCSEL indicates that there is an approximately quadratic dependence of its wavelength on the injection current. The proposal of this paper is to use a four-point peak detection method and the real-time calibration using the absorption lines of C_2H_2 to realize a uniform and predictable wavelength calculation. Owing to logarithmic amplifiers, the system has a large optical dynamic range. The results show that the system achieves an accuracy of 1.2 pm with a tuning range of 3 nm and a tuning rate of 1 kHz. It is demonstrated that this system is practical and effective by applied in the transformer temperature monitoring. Furthermore, the system can also be used to interrogate some dynamic signals such as acceleration, vibration, and shock.

Acknowledgment

This work was partly supported by the Science and Technology Development Plan of Shandong Province (Grant No. 2014GSF120017) and Science Fund for Young Scholars of Shandong Academy of Sciences (Grant No. 2013QN005).

References

[1] A. Zhang, S. Gao, G. Yan, and Y. Bai, "Advances in optical fiber Bragg grating sensor technologies," *Photonic Sensors*, 2012, 2(1): 1–13.

[2] J. Wang, T. Liu, G. Song, H. Xie, L. Li, X. Deng, *et al.*, "Fiber Bragg grating (FBG) sensors used in coal mines," *Photonic Sensors*, 2014, 4(2): 120–124.

[3] J. Song, Q. Jiang, Y. Huang, Y. Li, Y. Jia, X. Rong, *et al.*, "Research on pressure tactile sensing technology based on fiber Bragg grating array," *Photonic Sensors*, 2015, 5(3): 263–272.

[4] S. Ma, J. Guo, Y. Guo, J. Chao, and B. Zhang, "On-line monitoring system for downhole temperature and pressure," *Optical Engineering*, 2014, 53(8): 087102.

[5] V. R. Pachava, S. Kamineni, S. S. Madhuvarasu, K. Putha, and V. R. Mamidi, "FBG based high sensitive pressure sensor and its low-cost interrogation system with enhanced resolution," *Photonic Sensors*, 2015, 5(4): 1–9.

[6] H. Nasim and Y. Jamil, "Recent advancements in spectroscopy using tunable diode lasers," *Laser Physics Letters*, 2013, 10(4): 916–923.

[7] Y. Huang, T. Guo, C. Lu, and H. Y. Tam, "VCSEL-based tilted fiber grating vibration sensing system," *IEEE Photonics Technology Letters*, 2010, 22(1): 1235–1237.

[8] B. V. Hoe, G. Lee, E. Bosman, J. Missinne, and S. Kalathimekkad, "Ultra small integrated optical fiber sensing system," *Sensors*, 2012, 12(9): 12052–12069.

[9] T. Mizunami, S. Hirose, T. Yoshinaga, and K. I. Yamamoto, "Power-stabilized tunable narrow-band source using a VCSEL and an EDFA for FBG sensor interrogation," *Measurement Science and Technology*, 2013, 24(9): 094017–094023.

[10] A. Lytkine, W. Jäger, and J. Tulip, "Frequency tuning of long-wavelength VCSELs," *Spectrochimica Acta Part A: Molecular and Biomolecular Spectroscopy*, 2006, 63(5): 940–946.

[11] M. Fernandez-Vallejo and M. Lopez-Amo, "Optical fiber networks for remote fiber optic sensors," *Sensors*, 2012, 12(4): 3929–3951.

[12] P. Wang P, X. Han, S. Guan, H. Zhao, and M. Shao, "Research on peak-detection algorithm for high-precision demodulation system of fiber Bragg grating," *International Journal of Hybrid Information Technology*, 2014, 7(6): 337–344.

[13] M. Basu and S. K. Ghorai, "Sequential interrogation of multiple FBG sensors using LPG modulation and an artificial neural network," *Measurement Science and Technology*, 2015, 26(4):1–9.

[14] X. Zhang, R. Huang, W. Huang, S. Yao, D. Hou, and M. Zheng, "Real-time temperature monitoring system using FBG sensors on an oil-immersed power transformer," *High Voltage Engineering*, 2014, 40(2): 253–259.

[15] M. R. Meshkatoddini and S. Abbospour, "Aging study and lifetime estimation of transformer mineral oil," *American Journal of Engineering and Applied Sciences*, 2008, 1(4): 384–388.

23

Material Measurement Method Based on Femtosecond Laser Plasma Shock Wave

Dong ZHONG[*] and Zhongming LI

School of Electronic and Information, Hubei University of Science and Technology, Xianning, 437100, China

[*]Corresponding author: Dong ZHONG   E-mail: zhongdong0129@163.com

Abstract: The acoustic emission signal of laser plasma shock wave, which comes into being when femtosecond laser ablates pure Cu, Fe, and Al target material, has been detected by using the fiber Fabry-Perot (F-P) acoustic emission sensing probe. The spectrum characters of the acoustic emission signals for three kinds of materials have been analyzed and studied by using Fourier transform. The results show that the frequencies of the acoustic emission signals detected from the three kinds of materials are different. Meanwhile, the frequencies are almost identical for the same materials under different ablation energies and detection ranges. Certainly, the amplitudes of the spectral character of the three materials show a fixed pattern. The experimental results and methods suggest a potential application of the plasma shock wave on-line measurement based on the femtosecond laser ablating target by using the fiber F-P acoustic emission sensor probe.

Keywords: Optical fiber sensing; femtosecond laser; plasma shock wave; acoustic signal; material testing

1. Introduction

In recent years, the material detection methods consist of chemical composition analysis, photo acoustic spectrum analysis, performance test, and so on. Among them, photo acoustic spectroscopy (PAS) is comparatively a new material testing technology. It has made great progress since the 1970s [1]. Rosencwaig and Gersho [2] published one-dimensional photo-acoustic theory for condensed matter in 1976, which is known as the R-G theory. Based on the R-G theory, when using a gas microphone detecting photo-acoustic signal [3], the signal depends on the pressure of gas samples perturbing and disturbing on the interface accepted by the microphone.

The complexity of the Jackson-Ainer theory limits its practical application, and then Blonskij *et al.* [4] simplified the computing method of piezoelectric photo-acoustic signal. The photo acoustic spectrum detection technology has the advantages of high sensitivity, wide range of detectable spectrum, and non-contact character [5, 6]. It has been applied widely to the detection and analysis for materials. However, the photo acoustic spectrum technology usually adopts radiation light sources including ultraviolet, visible, and infrared laser [7–9], which can produce light saturation phenomenon easily, and it will led to a decrease in the signal-to-noise ratio and influence the effect of detection. The microphone and photo acoustic spectrum test system usually adopts the piezoelectric element (PZT) or uses microphone (MIC) [10]. PZT has the problem of electric charge [11], which would affect the sensitivity of system testing. The frequency range for the MIC is quite limited and is affected easily by noise, as we all know that the PAS technology has heating effect.

However, femtosecond laser has the advantage of feeble heating effect which is better in the detecting technology. Therefore, the photo acoustic spectroscopy detection theory remains to be further improved. Based on this, this paper uses optical fiber Fabry-Perot (F-P) acoustic emission sensor probe instead of PZT, which belongs to the fiber external cavity type sensor, and it is suitable for small amplitude signals' detection because of its high sensitivity, with advantages of the anti-interference and sensing probe's smaller type at the same time. It is better than the micro-plasma measurement. Considering the shortcoming of general radiation sources, femtosecond laser has been used as the light source [12–14] in this paper to analyze the change rules of plasma shock wave acoustic emission signal produced by femtosecond laser ablating pure Cu, Fe, and Al, thus providing a new method for testing materials.

2. Experiment

2.1 Characteristics of the F-P sensor

(1) The light beam in the F-P sensor is actually located at the center of the quartz diaphragm, and the size of the light spot is far smaller than that of diaphragm. The quartz diaphragm would de deformed under the air pressure P, and the maximum deflection is defined as y max. The deflection y_{max} at the center of a quartz membrane could be considered as the effective deflection, i.e. the amount of actual cavity length change $\Delta d = y_{max}$. When the signal is detected by the sensor, the diaphragm would oscillate at a very high speed, leading to the quick change in the cavity length. So the interference signal strength I_R varies with the cavity length d very fast.

$$y_{max} = \frac{3P(1-\mu^2)}{16E} \cdot \frac{a^4}{h^3} \qquad (1)$$

where h stands for the thickness of the diaphragm, a is the effective radius of the diaphragm, E is the Young modulus, and μ is the Poisson ratio of the diaphragm. The pressure sensitivity of quartz diaphragm at the center can be defined as

$$S_{max} = \frac{y_{max}}{P} = \frac{3(1-\mu^2)}{16E} \cdot \frac{a^4}{h^3}. \qquad (2)$$

(2) The modulation method of the optical fiber F-P sensor in the experiment is based on the double beam interference. When the wavelength of the tunable diode laser remains the same, the relation between the reflection intensity I_R and the cavity length is a cosine function as

$$I_R \approx I_0 \cdot 2R(1-\cos\varphi) = 2RI_0\left(1-\cos\frac{4\pi d}{\lambda}\right) \qquad (3)$$

where R stands for the surface reflectance, I_0 is the incident light intensity, and they are all constants. φ is the phase between two adjacent refracted beams.

(3) The free spectral range (FSR) of the F-P sensor reflects dense degree of the interference fringes. It could be calculated as follows:

$$\text{FSR} = |\lambda_{m1} - \lambda_{m2}|. \qquad (4)$$

The relation between the length of cavity and FSR could be estimated as

$$d = \frac{1}{2} \cdot \frac{\lambda_{m1}\lambda_{m2}}{|\lambda_{m2} - \lambda_{m1}|}. \qquad (5)$$

The length of cavity could be calculated when knowing λ_{m1} and λ_{m2}. From this, we know that the longer the length is, the smaller the FSR is, and the fringes would be more dense.

2.2 Principles of the detecting method

The optic signal generated by tunable semiconductor laser would pass into the circulator through the fiber and then into the sensors. The signal would be modulated when the sensor receives the acoustic signal of the plasma shock wave produced by femetosecond ablation targets. Then the modulated signal would be reflected back to the circulator. In this way, the signal owning sensing information is detected by the photoelectric conversion and analyzed by the data acquisition system. The modulated information wave could be timely observed in the computer finally. The experimental apparatus is shown in Fig. 1. The experimental femtosecond laser is a kind of femtosecond laser micro-processing system from Japanese Cyber Laser Company, of which the type is titanium gem LS-IF-FW-C-401. The related technology parameters are as follows: the wavelength is 780 nm, the pulse width is 180 fs, the average output power is 1.1 W, the pulse frequency is 1 kHz, the pulse energy is 1.1 mJ, the pulse stability is 1.5% rms, the three-dimensional work bench moving range (x, y, z) is ±100 mm $\times$ ±100 mm $\times$ ±25 mm, and the workbench mobile precision (x, y, z) is $1.0\,\mu m \times 1.0\,\mu m \times 0.5\,\mu m$. Laser light source is a tunable semiconductor laser light source produced by Santec, of which the model is tunable semiconductor laser TSL-510, with high stability. The wavelength (λ) of tunable semiconductor laser light source is 1545 nm, and the power (P) is 0.2 mW in this experiment. Data acquisition module is a kind of multi-channel data acquisition card made by NI company, with voltage range approximately 0.56 V–0.87 V, $f=600$ kHz, which ensures the signal without distortion. The photoelectric conversion module is developed independently by our laboratory, composed of photoelectric conversion circuit, signal amplification circuit, and filter circuit. The photoelectric conversion module is a kind of high sensitivity photoelectric signal processor to achieve photoelectric conversion.

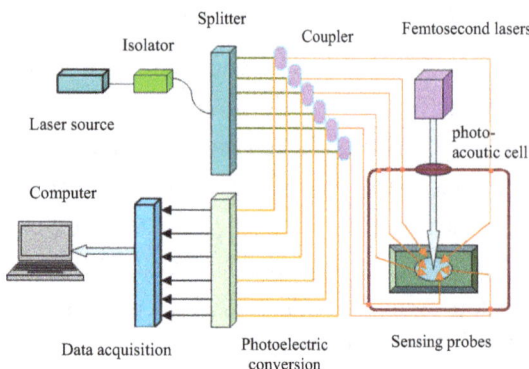

Fig. 1 Experimental device diagram.

In experiments, tunable semiconductor laser has been used as the light source in the measurement process. Femtosecond laser produces femtosecond laser beam with high power density and short pulse, then the beam irradiates pure Cu, Fe, and Al target to produce high voltage laser plasma shock wave, using the way of space division multiplexing in a sealed space. The detected signal passes through the coupler and the branching device into the photoelectric signal processing module, then data acquisition module. Finally, we analyze and research the collected signal by computer processing.

As shown in Fig. 2, the sensor has a structure of all-fiber with F-P acoustic emission sensor probe owning an adhesive microstructure. It has characteristics of resistance electromagnetic interference and a wide frequency band. As for the structure of the sensor probe, the capillary glass tube's outside diameter size is identical with the optical fiber. The capillary glass tube is spliced with the optical fiber. The thickness of the capillary glass tube is 50 μm, and the thin quartz crystal diaphragm is stuck in the surface of capillary glass tube. The quartz crystal diaphragm determines the sensibility of the sensor probe. The diameter of the sensor probe is only 125 μm. It has high sensitivity and accuracy, and can detect the time distribution and intensity space distribution of femtosecond laser plasma shock wave in the process of formation, development, and attenuation. It realizes accurate measurement of laser plasma shock wave with superiorities such as non-contact, high precision, high response speed, and high sensitivity, which can help us analyze the action mechanism between the femtosecond laser plasma shock wave and different materials.

Figure 3 shows the time domain diagram of the typical femtosecond laser plasma shock wave acoustic emission signal.

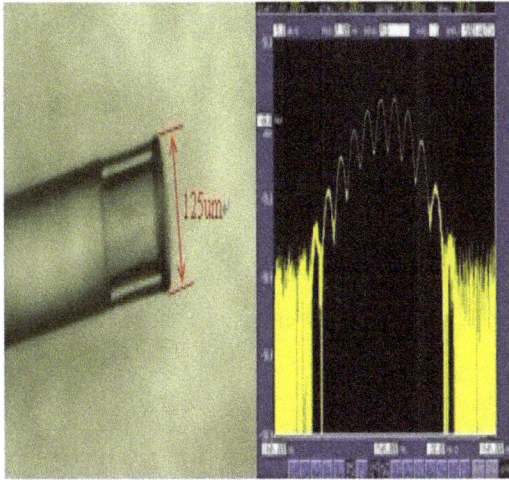

Fig. 2 Optical fiber F-P sensor probe and spectrum.

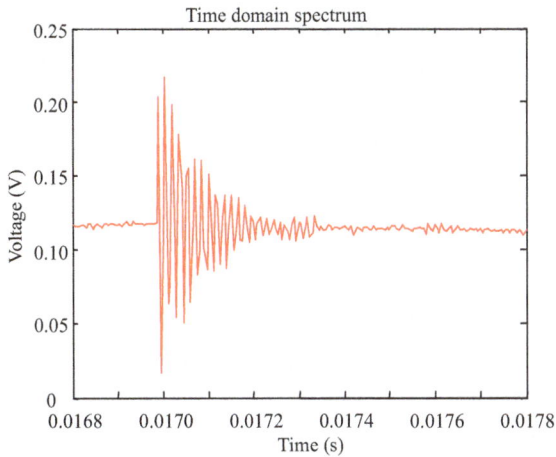

Fig. 3 Acoustic emission signal.

3. Results and discussion

To measure and analyze the characteristics of femtosecond laser plasma shock wave acoustic emission signals, we use different laser pulse energies including 300 mW, 400 mW, 500 mW, 600 mW, 700 mW, and 800 mW with different detection ranges including 1500 μm, 2000 μm, 3000 μm, 5000 μm, 6000 μm, and 8000 μm. What's more, the purity of Cu, Fe, and Al is 99.99%. Femtosecond laser ablates the pure Cu, Fe, and Al to produce plasma shock wave in a sealed space. Considering that the signal detected by F-P acoustic emission sensor has vibration attenuation which is mainly due to the forced vibration attenuation rules of the sensor piezoelectric membrane, the first peak of the signal frequency can be identified as the laser

plasma shock wave acoustic emission signal. We take the first peak as the frequency of acoustic emission signal of the plasma shock wave. The experimental results in Fig. 4 show the change rules of the plasma shock wave acoustic emission signal frequency peak value of pure Cu, Fe, and Al under different energies and detection ranges.

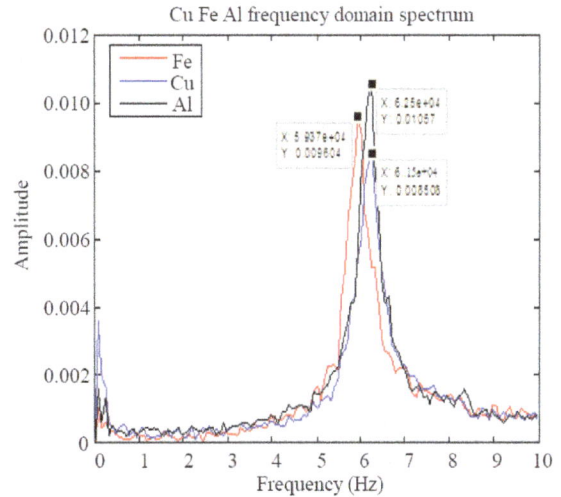

Fig. 4 Frequency change diagrams of pure Cu, Fe, and Al plasma shock wave acoustic emission signal with the change pulse energies and detection distances.

From the measurement results of the femtosecond laser ablating the pure Cu, Fe, and Al to produce plasma shock wave under changed femtosecond laser pulse energies and detection ranges in the sealed space, we get that the plasma shock wave acoustic emission signal frequency has a certain change rule. In Fig. 4, for the same target material, the plasma shock wave acoustic emission signal frequency remains the same with changed pulse energies and detection ranges. The acoustic emission signal frequency of the femtosecond laser ablating pure Cu is mainly concentrated in the frequency of 61.5 kHz, and the acoustic emission signal frequency of the femtosecond laser ablating pure Fe and Al are concentrated in the frequencies of 59.37 kHz and 62.5 kHz. Obviously, the acoustic emission signal frequency of the femtosecond laser ablating the pure Fe is lower than that of the pure Cu and Al in the sealed space, what's more it almost remains unchanged. Maede et al. [15] calculated the molten pool's natural oscillation frequency of the laser irradiation target, which was below 10 kHz. It is visible that the acoustic emission signal of the femtosecond laser plasma shock wave measured in this experiment has a high frequency. In the process of femtosecond laser ablating target material, the strong sound and light signals are similar with lightning. The low frequency features of thunder are related with air expansion rate, for this, we conclude that the detected frequency of the acoustic spectrum is related with the environment gas properties in the sealed space. Furthermore, in the process of interaction between laser and different materials, ionization properties and thermal properties are different. The differences of the doped metal ions and gas vibration intensity of electrons are mainly reflected on the diversities of sound pressures corresponding to different materials under the same experimental condition.

When the femtosecond laser radiation metal target is to meet or exceed the optical breakdown threshold, producing photo-acoustic signal could be expressed as

$$E_r = \xi \chi \qquad (6)$$

where E_r is the sound energy of the initial micro plasma, χ is the threshold energy of laser ablating, and ξ is the photo-acoustic conversion efficiency of pure metal materials. The value of ξ and χ are constant for the same metal material. The acoustic energy of the initial micro plasma for some kind of metal materials is constant. Therefore, the acoustic signal frequency of the initial plasma for the metal material is also constant. With the expansion of laser micro plasma, the micro plasma acoustic emission signal frequency for the same material should be certain after micro plasma energy attenuation in the same detection distance. These are in good agreement with the results from the experiment.

At the same time, there is the energy coupling of laser and metal in the process of laser acting on metal materials. Under the same laser conditions, the absorbing laser energy is different for the metals of different electrical conductivities. The absorption

coefficient (in power) [16] could be expressed as

$$A(\theta) = \frac{2\Delta\cos\theta}{(\omega_p / \omega)^2 \sqrt{(\omega_p / \omega)^2 - \cos\theta^2}} \times$$
$$\left[\frac{(\omega_p / \omega)^2 - \cos 2\theta}{(\omega_p / \omega)^2 \cos^2\theta - \cos 2\theta} \right] \qquad (7)$$

$$\Delta = (\omega\varepsilon_0 / \sigma)(\omega_p / \omega)4 \qquad (8)$$

where ω_p is the electron plasma frequency, ω is the frequency of the laser light, ε_0 is the free space permittivity, and σ is the electrical conductivity of the metallic target.

For $\Delta \ll 1$ and $\omega_p/\omega > 1$, where θ is the incident angle. As everyone knows, the numerical relationship of the electrical conductivity among these metal materials is $\sigma_{Fe} > \sigma_{Cu} > \sigma_{Al}$. It can derive the numerical relationship of the absorbed laser energy for Al, Cu, and Fe that $E_{Al} > E_{Cu} > E_{Fe}$. In addition, the target ablation is related to the first materials ionization. The lower the first ionization energy of target materials becomes, the easier the ablation is, which leads to a bigger micro plasma density. The first ionization energies (EL) for the Al, Cu, and Fe are different [EL(Al)=5.97 eV, EL(Cu)=7.72 eV, EL(Fe) =7.83 eV]. Considering the influence of this comprehensive action of the absorbed laser energy and the first ionization energy influence on the target ablation, it could be inferred that the numerical relationship of the micro plasma density for Al, Cu, and Fe is $\rho_{Al} > \rho_{Cu} > \rho_{Fe}$. As a result of the expansion and friction with background gas of the different density micro plasma, it would lead to different acoustic emission frequency signals at the same range. Figure 2 shows that the frequency constants for Al, Cu, and Fe target are 62.5 kHz, 61.5 kHz, and 59.37 kHz, respectively. It indicates that the greater the micro plasma density is, the higher the frequency of the acoustic emission signal is. It could be inferred that the acoustic emission frequency signal is related to the micro plasma density.

Our further study found that, through the analysis of the spectrum amplitude of plasma shock wave acoustic emission signal, Fe and Al can be separated. In Fig. 5, the spectrum amplitude of the acoustic emission signal has a variation trend when the femtosecond laser ablates the pure Fe and Al under different incident laser pulse energies and different distances.

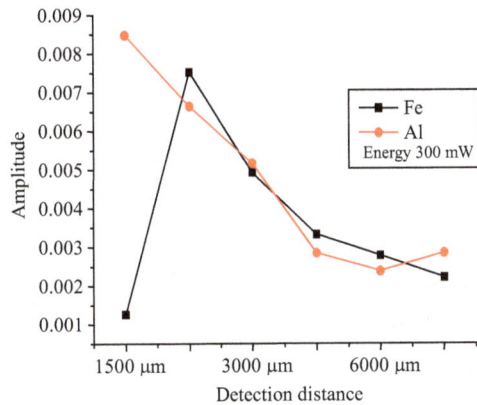

(a)

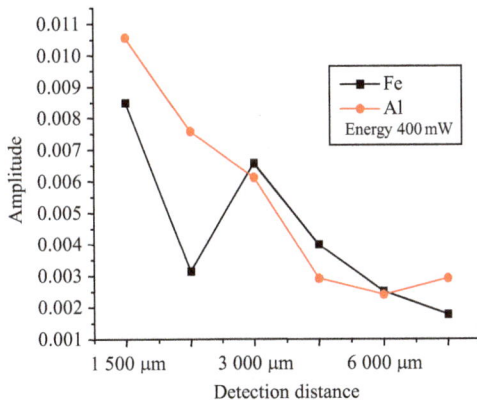

(b)

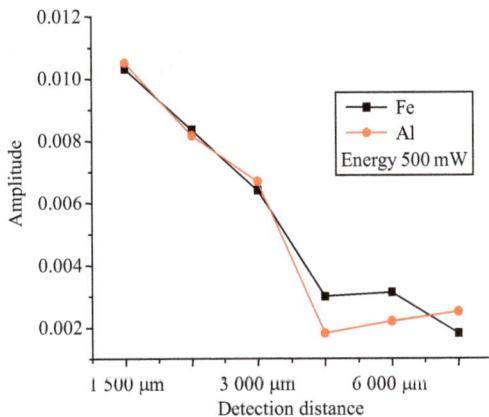

(c)

Fig. 5 Spectrum amplitude of the acoustic emission signal for Fe and Al at different energies: (a) energy 300 mW, (b) energy 400 mW, and (c) energy 500 mW.

From Fig. 5, we draw the conclusion that under the same incident laser pulse energy, the spectrum amplitude of the plasma shock wave acoustic emission signal produced by femtosecond laser ablating the pure Al has a gradually decreasing trend with the change in operating distance. But the spectrum amplitude has a certain fluctuation when ablating the pure Fe. Under different laser pulse energies, we can see that the spectrum amplitude of the pure Al is higher than that of pure Fe in the distances of 1500 μm and 8000 μm. So, we can distinguish two kinds of target material of Fe and Al according to the ablating characteristics and rules. Through additional experiments, we find that Cu and Al have the similar characteristics with Fe and Al.

From the above analysis, we can use this designed detection system, with an optical fiber F-P acoustic emission sensor probe, and a femtosecond laser as a radiation source to detect the frequency and the spectrum amplitude of the plasma shock wave acoustic emission signal produced by ablating pure Cu, Fe, and Al, thus determining the composition of the materials. This detection system and method can be applied to the detection of different materials' compositions and characteristics, which provides a new thought for the material test.

In order to further validate test results, the time-frequency signal processing methods have been explored, and the micro plasma shock wave transmit's frequency and energy trace information have been researched in this article. By researching the time-frequency distribution of the micro plasma shock, we can reveal the frequency, energy, and time of duration information. It will be of great importance to explore the femtosecond laser erosion crystalline material dynamics features.

Usually, the micro plasma shock wave acoustic emission signal meets the relation: $x(t) \in L^2(R)$, in which the $x(t)$ is limited energy. Meanwhile, we use the tradition Fourier transform method and combine with Parseval, then we can know the signal $x(t)$ in the time of total energy and total energy in frequency domain is equal.

$$E_x = \|x(t)\|^2 = \int_{-\infty}^{\infty}[x(t)]^2 dt = \frac{1}{2\pi}\int_{-\infty}^{+\infty}[X(j\omega)]^2 d\omega$$

$$(9)$$

where $X(j\omega)$ is the fast Fourier transform (FFT) of $x(t)$. FFT cannot determine different time of the frequency of the details, which can only obtain some basic information of energy distribution in the whole frequency range. It's hard to get the corresponding relation between energy and time. We also know that FFT don't have the ability to automatically adjust the time domain and frequency domain resolution.

In order to break through the limitations of the traditional analysis method, for non-stationary micro plasma shock wave in the acoustic emission signal processing, we put the window function $g(\tau)$ add to short time Fourier transform (STFT). Therefore, the time-frequency analysis can reflect the time-frequency local characteristics.

$$STFT_x(t,\omega) = \int x(\tau)g_{t,\omega}^*(\tau)e^{-j\omega\tau}d\tau$$
$$= \int x(\tau)g^*(t-\tau)e^{-j\omega\tau}d\tau$$
$$= \langle x(\tau), g(t-\tau)e^{-j\omega\tau} \rangle. \qquad (10)$$

$$g_{t,\omega}(\tau) = g_{t,\omega}(\tau-t)e^{j\omega t}, \quad \|g_{t,\omega}(\tau)\|=1. \qquad (11)$$

From STFT, it could be comprehended that $x(\tau)$ can be separated by the relevant translation function of $g(\tau)$ on the time axis. We can get the two-dimensional function STFT $x(t, \omega)$ by Fourier transform of these separated signals. In the time domain and frequency domain, the primary function $g(t-\tau)e^{-j\omega t}$ has the feature of limited support, so we can accomplish to achieve the position information of the time domain by inner product operation for (11). According to the window function $g(\tau)$ when we choose the narrow width $\Delta\tau$ and bandwidth $\Delta\nu$, the time-frequency localization can be achieved. Based on the basic theory of the uncertainty principle, the minimum values of $\Delta\tau$ and $\Delta\nu$ cannot be achieved at the

same time. Therefore, we can consider the influence of the factors on the signal processing according to the actual signal analysis. Since the signal $x(t)$ uses spectra, the time domain can be expressed as the amount type:

$$S_x(t,\omega) = |STFT_x(t,\omega)|^2$$
$$= \left|\int x(\tau)g^*(t-\tau)e^{-j\omega\tau}d\tau\right|^2. \qquad (12)$$

Hamming:

$$g(n) = 0.54 - 0.46\cos[2\pi n/(N-1)] \qquad (13)$$

where N is the number of frequency points, and n is the window function points. We want to get time domain energy distribution of acoustic emission signal by using (11), (12), and (13).

According to this method, we get much information through femtosecond laser ablation of copper alloy (90:10), aluminum, and iron copper alloy (93:2:5) in the process of experiment. The typical crystal materials produce the groups of micro plasma shock wave acoustic emission signal as a typical signal. Based on STFT, the time-frequency analysis technology of femtosecond laser ablation crystal material micro plasma shock wave produced by the acoustic emission signal is analyzed and studied.

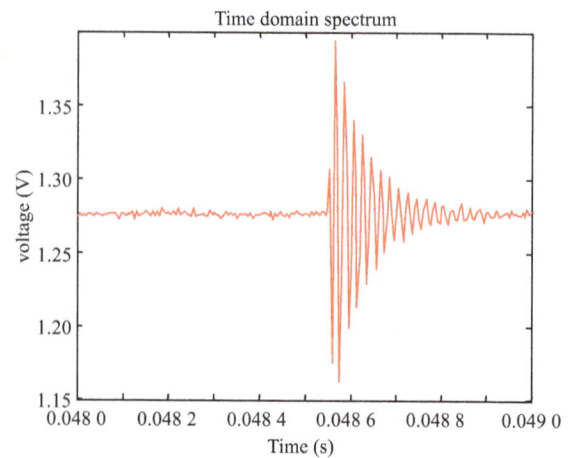

Fig. 6 Typical micro plasma shock wave acoustic emission signal by femtosecond laser ablation of aluminium copper alloy (90:10).

In order to further explore the femtosecond laser ablation characteristics of different crystal materials, under the same experimental conditions, the femtosecond laser ablation of aluminum copper alloy micro plasma shock wave produced by the characteristics of acoustic emission signals is studied. Femtosecond laser ablation of aluminum copper alloy (90:10) produced by the micro plasma acoustic emission signal is shown in Fig. 6.

In the same way, the femtosecond laser ablation of aluminum copper alloy (90:10) produced by acoustic emission signal time-frequency distribution is shown in Fig. 7. The study finds that the aluminum copper alloy (90:10) micro plasma shock wave acoustic emission signals and pure aluminium micro plasma shock wave acoustic emission signal have a similar change rule. But the micro plasma shock wave acoustic emission signal is only one obvious stage, which is in 0 kHz–4 kHz and 40 kHz–60 kHz range. We can see that the micro plasma shock wave acoustic emission signal frequency range of aluminum-copper alloy (90:10) is narrower than that of pure aluminum, meanwhile, the two kinds of acoustic emission signals have different energy densities.

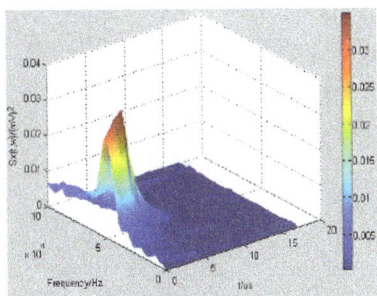

(c)

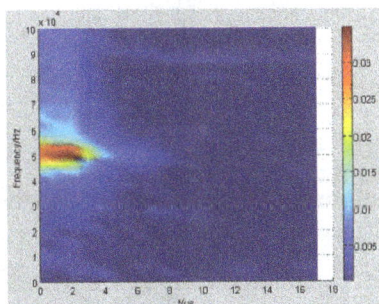

(d)

Fig. 7 Micro plasma shock wave acoustic emission signal time-frequency distribution by femtosecond laser ablation of aluminium copper alloy (90:10): (a) graphic model, (b) plan view (x-y), (c) front view (x-z), and (d) side view (y-z).

In order to explore the characteristics of a wider variety of crystal materials, we have researched the feature of the micro plasma shock acoustic emission signal by femtosecond laser ablating aluminum iron copper alloy (93:2:5) under the same experimental conditions, and the results have been shown in Figs. 8 and 9.

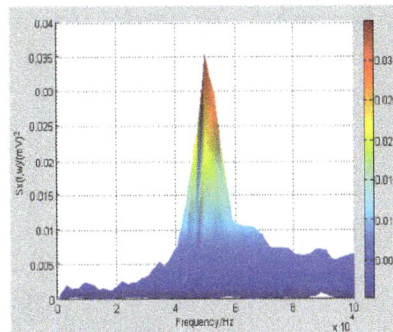

(a)

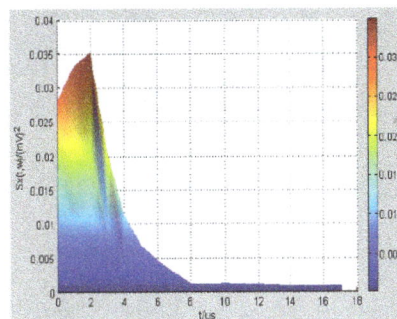

(b)

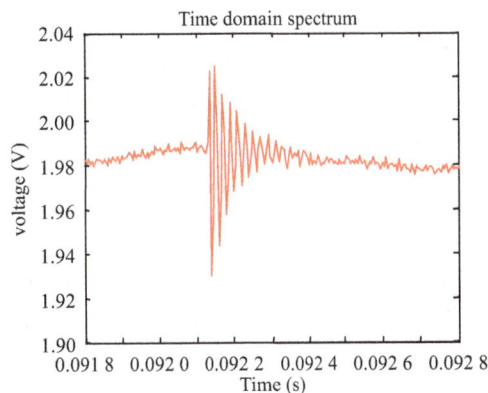

Fig. 8 Typical micro plasma shock wave acoustic emission signal by femtosecond laser ablation of aluminium iron copper alloy (93:2:5).

After the same method of micro plasma shock wave acoustic emission signal processing, the result has been shown in Fig. 9. Aluminum iron copper alloy micro plasma shock wave acoustic emission signals in 0 kHz – 4 kHz frequency composition is rich, but the frequency range is relatively fuzzy, and we can't get the precise frequency range from picture that the signal distributes the whole area.

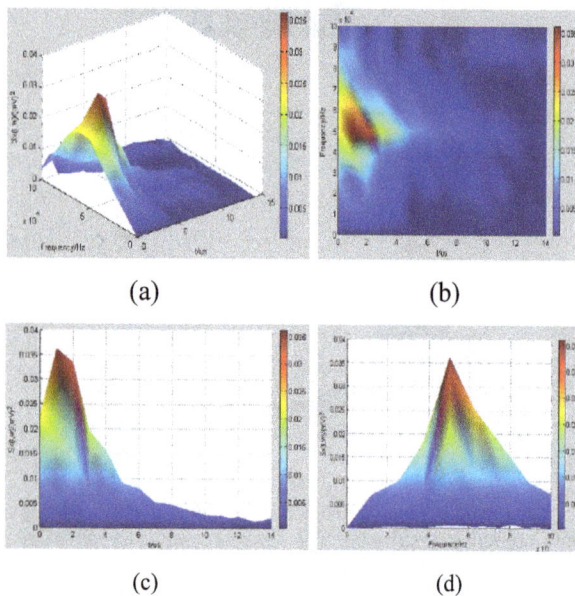

(a)　　　　　(b)

(c)　　　　　(d)

Fig. 9 Micro plasma shock wave acoustic emission signal time-frequency distribution by femtosecond laser ablation of aluminium iron copper alloy (93:2:5): (a) graphic model, (b) plan view (x-y), (c) front view (x-z), and (d) side view (y-z).

Through time-frequency analysis based on STFT for the femtosecond laser ablating different crystalline materials, we can get the similar characteristics. For example, the spectrum distribution of micro plasma shock wave acoustic emission signal is similar, these signals have close frequency ranges and certain frequency peaks. The different crystal materials have different peak frequencies. We also can conclude that the energy of micro plasma shock wave acoustic emission signal is concentrated in the initial stage. The energy of high frequency component is higher than that of the low frequency component.

From Figs.7 and 9, we can get the STFT time-frequency distribution formula (11), (12), and (13) of the micro plasma shock wave acoustic emission signal $x(t)$ .We can research the

characteristics of the micro plasma shock acoustic emission signal, which is produced by the femtosecond laser ablation crystal materials. The time-frequency analysis techniques have been used to study micro plasma shock wave acoustic emission signal in the paper, which can provide a new research method for the interaction between the femtosecond laser and material. It will play a guiding role for understanding the interaction mechanism between the femtosecond laser and matter.

4. Conclusions

In conclusion, the designed system and the detection method can provide a significant guide for the material testing. For the reason that the interaction mechanism of femtosecond laser and materials is not clear, it needs further research. Ideas and methods of this paper can also provide a further research direction for the interaction between the femtosecond laser and materials.

The research finds that femtosecond laser ablating different crystal materials micro plasma shock wave produced by the acoustic emission signal frequency spectrum distribution of the overall trend is similar. For each kind of crystal materials, under the irradiation of a femtosecond laser energy, producing micro plasma shock wave spectrum distribution of acoustic emission signals remains the same, each kind of materials remains the same, the characteristic frequency of peak signal strength decreases gradually along with an increase in the detection range, the signal intensity increases with an increase in energy, and signal intensity varies for the different crystalline materials.

Acknowledgment

The authors gratefully acknowledge the financial support for this work provided by the Dr. Start-up Fund of Hubei University of Science and Technology under Grant No. BK1524 and the Science Research Project in Hubei Province

Department of Education under Grant No. B2015077, the National Natural Science Foundation of China (NSFC) under Grant No. 61575148, the Colleges and Universities of Hubei Province Innovation and Entrepreneurship Training Plan under Grant No. 201510927017, No. 201510927018, and the Teaching Reform Program of Hubei University of Science and Technology under Grant No. 2015-XA-007.

References

[1] S. I. Anisimov, B. L. Kapeliovich, and T. L. Perelman, "Electron emission from metal surfaces exposed to ultrashort laser pules," *Zhurnal Eksperimentalnoi I Teroreticheskoi Fiziki*, 1974, 66(776): 776–781.

[2] A. Rosencwaig and A. Gersho, "Theory of the photo acoustic effect with solids," *Photoacoustic Effect: Vieweg*, 1984, 1(6): 631–636.

[3] J. S. Lv and H. Qi. "Multi-wavelength narrow line width fiber laser based on distributed feed back fiber lasers," *Photonic Sensors*, 2016, 6(3): 256–260.

[4] V. Blonskij, V. A. Thaoryk, and M. L. Shendeleva, "Thermal diffusivity of solids determination by photo acoustic piezoelectric technique," *Journal of Applied Physics*, 2003, 93(1): 790.

[5] P. Sprangle, J. R. Penano, B. Hafizi, R. F. Hubbard, A. Ting, D. F. Gordon, *et al.*, "Propagation of ultra-short, intense laser pulses in air," *Physics of Plasmas*, 2004, 11(5): 2865–2874.

[6] J. K. Chen, J. E. Beraun, L. E. Grimes, and D. Y. Tzou, "Modeling of femtosecond laser-induced non-equibrium deformation in metal films," *International Journal of Solids and Structures*, 2002, 39(12): 3199–3216.

[7] W. P. Leemans, B. Nagler, A. J. Gonsalves, C. Toth, K. Nakamura, C. G. R. Geddes, *et al.*, "GeV electron beams from a centimetre-scale accelerator," *Nature Physics*, 2006, 2(10): 696–699.

[8] S. V. Garnov, V. V. Bukin, A. A. Malyutin, and V. V. Strelkov, "Ultra fast space time and spectrum time resolved diagnostics of multicharged femtosecond laser micro plasma," in *AIP Conference Proceedings*, Hawaii, vol. 1153, pp. 37–48, 2009.

[9] H. Ditlbacher, J. R. Krenn, G. Schider, A. Leitner, and F. Aussenegg, "Two-dimensional optics with surface plasmon polaritons," *Applied Physics Letters*, 2002, 81(10): 1762–1764.

[10] D. Zhong and X. Tong. "Application research on hydraulic coke cutting monitoring system based on optical fiber sensing technology," *Photonic Sensors*, 2014, 4(2): 147–151

[11] A. Y. Basharin, V. S. Dozhdikov, V. T. Dubinchuk, A. V. Kirillin, I. Y. Lysenko, and M. A. Turchaninov, "Phases formed during rapid quenching of liquid carbon," *Technical Physics Letters*, 2009, 35(5): 428–431.

[12] J. Chen, G. Conache, M. Pistol, S. Gray, M. T. Borgström, H. Xu, *et al.*, "Probing strain in bent semiconductor nanowires with Raman spectroscopy," *Nano Letters*, 2010, 10(10): 1280–1286.

[13] X. L. Tong, D. S. Jiang, W. B. Hu, Z. M. Liu, and M. Z. Luo, "The comparison between CdS thin films grown on Si (111) substrate and quartz substrate by femtosecond pulsed laser deposition," *Applied Physics A*, 2006, 84(1–2): 143–148.

[14] M. Beresna, P. G. Kazansky, Y. Svirko, M. Barkauskas, and R. Danielius, "High average power second harmonic generation in air," *Applied Physics Letters*, 2009, 95(12): 121502-1–121502-3.

[15] T. Maede, E. Ohmura, and I. Miyamoto, "Analysis of key hole behavior in laser welding," in *Proceeding of 6th International Symposium of Japan Welding Society*, Nagoya, pp. 104–105, 1998.

[16] M. Nikoufard, M. K. Alamouti, and A. Adel, "Ultra-compact photonic crystal based water temperature-sensor," *Photonic Sensors*, 2016, 6(3): 274–278.

24

Cost-Effective Fiber Multiplexing System Based on Low Coherence Interferometers and Application to Temperature Measurement

Meng JIANG[*], Zhongze ZHAO, Kun LI, Zeming WANG, Yage ZHAN, Hongying ZHOU, and Fu YANG

College of Science Donghua University of China, Shanghai, 201620, China

[*]Corresponding author: Meng JIANG   E-mail: jiangmeng@dhu.edu.cn

Abstract: Based on the low-coherence interferometric principles, a cost-effective all-fiber Mach-Zehnder multiplexing system is proposed and demonstrated. The system consists of two interferometers: sensing interferometer and demodulation interferometer. By scanning an optical tunable delay line back and forth constantly with a stable speed, sensing fibers with different optical paths can be temporal interrogated. The system is experimentally proved to have a high performance with a good stability and low system noises. The multiplexing capacity of the system is also investigated. An experiment of measuring the surrounding temperature is carried out. A sensitivity of $12\,\mu m/^{\circ}C$ is achieved within the range of $20\,^{\circ}C$ to $80\,^{\circ}C$. This low cost fiber multiplexing system has a potential application in the remote monitoring of temperature and strain in building structures, such as bridges and towers.

Keywords: Mach-Zehnder interferometer; low-coherence; multiplexing fiber system

1. Introduction

Fiber optical sensors have been widely researched in the fields of physics, chemistry, bio-medicine, and industry [1–5] in recent decades because of their high performance compared with the traditional electronic sensors. With the advantages of anti-corrosion and anti-electromagnetic interference, a variety of new fiber optical sensors appeared in the world's laboratories [6–12]. Optical fiber interferometry has been wildly used to produce optical sensors with extremely high resolution [13, 14]. Low coherence interferometry (LCI), white light interferometry, has been widely developed as a

multiplexing sensing system with high sensitivity, large measurement range, high resolution, and low cost. One of the medical applications of LCI is optical coherence tomography (OCT) [15], in which it has been proved that the axial resolution of OCT improves with the bandwidth. The LCI technique can also be used to simultaneously resolve the mode profile and to measure the intermodal dispersion of guided modes of a few-mode fiber [16]. The interference cross-correlation function is only significant over an optical path difference approximating to the coherence length of the light source. The zero-order interference fringe can be identified for absolute position of optical path difference. The interferogram is obtained by a temporal scan of the group delay in one arm of the demodulation interferometer. By measuring the phase delay in one of the sensing interferometer's arms, the corresponding change in the interference signal can be obtained, and the measurand can be deduced. Most importantly, LCI is a multiplexing system with high capacity and consequent demodulation mechanism [9–11]. LCI can integrate multiplex sensors onto one single optical signal without requiring the use of other complex time or frequency multiplexing techniques. The sensing interferometer which can transfer sensing signal to phase delay can be multiplexed and interrogated by the LCI system.

In this paper, we demonstrate a cost-effective all-fiber multiplexing sensing scheme based on LCI [12], which measures the absolute optical path difference between reference fiber and sensing fiber. The interrogation of the system is realized by scanning an optical tunable delay line with a stable speed. This multiplexing system provides high performance with a good stability and low system noises. The multiplexing capacity is also investigated here. An experiment of measuring the surrounding temperature is carried out. A sensitivity of $12\,\mu m/°C$ is achieved within the range of 20 °C to 80 °C by our system.

2. Experimental principles

Low-coherence measurement method has been elaborated in classical optics [17]. It utilizes broad-band light source (low-coherence light source) such as a laser diode (LD) or a light-emitting diode (LED), combined with Mach-Zehnder, Michelson or other interferometer, to achieve a high precision measurement.

2.1 Low-coherence interferometer

Low coherence interferometry is simply formed by two interferometers: sensing interferometer and demodulation interfereometer. The structure of low-coherence interferometer based on Mach-Zehnder interferometer is shown in Fig. 1.

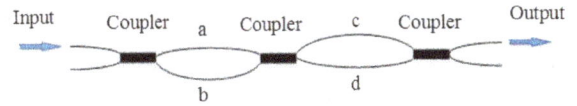

Fig. 1　Structure of low-coherence Mach-Zehnder interferometer

Light input is divided into two beams equally by a 2×2 fiber coupler and propagates through "a" and "b" arms (sensing arm and reference arm), respectively. The electric components of the two beams can be expressed as follows [18]:

$$E_a(\omega t,\varphi)=A_a(\omega\tau)\exp(i\varphi) \quad (1)$$
$$E_b(\omega t,\varphi)=A_b(\omega\tau)\exp\left[i(\varphi+k\delta_1)\right] \quad (2)$$

where A is the electric amplitude, φ is the phase angle, δ_1 is the optical path difference (OPD) between "a" and "b" arms, κ is the wave vector, and i is the imaginary part. After the two beams are coupled into "c" and "d" arms, there will be four beams:

$$E_{ad}=A_{ad}\exp(i\varphi) \quad (3)$$
$$E_{bd}=A_{bd}\exp\left[i(\varphi+k\delta_1)\right] \quad (4)$$
$$E_{ac}=A_{ac}\exp\left[i(\varphi+k\delta_2)\right] \quad (5)$$
$$E_{bc}=A_{bc}\exp\left[i(\varphi+k(\delta_1+\delta_2))\right] \quad (6)$$

where δ_2 is the OPD introduced between "c" and "d" arms. Once the OPD between E_{bd} and E_{ac} $\Delta\delta=|\delta_2-\delta_1|$ is less than the coherence length of the light source, the interference fringes can be obtained at the output of the fiber.

In our experimental setup, the OPD δ_1 will be influenced by the under test parameters. The OPD δ_2 is scanning to compensate the OPD δ_1. In our system, a motorized optical tunable delay line (OTDL, General Photonics, MDL-002) is embedded in the fiber "d" arm to adjust the δ_2.

2.2 Optical fiber characteristics for temperature

The group delay of the optical pulse propagating in the optical fiber can be represented by [19]

$$\tau=\frac{LN}{c} \quad (7)$$

where L is the length of optical fiber, N is the effective refractive-index of the fiber core, and c is the velocity of light in vacuum.

Hence, the fiber temperature delay drift constant is defined by

$$K_f = \frac{d\tau}{dT}\frac{1}{L} = \frac{1}{C}\left(\frac{dN}{dT} + \frac{N}{L}\frac{dL}{dT}\right). \qquad (8)$$

Substituting the thermal expansion coefficient ($\approx 5.5 \times 10^{-7}$ /℃) and the thermo-optical coefficient ($\approx 1 \times 10^{-5}$ /℃) for silica into (8), the K_f of fiber is approximately 0.036 ps/(m·℃). When surrounding temperature of a certain length of fiber ('a' arm in Fig. 1) changes, the OPD δ_1 will be changed relatively. The interference signal will emerge when the OPD δ_2 (introduced by the scanning OTDL) can compensate the changed δ_1. By observing the position drift of the interference fringes, the temperature variation can be detected.

3. Experimental setup

The structure of our multiplexing sensing system is shown in Fig. 2. An amplified spontaneous emission (ASE) light source with the central wavelength of 1567 nm and spectral bandwidth of 73 nm is used as a low-coherence light source here. The system is composed of two Mach-Zehnder interferometers: the first one is multi-channel for multiplexing sensors; the second interferometer which contains an OTDL is used to demodulate all sensing signals. The polarization controller in the demodulation interferometer is used to minimize the polarization influence to obtain a clearly interference fringes.

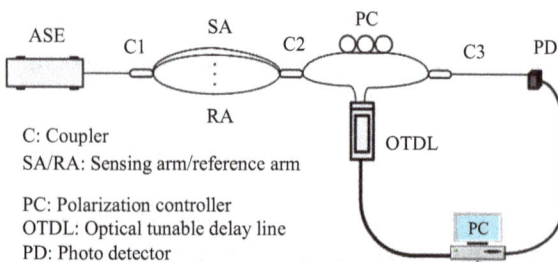

C: Coupler
SA/RA: Sensing arm/reference arm
PC: Polarization controller
OTDL: Optical tunable delay line
PD: Photo detector

Fig. 2 Structure of the multiplexing sensing system.

The output signal is detected by an InGaAs-based photodetector which is connected by a data acquisition (DAQ) card installed in the computer. The OTDL can scan back and forth within a range of

0 to 330 ps, corresponding to an optical path of 0 to 99 mm. In our system setup, the OTDL and DAQ card are computer-controlled by a Labview program. We connect a reference arm and two sensing arms in the experiment for demonstration. The length of reference fiber is set as 1970 mm, while the two sensing arms are 1986 mm and 1990 mm, respectively.

3.1 Demodulate scheme

When the OTDL scans for one single trip, four interference fringes would appear in the entire interferogram, as shown in Fig. 3.

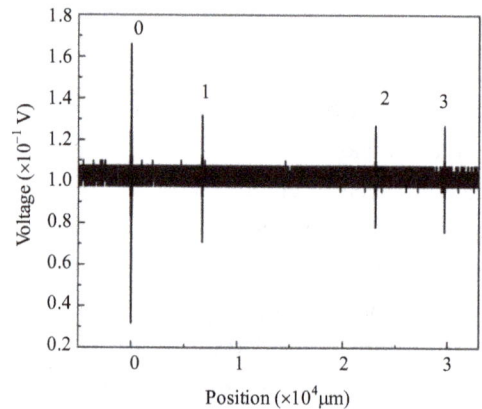

Fig. 3 Measured interferogram of the multiplexing system when it has two fiber sensors (the scanning speed is 16 ps/s).

The highest interference fringe is the zero-order interfering where the OPD δ_2 approaches zero. The position of this fringe can be the reference for measure "2" is generated when the OPD introduced by OTDL can compensate the OPD between the reference arm and first fiber sensing arm, while fringe "3" is generated when the OPD between the reference arm and second fiber sensing arm is compensated. Fringe "1" is generated when the OPD of two sensing arms is compensated, which could be regarded as a "cross-talk" of the system here.

The OPD introduced by the sensing interferometers should be less than the range of the OTDL, so that the fringes can be obtained during the OTDL scans. We should note that the "cross-talk" fringes (like fringe "1") caused between sensors should be designed away from the sensing fringes.

Therefore, we set the OPD between the first fiber sensor and reference as half of the OTDL's scanning range. And other fiber sensors have longer optical paths than the first one. In this way, all of the "cross-talk" fringes could appear between the fringe "0" and the fringe "2", so that one can distinguish the sensing fringes from "cross-talk" fringes easily.

As illustrated in Fig. 4, by employing a DAQ card and a program written by Labview, the peak position of the zero-order interference fringe envelope can be accurately located. So do the fringes "2" and "3". Figure 4 also indicates that the scanning speed of the OTDL is uniform, which is important in extracting the interference positions. Table 1 gives the four-peak points positions of fringes envelope in the interferogram in Fig. 3.

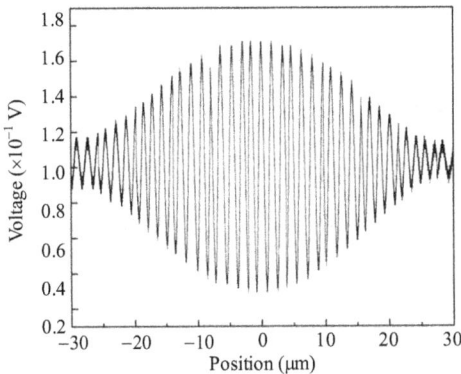

Fig. 4 Detailed fringes of the zero-order interferogram (the scanning speed is 0.01 ps/s).

Table 1 Peak positions of interference fringes envelope in Fig. 3.

Fringe	Peak position (μm) (20℃)
0	0
1	6 703
2	22 195
3	29 714

To prove the stability of the system, we carry out the repeatability test on the position of the fringe "2" under certain environment at room temperature (22.2±0.2 ℃). The peak point positions of the fringe "2" are given in Table 2. The stability of the system is proved to be good. Since the accuracy of the OTDL is 0.01 ps, the accuracy of our system is 3 μm correspondingly.

Table 2 Peak positions of fringe "2" under stable environment.

Number	Temperature (℃)	Peak position (μm)
1	22.0	22 199
2	22.4	22 204
3	22.3	22 201
4	22.3	22 201
5	22.3	22 201
6	22.3	22 201
7	22.4	22 204
8	22.4	22 204
9	22.4	22204
10	22.4	22204

3.2 Temperature measurement

As analyzed in Section 2.2, the system can be utilized to detect the surrounding temperature. Theoretically, parameters which can influence the phase delay of the optical fiber can be detected by our system, such as strain, vibration, and stretching.

In the experiment, a single mode fiber with 0.9-meter length of the first sensing fiber is heated by a temperature oven. The scanning speed of the OTDL is 3 μm/s, and the sample rate of the DAQ card is 8000 Hz. The position drift of the peak "2" with the temperature changes is illustrated in Fig. 5. The scanning speed of the OTDL is set as 8 ps/s so that all the interference fringes could be discerned on the screen of the computer, and it will take about forty seconds to demodulate all sensors. The sensitivity of the temperature is 12 μm/℃, with the range of 20 ℃ to 80 ℃. The group delay of 12 μm is about 0.04 ps in the OTDL, which corresponds to the optical fiber of 0.03 ps/(m·℃). So the measurement results match with the theoretical sensitivity in Section 2.2. Since the accuracy of the system is 3 μm, the resolution of this experiment is 0.25 ℃. One can get higher sensitivity and resolution by simply lengthening the sensing fiber under measured temperature.

The capacity of this multiplexing system heavily relies on the scanning range of the OTDL, which is 0–330 ps in our system. So the OTDL could introduce 0–99 mm optical path in free space into the Mach-Zehnder interferometer. In the experiment, temperature change of 100 ℃ could cause a 1.2-mm position shift of the interference fringes. Furthermore, the cross-talks made by the sensor array should be separated from the sensing signals. In our system, the cross-talks are designed close to the zero order interference fringes by means of designing the OPD between every two sensors less than the OPD between the reference optical fiber and the minimal optical sensor:

$$OPD_{0,1} \gg OPD_{0,N} \qquad (9)$$

where "0" means the reference arm, and "1" and "N" mean the sensors which have the minimal and maximal optical paths.

Therefore, only half of the scanning range of OTDL is used for sensing interference fringes, thus the capacity of our system could be

$$n \ll (99/2)/1.2 \approx 40 \qquad (10)$$

where "n" represents the number of sensors.

The capability of this system can be further enlarged by utilizing a longer scanning range OTDL, which is commercial available. Besides temperature, strain on the fiber sensor can also involve additionally optical phase change in the system. If we package each fiber sensor properly to make it only sensitive to temperature or strain, our system can be used to demodulate all the sensors to measure both temperature and strain.

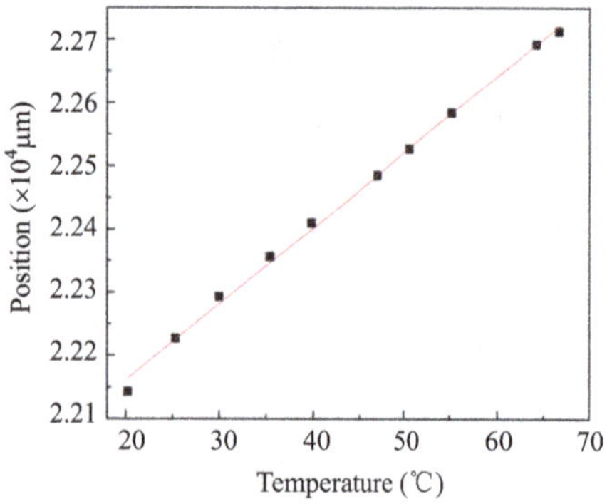

Fig. 5 Position drifts of the interference fringe envelope peak "2" versus temperature.

4. Conclusions

In this paper, an optical fiber multiplexing sensing system based on LCI is built up. By scanning an optical tunable delay line back and forth constantly with a stable speed, fiber sensors with different optical paths can be temporal interrogated. This multiplexing system shows high performance with a good stability in the experiment. The "cross-talk" fringes between multiplexed sensors can be designed to be eliminated from the sensing signals. The capacity of the system is also discussed. An experiment of measuring the surrounding temperature is carried out. A sensitivity of $12\,\mu m/°C$ is achieved within the range of $20\,°C$ to $80\,°C$ by our system, which matches well with theoretical results.

Our proposed multiplexing sensing system will be useful for the remote monitoring of the temperature and strain. By incorporating fiber optic sensor arrays into structures such as bridges and buildings, smart structures can be realized with relatively low cost.

Acknowledgment

This work was supported by Pujiang Program (No. 14PJ1400100) and the Fundamental Research Funds for the Central Universities of China (No. 2232014D3-28).

References

[1] N. B. Kosa, "Key issues in selecting plastics optical fibers used in novel medical sensors," *Proc. of SPIE*, 1991, 1592: 114–121.

[2] T. G. Gaillorenze, "Optical fiber sensor technology," *IEEE Journal of Quantum Electronics*, 1982, 18(4): 626–665.

[3] R. C. Spooncer, "*Fiber optics in instrumentation*," in Handbook of Measurement Science, Ed: P. H. Sydenham & R. Thorn, Chichester. UK: Wiley, 1992: 1691–1720.

[4] H. Chen and Y. Liang, "Analysis of tunable asymmetric fiber F-P cavity for fiber strain sensor edge-filter demodulation," *Photonic Sensors*, 2014, 4(4): 338–343.

[5] Q. Li and H. Chen, "Design of fiber magnetic field sensor based on fiber Bragg grating Fabry-Perot cavity ring-down spectroscopy," *Photonic Sensors*, 2015, 5(2): 189–192.

[6] K. T. V. Grattan, "Fiber optic sensors-the way forward," *Measurement: Journal of the International Measurement Confederation*, 1987, 5: 122–134.

[7] E. M. Dianov, S. A. Vasiliev, A. S. Kurkov, O. I. Medvedkov, and V. N. Protopopov, "In-fiber Mach-Zehnder interferometer based on a pair of long-period gratings," in *22th European Conference on Optical Communication*, Oslo, 1996.

[8] D. P. Hand and P. St J. Russell, "Photo induced refractive-index changes in germanosilicate fiber," *Optics Letters*, 1990, 15(2): 102–104.

[9] J. L. Brooks, R. H. Wentworth, R. C. Youngquist, M. Tur, B. Y. Kim, and H. J. Shaw, "Coherence multiplexing of fiber-optic interferometricsensors," *IEEE Journal of Lightwave Technology*, 1985, 3(5): 1062–1072.

[10] W. V. Sorin and D. M. Baney, "Multiplexed sensing using optical low-coherence reflectometry," *IEEE Photonics Technology Letters*, 1995, 7(8): 917–919.

[11] L. B. Yuan and J. Yang, "Two-loop-based low-coherence multiplexing fiber optic sensor network with a Michealson optical path demodulator," *Optics Letters*, 2005, 30(6): 601–603.

[12] L. Jin, W. G. Zhang, Q. C. Tu, and X. Y. Dong, "Applications of interferential technique in designing optics fiber sensors," *Laser and Optoelectronics Exhibition*, 2004, 41(7): 54–55.

[13] C. D. Butter and G. B. Hocker, "Fiber optic strain gauge," *Applied Optics*, 1978, 17(18): 2867–2869.

[14] S. R. Kidd, P. G. Sinha, J. S. Barton, and J. D. C. Jones, "Interferometric fiber sensors for measurement of surface heat transfer rates on turbine blades," *Optics and Lasers in Engineering*, 1992, 16(2–3): 207–221.

[15] P. Jansz, S. Richardson, G. Wild, and S. Hinckley, "Modeling of low coherence interferometry using broadband multi-Gaussian light sources," *Photonic Sensors*, 2012, 2(3): 247–258.

[16] Y. Z. Ma, Y. Sych, G. Onishchukov, S. Ramachandran, U. Peschel, B. Schmauss, *et al.*, "Fiber-modes and fiber-anisotropy characterization using low-coherence interferometry," *Applied Physics B*, 2009, 96(2): 345–353.

[17] M. Born and E. Wolf, "*Principles of optics*. New York: Pergamon Press, 1986.

[18] K. T. V. Grattan and B. T. Meggitt, *Optoelectronics, imaging and sensing series: optical fiber sensor technology volume 2 devices and technology.* London: Chapman & Hall, 1998: 169–171.

[19] X. Z. Hu, *Fiber optic and fiber cable.* Beijing: Publishing House of Electronics Industry, 2007: 24–243.

25

Investigation into the Electromagnetic Impulses From Long-Pulse Laser Illuminating Solid Targets Inside a Laser Facility

Tao YI[1], Jinwen YANG[1,2], Ming YANG[2], Chuanke WANG[1], Weiming YANG[1], Tingshuai LI[2*], Shenye LIU[1], Shaoen JIANG[1], Yongkun DING[1], and Shaoqiu XIAO[3]

[1]*Laser Fusion Research Center, Chinese Academy of Engineering Physics, Mianyang, 621900, China*

[2]*School of Energy Science and Engineering, University of Electronic Science and Technology of China, Chengdu, 611731, China*

[3]*School of Physical Electronics, University of Electronic Science and Technology of China, Chengdu, 611731, China*

*Corresponding author: Tingshuai LI   E-mail: litingshuai@uestc.edu.cn

Abstract: Emission of the electromagnetic pulses (EMP) due to laser-target interaction in laser facility had been evaluated using a cone antenna in this work. The microwave in frequencies ranging from several hundreds of MHz to 2 GHz was recorded when long-pulse lasers with several thousands of joules illuminated the solid targets, meanwhile the voltage signals from 1 V to 4 V were captured as functions of laser energy and backlight laser, where the corresponding electric field strengths were obtained by simulating the cone antenna in combination with conducting a mathematical process (Tiknohov Regularization with L curve). All the typical coupled voltage oscillations displayed multiple peaks and had duration of up to 80 ns before decaying into noise and mechanisms of the EMP generation was schematically interpreted in basis of the practical measuring environments. The resultant data were expected to offer basic know-how to achieve inertial confinement fusion.

Keywords: Electromagnetic impulse; laser; voltage; simulation; decay

1. Introduction

Although various radiations tend to be generated when some particles strike solid targets like Cu [1], the concentration of extremely powerful laser beams onto a solid target would induce significant electromagnetic radiation, which can not be only used for scientific research concerning the physics of extreme energy densities and pressures, but also provide a feasible path to achieve inertial confinement fusion (ICF) and a promising way to

generate carbon-free sustainable energy [2]. Fast heating of ultrahigh-density plasma was reported to be an effective step towards laser fusion ignition [3]. However, an intensive electromagnetic pulse (EMP) emitted within the target chamber of a high power laser are well-known issues that cause malfunction and damage of some hardware, internal, and external diagnostics for the laser fusion. The EMP is possibly originated by the escaping electrons from the target surface during the laser-target interaction process [4], in which the amplitudes of EMP are closely related to the target sizes and laser power density. It is thus significantly imperative to evaluate the EMP inside and outside laser chamber before some necessary electromagnetic shielding fashions can be proposed to protect all the diagnostic setups.

Brown *et al.* [5] at Lawrence Livermore National Laboratory installed multiple B- and D-dot sensors in the Titan short-pulse laser facility (ps) to measure the levels of electromagnetic fields, by which the electric field strength from the sensors locating about 32 cm from target chamber center was estimated to be 167 kV/m. During this process, the ps and ns laser pulses generated quite different ion emission current densities, leading to up to 100 MeV by sub-picosecond laser plasma [6]. On the other hand, Brown *et al.* [7] also reported that the EMP signals collected at Lawrence Livermore National Laboratory mainly covered the frequency domains from several hundreds of MHz to several GHz. Chen *et al.* [8] reported the microwave emission generated by a laser pulse incident on a metallic disc surface had a frequency ranges from 0.5 GHz to 4 GHz. Moreover, the fs laser pulse inducing EMP can be enhanced by a long (ns) laser pulse due to the formation of pre-plasma [9].

The generation of EMP by laser illuminating solid targets is closely bound up with the plasmas induced during this process. A long-pulse laser of high power density focuses on the solid target surface, rapidly producing plenty of plasma and the time needed is far shorter than that of the laser pulse. Moreover, for the long-pulse laser without high power density ($< 10^{15}$ W/cm^2), the electrons among plasmas take swift quiver, by which the adsorbed energy by electrons is converted to thermal energy of the random thermal motional plasma [10]. The self-radiation of plasma can bring out X-ray, electrons, ions, and neutrons, but there is still lack of deep understanding on the specific mechanism for the EMP generation. To further study the EMP features, a Faraday cup was developed [11].

This work aims to evaluate the EMP emissions from an ICF integrated diagnostic system when a long pulse (ns) laser of several kJ level is utilized to ignite the target and it is expected to provide possible solutions to electromagnetic interference (EMI) problems and also lay a preliminary foundation to achieve fusion. The microwave can potentially damage various interior electrics including energy balance components, X-ray cameras, and optical parts.

2. Experiment

All the measurements were performed outside a 9-beam Nd-glass laser facility as sketched in Fig. 1, where various diagnostic setups were installed inside and outside a 2.7-meter diameter sphere chamber. The 9-beam laser included 8-beam laser from the northern hemisphere (4 beams) and the southern hemisphere (4 beams), plus a backlight beam normally from top of the north-south line 45 degrees. The detailed information on laser energy, focal spot size, backlight energy, and laser pulse width used in this measurement is listed in Table 1, in which an estimated peak power intensity is around 10^{15} W/cm^2, and the amplitude of emitted microwave are compared and analyzed as functions of these parameters. The solid targets used in this study were typical dual-media targets of Rayleigh-Taylor instability. The chamber vacuum remained less than 5×10^{-3} Pa.

Table 1 Detailed information of laser used to illuminate solid targets.

	Laser energy (kJ)	Focal spot size (μm)	Back ground light (kJ)	Laser pulse width (ns)
L1$^\#$	5.6	500	1	3
L2$^\#$	0	500	1	3
L3$^\#$	0	500	2.7	3
L4$^\#$	3.35	500	0	1(0.2)
L5$^\#$	3.2	500	2.7	1(3)
L6$^\#$	5.6	500	1.8	2
L7$^\#$	5.6	500	2.6	2

A conical antenna was designed and made from copper coated nickel to collect the EMP signals, linked by a standard 50-coaxial cable to a high-speed real-time oscilloscope as shown in Fig. 1. The antenna was placed close to a glassy flange fixed in the chamber wall to reduce effects from the hardware. The antenna had been both calibrated and modeled from 0.8 GHz to 5 GHz, by which the peak gain of antenna in the frequency ranges of 1 GHz to 5 GHz and distribution of electrical field obtained at 1 GHz are displayed in Fig. 2.

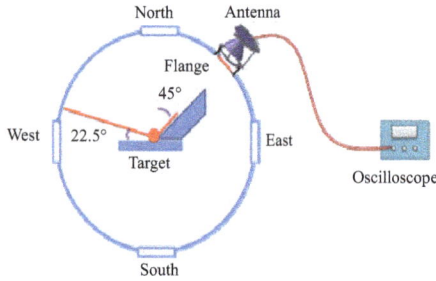

Fig. 1 Schematic representation of the experimental setup.

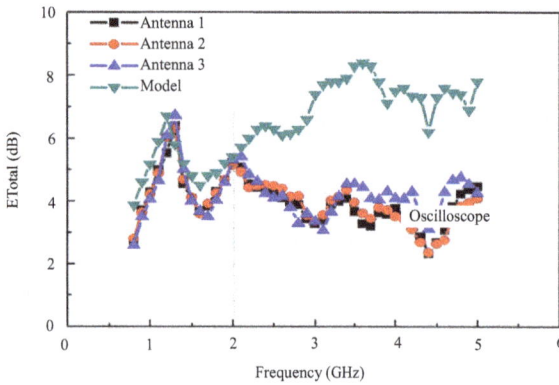

(a)

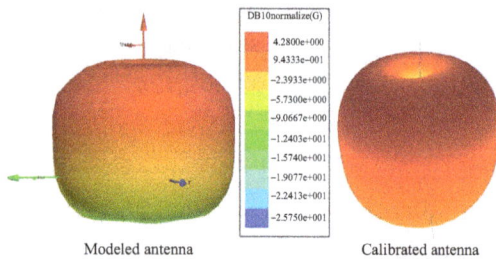

(b)

Fig. 2 Parameters for antenna used: (a) peak gains of the used antenna at different frequencies obtained by calibration and simulation and (b) calibrated and simulated electric field distributions along the antenna at 1 GHz.

The transfer function $\mathbf{H}(\omega)$ is of importance because it bridges the voltage signal and electromagnetic field, which was obtained by simulating the antenna. The effects of measurement system on low-frequency domains caused non-negligible noises. In order to eliminate this noise, Tikhonov regularization with L-curve analysis

was selected to address the distorted signals. The measured voltage $\mathbf{V}(\omega)$ can be related to the electrical field $\mathbf{E}(\omega)$ by a transfer function of $\mathbf{H}(\omega)$ as expressed by (1), in equal to a time-domain formulation in (2), which can be mathematically discrete into $\mathbf{V} = \mathbf{HE}$, where $\mathbf{V}$ and $\mathbf{E}$ denote m-directional vector and

$$\mathbf{V}(\omega) = \mathbf{H}(\omega)\mathbf{E}(\omega) \qquad (1)$$

$$\mathbf{V}(t) = \mathbf{H}(t)\mathbf{E}(t) \qquad (2)$$

where $\mathbf{H}$ is an $m \times m$ matrix. We then use the zero order of Tikhonov regularization to figure out the extremum $\min_{e \in R^n}\left(\|He - v\|^2 + \mu\|e\|^2\right)$, where $\|\cdot\|$ denotes the Euclidean norm, and μ is a parameter that needs to be cleared. The solution to the above problem is

$$e_{\mu,i} = H_i^* v_i / H_i^* H_i - \mu, \quad i = 1, 2, \cdots, m.$$

In this case, the L curve analysis is employed to confirm the parameter μ, and then the electrical field can be obtained by reversely undergoing these procedures.

3. Results and Discussion

Figure 3 presents some typical voltage wave forms that have been recorded and labeled from L1[#] to L7[#] with amplitudes ranging from 1.5 V to 6.6 V. Note that the measurements are all conducted outside the chamber, which means the resultant EMP would be greatly reduced when it penetrates the thick quartz glass (to ensure good leak-tightness), making the signals not that strong. Each EMP has transient duration of up to 80 ns before the signal attenuation decays into noise.

However, all the wave forms show that there are some unfixable periods for EMP peaks, where oscillation of the waveforms can be observed with time course, and the time intervals of some evident peaks are assessed from several ns to dozens of ns. The wave oscillation lasts nearly 20 ns without any decaying with regard to L1[#], which implies there are at least two coupled modes of different frequencies when 8-beams laser and the backlight energy are totally accounted to 6.6 kJ (5.6 kJ+1 kJ).

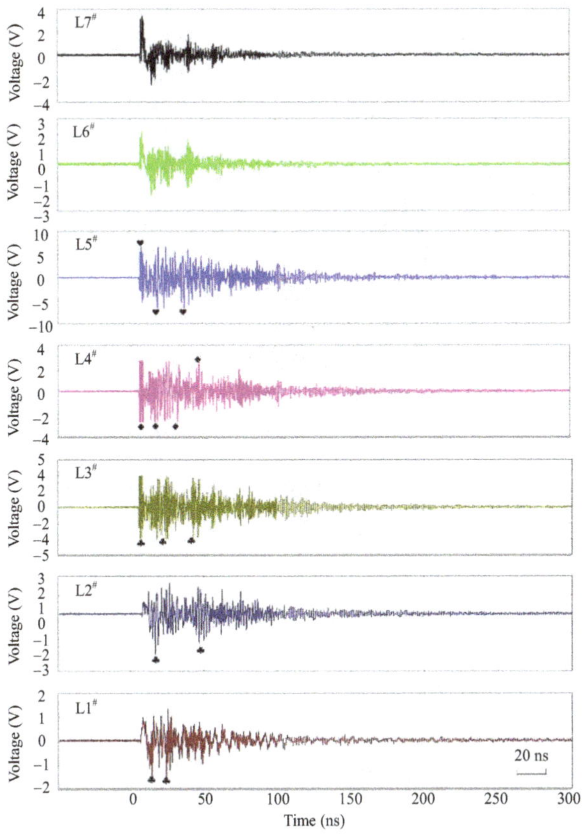

Fig. 3 Typical coupled voltage outputs recorded by the cone antenna at varying laser input and laser width.

To emphasize the effects of backlight laser beam, L2$^{\#}$ is carried out by adjusting the backlight to be 1 kJ and leaving the 8-beams laser. There are two separate peaks at 15 ns (2.1 V) and 50 ns (1.5 V), which are much bigger than that of L1$^{\#}$. To increase the backlight to 2.7 kJ as depicted in L3$^{\#}$, the peak voltage is enhanced to 4.0 V, and the couple modes of at least four different frequencies occur in this process, indicating the backlight shot can not only determine the microwave intensity, but also create more different radiating modes.

On the other hand, the effects of laser width on the formation of electromagnetic impulse are taken into account. With respect to L4$^{\#}$, we use 4-beams 1 ns pulses (3.2 kJ) and 1-beam 0.2 ns pluses (0.15 kJ) to react with the solid target. Several peaks with amplitudes of 2.6 V are obtained, and a series of strong oscillations are caused by the shorter pluses lasting for about 40 ns and then decayed into noise. We maintain the 1 ns pulses and meanwhile add

2.7 kJ backlight laser (3 ns) in L5$^{\#}$ to see the changes. A pronounced peak of 6.1 V is achieved, and three evident modes are coupled during this process, suggesting that the backlight shot could be related to the total radiation energy. The resultant waveforms in L6$^{\#}$ and L7$^{\#}$ are carried out to further confirm the conclusion that the intensity of backlight beam is critical to the strength of microwave.

As shown in L3$^{\#}$ (4.0 V) and L5$^{\#}$ (6.1 V), an increase in total energy from the 8-beams laser with shorter pulse tends to induce extra radiation upon the reactions between laser and target, differing to the situation in L1$^{\#}$ and L2$^{\#}$, where the 9-beams laser has identical width (3 ns) and aroused similar voltage amplitudes. The corresponding frequency wave profiles obtained by conducting the fast frontier transform (FFT) are shown in Fig. 4, from which it can be seen that the frequency bandwidth resides at around 2 GHz or lower. It is also noted that the emissions have characteristic peak structures with several peaks and all the peaks located below approximately 2 GHz possibly due to the glass flange act as an attenuator or filer, which is more evident when the antenna is placed outside the metal shell as measured in this study (the highest voltage peak was only 0.4 V).

The corresponding electric field (E-field) strength was obtained by combining the antenna simulation and mathematical method as depicted in Fig. 5. E-field signals reached from over 200 V/m to 600 V/m, which indicates some appropriate shielding methods are particularly required to protect various electronics including the charged coupled device (CCD), dilation X-ray imager (DIXI), and the filter-fluorescer experiment. The results in this experiment are able to provide a suitable standard on how to design additional shielding for all neighboring diagnostic instruments since the sources of EMP cannot be completely eliminated, even we tried the best to control or lower the EMP origins.

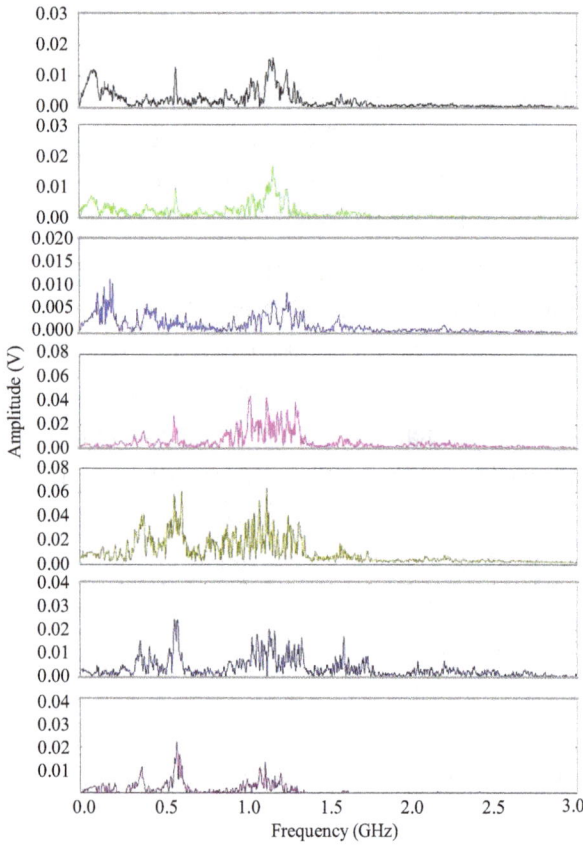

Fig. 4 Fast Fourier transform (FFT) of the corresponding voltage signals shown in Fig. 3.

(1.35 m). Assuming the speed of microwave inside the chamber (vacuum) is the light speed, the diameter corresponds to 9 ns.

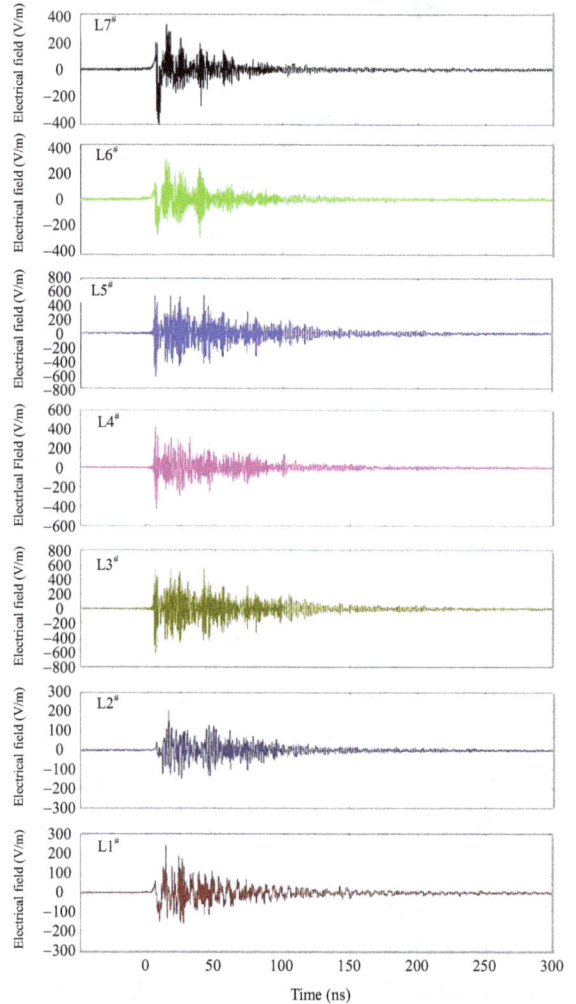

Fig. 5 Electric field strength obtained by processing voltage signal data [with $\mathbf{E}(\omega) = \mathbf{V}(\omega)/\mathbf{H}(\omega)$].

The patterns of EMP transportation are possibly responsible for the coupled voltage output in this measurement, which mainly stem from the laser-target interaction and diagnostic-generated EMI [12], where some preliminary results of EMI induced by various diagnostic setups are evaluated. More tests needed to be elaborated and performed to confirm the specific origins of EMI since the testing environment is extremely complicated.

To understand the multiple peaks in this study, the EMP generated from laser-target interaction randomly propagates inside the chamber as sketched in Fig. 6. There are probably four different coupled modes that are manifested in all testing results. The first peak may come from direct laser-target interaction labeled by the arrow as shown in Fig. 6(a) because of the largest intensity and the earliest time arriving at the antenna. In this case, the propagating length before hitting the antenna equals approximately to the radius of this sphere chamber

For another thing, the microwave is originated from reflection of inner wall as represented by the arrows. The second peak in L2$^{\#}$ waveform delays over 20 ns (6 meters), which could be attributed to several times of reflections as depicted in Fig. 6(b). Note that the second peak in L2$^{\#}$ is weaker than the first one, which is mainly due to energy loss caused by the chamber wall adsorption and scattering. The effects of interior diagnostic instruments are also taken into account when their electric components are very sensitive to electromagnetic pulses. Since the radiated electric can also generate huge EMP, it is expected to further intensify the reflected waves as demonstrated in Fig. 6(c). Some typical waveforms including L3#, L4#, and L5# can be

presumably explained by this coupled mode as they all have multiple peaks of similar amplitudes.

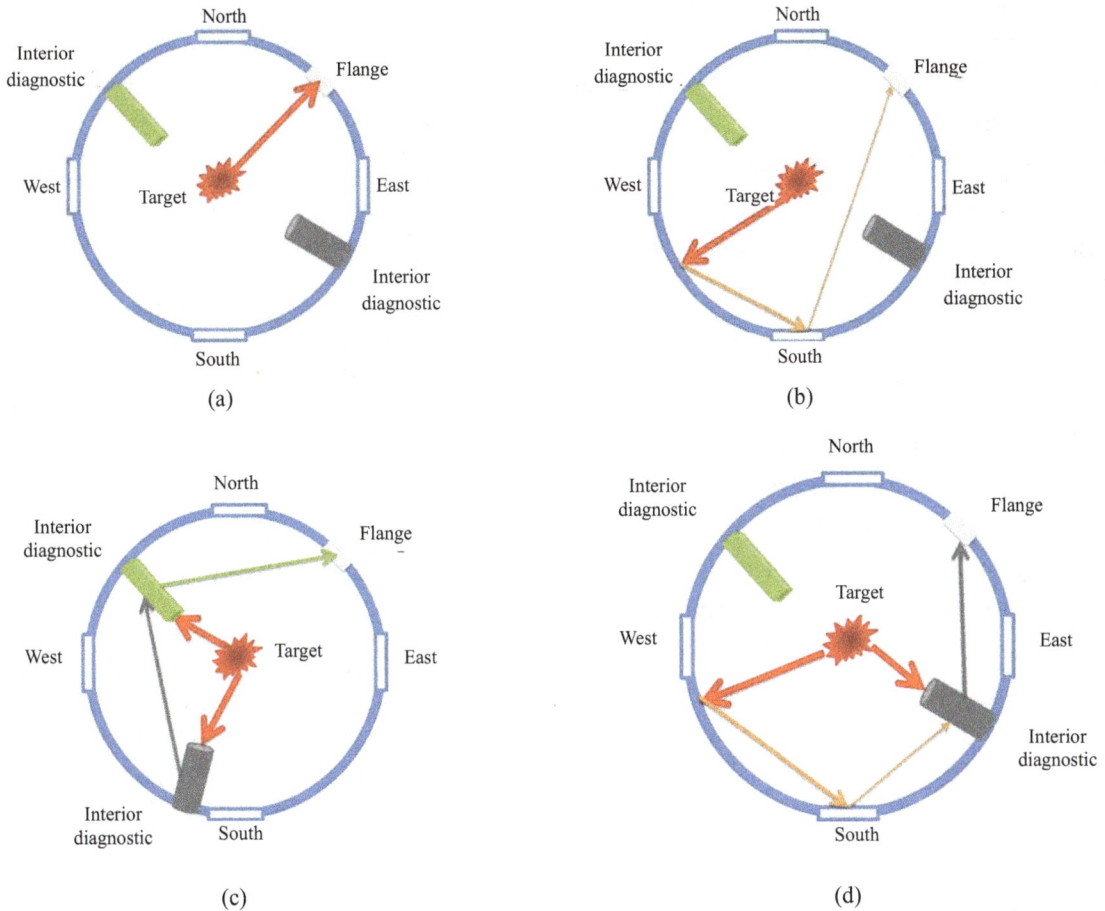

Fig. 6 Mechanism for EMP generation with a multiple peaks: (a) modes for EMP induced by a high-power laser directly interacting with solid target, (b) reflection from the inner chamber wall, (c) radiations from the interior diagnostics, and (d) adsorption and scattering of the wall in combination with inside electric-generated.

Attentions should be paid to some peaks with time intervals of 10 ns to 20 ns (equals to 3-m to 6-m length) owing to the inside diameter of sphere chamber being totally 2.7 m, meaning that the possible mode in Fig. 6(d) can maintain the wave strength and appear in a delayed way, in which multiple scattering (long distance transmission) on the inner wall surface interacts with some interior electronics to preserve or improve the intensity (L1#) to a high level.

To understand the relationship between the EMP amplitude and laser energy, it should relate the entry laser to escaping electron plasma density, which can be expressed in (3), where λ is the laser

wavelength and m_e is the electron mass.

$$n_c = \frac{m_e}{4\pi e^2}\left(\frac{2\pi c}{\lambda}\right)^2. \tag{3}$$

It can be seen that the electron number is closely pertinent to the laser wavelength. When a beam of backlight laser is added onto target surface, the escaping electrons must be increased and the spatial distribution can also be multiplied by characterizing the components that electrons strike [7], leading to the electrons escape in multi-directional way.

4. Conclusions

EMP generated by some powerful lasers interacting with solid targets was measured by using a cone antenna. The amplitudes of EMP were analyzed as functions of laser energy and bandwidth. All the waveforms in this experiment had multiple peaks and possessed duration of dozens of ns before they attenuated into noise. Several coupled modes were established to interpret these specific temporal electric signal oscillograms, indicating that the EMP was not only stemming from direct laser-target interaction, but also was induced by the interior diagnostics by inner wall reflection.

Acknowledgment

We would like to thank Dr. Peng Wang at University of Electric Science and Technology of China (UESTC) and Dr. Ziyu Chen at China Academy of Engineering Physics for their help in processing the data and understanding the potential mechanisms. This work was financially supported by the Fundamental Research Funds for the Central Universities (No. ZYGX2015J108) and National Natural Science Foundation of China (Nos. 11575166 and 5158 1140).

References

[1] R. Qiu, Y. Y. Liu, W. Q. Li, Y. X. Pan, S. L. Wang, Q. Li, *et al.*, " Measurement and validation of the cross section in the FLUKA code for the production of Zn-63 and Zn-65 in Cu targets for low-energy proton accelerators," *Nuclear Science and Techniques*, 2014, 25(1): 6−10.

[2] T. M. Tzschentke and W. J. Schmidt, "Ignition on the national ignition facility: a path towards inertial fusion energy," *Nuclear Fusion*, 2009, 49(10): 593−598.

[3] R. Kodama，P. A. Norreys, K. Mima, A. E. Dangor, R. G. Evans, H. Fujita, *et al.*, "Fast heating of ultrahigh-density plasma as a step towards laser fusion ignition," *Nature*, 2001, 412(6849): 798−800.

[4] M. J. Mead, D. Neely, J. Gauoin, R. Heathcote, and P. Patel, "Electromagnetic pulse generation within a petawatt laser target chamber," *Review of Scientific Instruments*, 2004, 75(10): 4225−4227.

[5] C. G. Brown, E. Bond, T. Clancy, S. Dangi, D. C. Eder, W. Ferguson, *et al.*, "Assessment and mitigation of electromagnetic pulse (EMP) impacts at short-pulse laser facilities," *Journal of Physics: Conference Series*, 2010, 244(3): 681−687.

[6] H. Hora, J. B. Heinrich, F. P. Boody, R. Höpfl, K. Jungwirth, B. Králikova, *et al.*, "Effects of ps and ns laser pulses for giant ion source," *Optics Communications*, 2002, 207(1−6): 333−308.

[7] C. G. Brown, A. Throop, D. Eder, and J. Kimbrough, "Electromagnetic pulses at short-pulse laser facilities," *Journal of Physics: Conference Series*, 2008, 112(3): 1231−1236.

[8] Z. Y. Chen, J. F. Li, J. Li, and Q. X. Peng, "Microwave radiation mechanism in a pulse-laser-irradiated Cu foil target revisited," *Physica Scripta*, 2011, 83(5): 434−440.

[9] S. Varma, J. Spicer, B. Brawley, and J. Miragliotta, "Plasma enhancement of femtosecond laser-induced electromagnetic pulses at metal and dielectric surfaces," *Optical Engineering*, 2014, 53(5): 688−696.

[10] S. Eliezer, *The interaction of high-power lasers with plasmas*. Boca Raton: CRC Press, 2002.

[11] J. Prokupek, J. Kaufman, D. Margarone, M. Krůs, A. Velyhan, J. Krása, *et al.*, "Development and first experimental tests of Faraday cup array," *Review of Scientific Instruments*, 2014, 85(1): 013302−1−013302−6.

[12] C. G. Brown, J. Ayers, B. Felker, W. Ferguson, J. P. Holder, S. R. Nagel, *et al.* "Assessment and mitigation of diagnostic-generated electromagnetic interference at the National Ignition Facility," *Review of Scientific Instruments*, 2012, 83(10): 10D729−1−10D729−3.

26

Variable Configuration Fiber Optic Laser Doppler Vibrometer System

Julio E. POSADA-ROMAN[1], David A. JACKSON[2*], and Jose A. GARCIA-SOUTO[1]

[1]*Department of Electronics Technology, GOTL University Carlos III de Madrid, Butarque 15, 28911, Leganés, Madrid, Spain*

[2]*Applied Optics Group, School of Physical Sciences, University of Canterbury, Kent, CT2 7NH, UK*

[*]Corresponding author: David A. JACKSON   E-mail: d.a.jackson@kent.ac.uk

Abstract: A multichannel heterodyne fiber optic vibrometer is demonstrated which can be operated at ranges in excess of 50 m. The system is designed to measure periodic signals, impacts, rotation, 3D strain, and vibration mapping. The displacement resolution of each channel exceeds 1 nm. The outputs from all channels are simultaneous, and the number of channels can be increased by using optical switches.

Keywords: Laser Doppler; vibration; digital processing; nanometer resolution; multiplexing; optical fiber

1. Introduction

Laser Doppler vibrometers (LDV) based on a Mach-Zehnder interferometer (MZI) incorporating an external air path with a Bragg cell frequency modulator have been available for several decades. Displacement resolution of $\sim 1*10^{-9}$ m has been achieved with these devices typically illuminated with relatively low power single frequency He-Ne laser. One of the major factors responsible for the high spatial resolution is the very low phase noise and nearly perfect Gaussian beam exhibited by these lasers. A problem with conventional LDVs is that a direct line of sight between the instrument and the target is required making it difficult to make measurements in confined or restricted areas. It is also not possible to make measurements at more than one location at a time with these LDVs. To overcome these problems, we have designed and

implemented a multichannel fiber optic based laser Doppler heterodyne vibrometer (FLDV) which can be operated at ranges from $\sim$ cm to more than 50 m. The vibrometer can be used to measure signals such as displacement impacts, rotation, strain and torsional vibrations. The displacement resolution of each channel exceeds 1 nm at frequencies up to 100 kHz (limited by the shakers available). The outputs from all channels are simultaneous; the number of channels can be increased using optical switches.

2. Conventional laser Doppler vibrometer

A conventional LDV is shown in Fig. 1. Light from a single frequency He-Ne laser is injected into a cube beam splitter (CBS1) where it is amplitude divided. One of the output laser beams propagates to CBS3 and is focused by the lens onto the target. The back reflected beam from the target is transferred to CBS4 via CBS3. The other output beam from CBS1 is reflected at CBS2 transferring it to the Bragg cell where it undergoes a frequency shift of typically 40 MHz. These beams interfere in CBS4 generating an optical interference signal at 40 MHz which is transposed to an electrical signal by the photodiodes. The outputs from the detectors can be combined differentially to minimize noise. When the target is subject to vibration, side bands are generated symmetrically around 40 MHz which are described by Bessel functions of the first kind. This signal is processed electronically to recover the frequency and amplitude of vibration of the target.

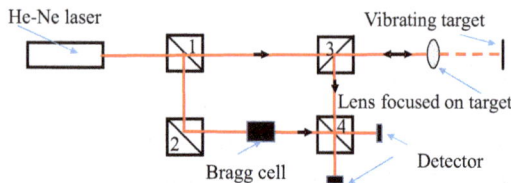

Fig. 1 Conventional laser Doppler vibrometer.

3. Fiber optic laser Doppler vibrometer

Figure 2 shows an implementation of an optical FLDV. Light from a single frequency 1.5 micron laser diode with low phase and intensity noise is pigtailed to low loss single mode optical fiber producing a Gaussian beam at the distal end of the fiber. This guided beam is injected into the first 3 dB couple as shown in Fig. 2. One output from the coupler is connected to a circulator which sends the fiber guided light to a target via a variable focus collimator generating a tightly focused Gaussian disc. The back reflected light from the target is coupled back into the fiber and hence the circulator, and transferred to the input of the second coupler. The other optical output from the first coupler is frequency shifted by the 40 MHz Bragg cell (as discussed above) and injected into the other input port of the second 3 dB coupler. Optical interference occurs at this coupler and is detected by a low noise 125 MHz optical detector [1]. Additionally, an inline variable attenuator in the MZI arm containing the Bragg cell is used to prevent saturation of the detector.

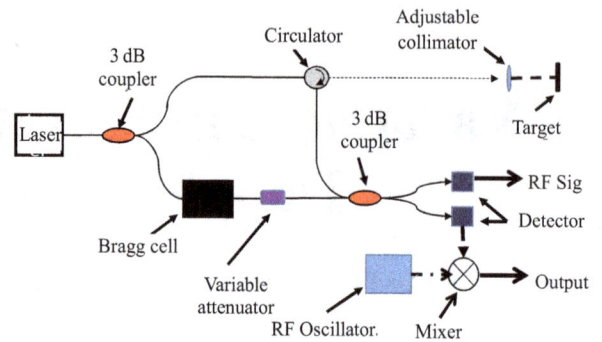

Fig. 2 Fiber laser Doppler vibrometer including a mixer to down shift recovered signal.

Optical fiber components can be fused together with extremely low optical loss; in addition, no optical alignment is required except the collimator with the target. As the FLDV is a complex MZI, we need to consider the impact on the fringe visibility of large optical path differences (OPDs). Ideally, the optical path length (OPL) of the sensing arm should equal the OPL of the reference arm. This would be very difficult to achieve in practice, however this requirement can be ameliorated to a large extent as the laser source used has an extremely narrow linewidth of less than 1 MHz. This corresponds to a coherence length of ~200 m, implying that it is unnecessary to closely match the OPDs of the sensor and reference paths.

As discussed above, the spectrum of the light backscattered from a vibrating target will generate a phase modulated signal at 40 MHz, which is described by Bessel functions [3]. The spectrum will be much more complicated for random modulation of the target. The signal can be analyzed with an electronic spectrum analyzer or a "phase locked loop" which can give valuable but limited information. A digital processor offers the best solution as it is able to recover the amplitude, frequency, and phase of the signal with higher accuracy. Rather than directly processing the signal at 40 MHz (Fig. 3(a)), it can be transposed to a lower frequency using an electronic mixer and a second high frequency oscillator at 39.9 MHz as shown in Fig. 2. An example of the resultant down shifted carrier is shown in Fig. 3(b).

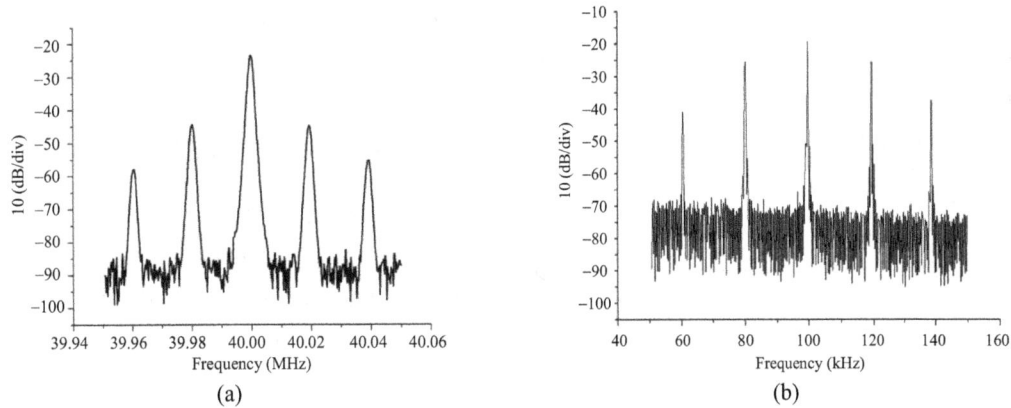

(a)

(b)

Fig. 3 Spectra of the optical carrier: (a) carrier at 40 MHz modulated at a frequency of 20 kHz and (b) the carrier frequency is 100 kHz modulated at 20 kHz.

4. Digital high speed processing system for heterodyne FLDV and calibration

4.1 Digital demodulator

Figure 4 shows the basic scheme proposed for the heterodyne FLDV signal recovery. The DAQ and signal processing implement an I/Q phase demodulation instead of a frequency demodulation [2]. It is completely digital and is implemented in LabVIEW. This reduces the size of the system and facilitates the implementation of a multichannel FLDV due to the software scalability.

By means of the I/Q demodulation, the sine and cosine of the carrier are obtained, and the phase angle is recovered using the arctangent function. The heterodyne detection together with the digital demodulation provides a wider measurement range because a digital phase accumulator can be implemented (Fig. 4) for fringe counting which extends the measurement range further than 2π radians.

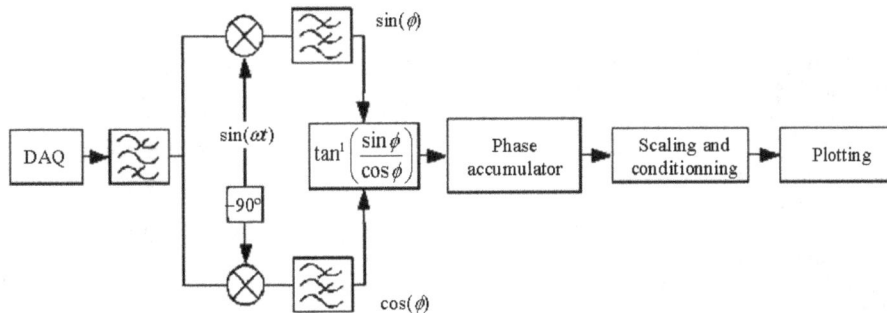

Fig. 4 Schematic of digital processing.

4.2 Performance of the heterodyne FLDV

Initial studies were performed to assess the performance of the 4-channel heterodyne FLDV. A simple method that exploits the Bessel function of first kind can be used for this propose [3]. It is based on the measurement of the ratio of

$$J_0(m) = J_1(m) \quad (1)$$
$$m = 2kx \quad (2)$$

where $J_n(m)$ is a Bessel function of the first kind order n and modulation index m, $k=2\pi/\lambda$, x is the displacement, and λ is the wavelength of the laser. A spectrum analyzer was used to monitor the carrier, and the harmonics produced by the modulation of a shaker used as a vibrating target for the calibration. The amplitude of the vibration was set to obtain the particular condition when $J_0 = J_1$. This calibration point corresponds to a modulation index $m = 2.4$ and can be used in (2) in order to find the magnitude of the target displacement. The carrier signal was observed with the spectrum analyzer for a stationary target (without vibration), when it is vibrating at 20 kHz with an amplitude corresponding to the calibration point where $J_0 = J_1$ are shown in Figs. 5(a) and 5(b), respectively.

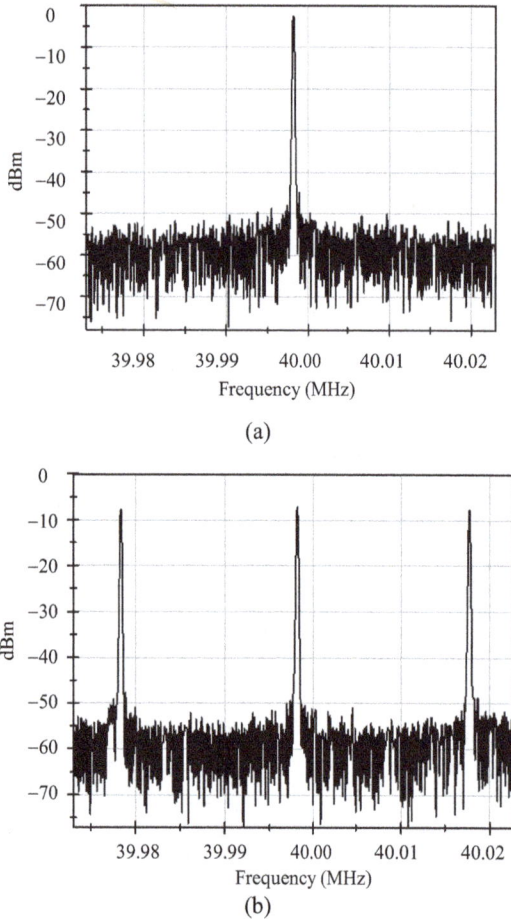

(a)

(b)

Fig. 5 Spectra recovered from a vibrating target: (a) spectrum of the un-modulated carrier and (b) spectrum of the carrier modulated at $J_0 = J_1$ (RBW=100 Hz).

The data shown in Fig. 6 are the typical amplitude noise (in microns) observed at the output

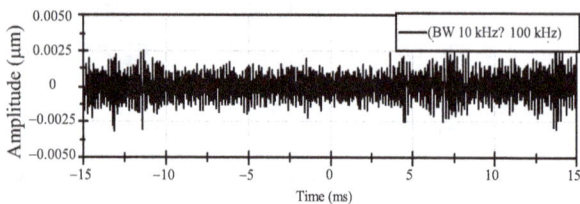

Fig. 6 Typical amplitude noise observed in a bandwidth of 90 kHz when the system is illuminated with a single frequency distributed feedback laser (DFB).

of the digital I/Q demodulator when the target is stationary. The amplitude of the noise measured in a bandwidth of 90 kHz was 0.7 nm (RMS), i.e. 2 pm/√Hz.

5. Multi-channel FLDV system

In this section, we consider optical fiber topologies for multichannel probes which allow simultaneous measurement for all channels [4–6]. The description of the mode of operation is for 4 channels but 8 channels or more are possible dependent on the source power available. Optical amplifiers can be used to increase the number of channels which could be deployed, providing the coherence of the illuminating laser is not affected. In addition, optical switches can be used at other channels as described above.

The fiber topology of the 4-channel system is based on the 4-channel multiplexing unit shown in Fig. 7. The input beam from the single frequency laser diode is initially divided by a 3 dB coupler. One of the outputs is transferred to a 4-channel fiber optic power divider. Each of the 4 emanating fiber guided beams propagates through separate fiber circulators and then via fiber links to adjustable fiber collimators that are focused on 4-individual targets. Light back scattered/reflected from the targets is re-injected back into the collimators and propagates to the inputs of a set of four 3 dB couplers via the third port of each circulator.

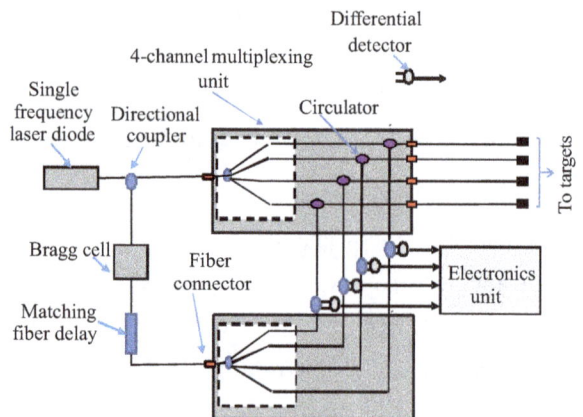

Fig. 7 Scheme of a four channel FLDV.

The other beam ejected from the 3 dB coupler via a delay and Bragg cell is divided into four guided beams by a 1×4 channel power divider. The 4 beams propagate to the 4 input ports of the set of 3 dB couplers where optical interference occurs. The four interferometric signals are detected with (differential) optical detectors and sent to the digital processing unit.

5.1 Experimental setup of vibration tests

In order to demonstrate that data could be recovered simultaneously from four sensors, targets T1 and T2 were placed on a carbon fiber plate driven by a B&K shaker shown in Fig. 8(a). The shaker was driven by a purpose-built high power amplifier. Above the target plate, there were 2 mirror mounts in which fiber collimator probes Ch1 and

Ch2 were mounted. The output beams were well focused on the targets. Figure 8(b) shows the arrangement of Targets T3 and T4. T3 was mounted directly on the shaft of a LING dynamics shaker, and T4 was on the end of a coaxial PZT. As in Ch1 and Ch2, the collimators were mounted in mirror mounts. The targets were chosen to simulate different target locations and vibration levels which could be encountered.

5.2 Experimental results with driven vibration sensors

Experiments were performed with the shakers being driven at different frequencies and amplitudes. Simultaneous measurements of vibration with the 4 sensors were taken in the experimental arrangement described in Section 5.1. The demodulated signals were presented as amplitude graphs in Fig. 9. Although Targets T1 and T2 were mounted on the same carbon plate, they were sufficiently far apart that there was a phase difference. The shaker of T1 and T2 was driven at 2 kHz, T3 was driven at 10 kHz, and the PZT connected to T4 was driven at 60 kHz. An optical power of ~2 mW provided by the DFB laser was enough to illuminate the 4-channel FLDV.

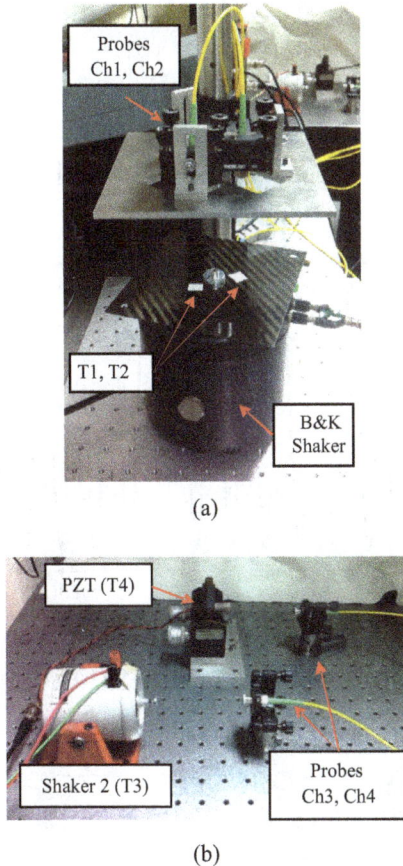

(a)

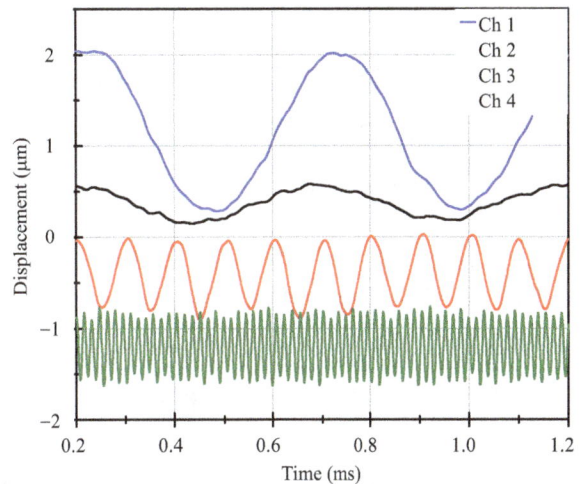

(b)

Fig. 8 Photograph in Fig. 8 shows the probes and targets: (a) Targets T1 and T2 are mounted on a carbon fiber plate with Probes Ch1 and Ch2 mounted vertically above and (b) Targets T3 and T4 are mounted vertically illuminated with horizontal beams from Probes Ch3 and Ch4, respectively.

Fig. 9 Demodulated output signals from the 4 targets (the top 2 traces were generated when the B&K shaker was driven at 2 kHz with amplitudes of 0.8 microns (Ch1) and ~0.25 (Ch2) microns, respectively; Channel 3 (third trace from top) was generated when shaker 3 was driven at 10kHz with amplitude of 0.4 microns and Channel 4 (bottom trace) at 60 kHz with amplitude of ~ 0.3 microns).

5.3 Large distance applications

In the multiplexing schemes described above (Fig. 7), the environments of the reference and signal arms are similar, however in field measurements, this condition cannot be maintained especially if large distances are involved. It is therefore necessary to match the operating conditions for the reference and sensor fibers.

This can be accomplished using the multiplexing scheme in Fig. 7 modified as shown in Fig. 10 such that at the output of the system the OPDs for "sensor reference pairs" are approximately equal (to better than ~10 cm). The reference and sensor probe transceiver leads are coupled together in a sheath to minimize the differential phase shifts induced by environmental effects in the links. The output from the sensing link is transferred to the target through a collimator. The distal face of the reference fiber has a reflective coating to ensure good fringe visibility. Typical results for a target at a 50 m measurement range vibrating at 60 kHz (amplitude of displacement is 70 nm) are shown in Figs. 11 and 12.

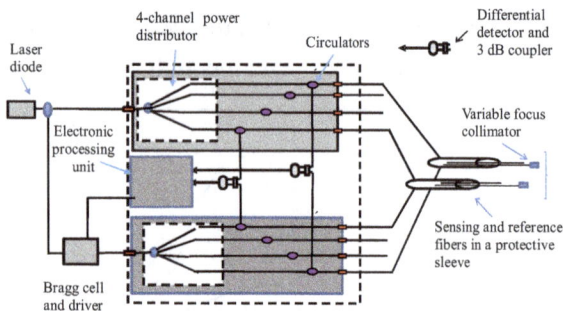

Fig. 10 Modifications of the system shown in Fig. 7 to enable large distance applications.

5.4 Increased number of channels

There are a large number of applications which require measurements to be made at more than one location where it is not essential that all the data are acquired at the same time. One low cost option is to incorporate an electrically controlled fiber bidirectional optic switch into a basic FDLV as shown in Fig. 13. Typically, the power loss for any switched channel is ~1.5 dB, which implies the

resolution and signal strength will only be marginal affected. A realistic number achievable by this approach is 4–8 channels. Although this is an attractive solution, the measurements at each channel are not simultaneous for example a commercial 4-channel switch takes ~2 ms per channel and 8 ms to make 4 sequential measurements at 4 locations. Experiments using the switch on one of the outputs from the four channels demonstrated no loss of performance.

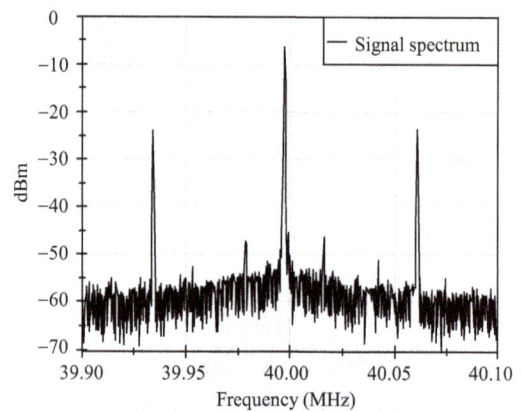

Fig. 11 Spectrum of the carrier modulated with a 60 kHz signal (PZT) measured at a range of 50 m.

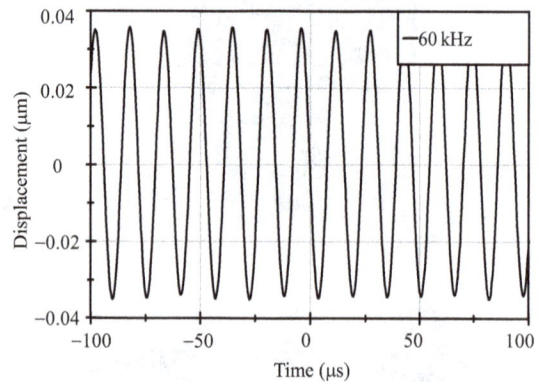

Fig. 12 Digitally demodulated signal for target at 50 m away.

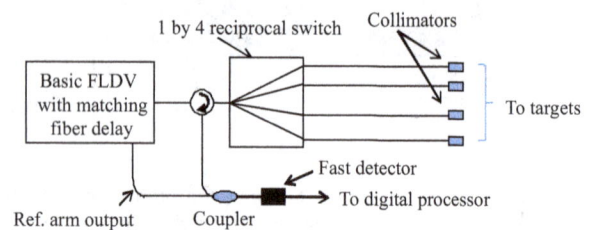

Fig. 13 Implementation of a bidirectional optical switch in the FLDV for increasing the number of channels.

6. Common path topologies

As part of this project, we explored the possibility of using a common path fiber optic topology (CPT) to form a sensor capable of making vibration measurements due to the potential simplicity of the instrumentation and low cost. A basic simple common path fiber sensor can be realized by forming an optical cavity between the distal end of a normally cleaved (transceiver fiber) and a remote target [7]. Figure 14 shows a non-contact common path sensor system. The OPD of the cavity must be less than the coherence length of the source. The input laser beam is transmitted to the target via a circulator and focused on the target and returns to the detector by the circulator. As can be seen in Fig. 14, the transceiver fiber acts as both, the reference and signal fibers for the sensor cavity. Therefore, environmental noise has a minimal impact on the system. This is especially the case for long-range measurements, as mentioned above in Section 5.3, and represents one of the main advantages of common path topology.

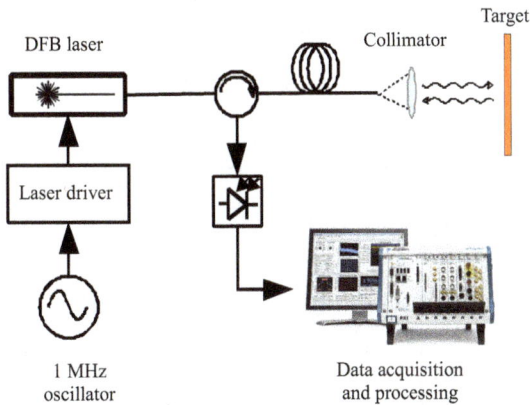

Fig. 14 Basic scheme of non-contact CPT vibrometer.

For the implementation of the CPT LDV, the cavity OPD is set at 4 cm. As in Section 6, the number of probes can be increased to 4 channels by using the multiplexing unit (Fig. 7), again a switch can also be used to increase further the number of probes.

Although the common path sensor has been used for many applications [7–9], to our knowledge there

are no reports in the literature of multiple channel common path fiber vibrometers, possibly due to problems associated with processing the data. Possible approaches for signal recovery include (a) phase generated carrier (PGC) [10], (b) channelled spectrum analysis [11], and (c) PGC with the fringe counter for every $\pm 2\pi$ phase change [2, 12]. In the work reported here, the PGC with the fringe counter approach was used. The complete system is shown in Fig. 14 and configured for testing a CPT vibrometer. A laser diode was used to illuminate the system, which was chirped over one free spectral range of the sensor cavity. This condition can be written [6] as

$$\Delta\phi_L = \frac{2\pi \cdot \Delta v \cdot \Delta L}{C} \quad (3)$$

where $\Delta\phi_L$ is the phase change induced in the cavity, Δv is the absolute frequency shift of the laser, and ΔL is the cavity length. It can be seen from (3) that modulation of the laser optical frequency produces phase changes of the interferometric signal. PGC approach exploits this characteristic in order to generate a carrier through these induced phase changes. Typically, a laser with an extremely low phase noise combined with frequency modulation capabilities is required. By means of PGC, quadrature components of the measurement phase shift can be recovered at an specific modulation depth (when Bessel functions are $J_1=J_2$ or $J_2=J_3$, etc.). This condition of the modulation depth can be established through a closed loop that controls the amplitude of the modulating signal oscillator.

Experiments were performed using a LabVIEW controlled PXI DAQ and processing system. This can be implemented by different approaches: PGC, the fast Fourier transform, and the fringe counting, but in these experiments, the PGC approach was used. The modulation signal used to chirp the DFB laser was a sinusoid of 1 MHz frequency. Figure 15(a) shows the results of a vibration at 1 kHz measured with the CPT LDV, and Fig. 15(b) shows the Lissajous curve of the quadrature components obtained from the PGC signal.

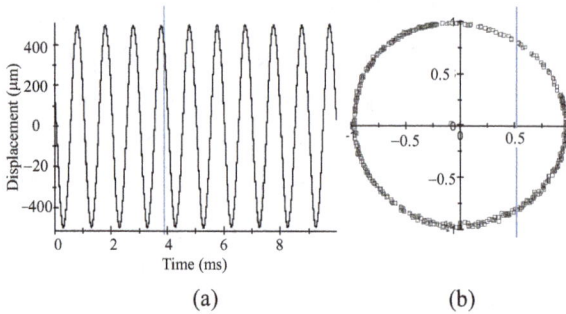

(a)   (b)

Fig. 15 Spectrum recovered with the CPT LDV and its associated Lissajous curve: (a) measurement of a target vibrating at 1 kHz and (b) Lissajous curve of the quadrature components of the signal recovered with digital processing of the PGC.

7. Measurement of rotating targets

FLDV can be used for measuring torsional vibrations, which is of interest in various applications, such as monitoring the instantaneous rotational motion of key components e.g. in car and airplane engines. Torque measurements in shafts can be also done with FLDV. This is a very interesting application for industry, but as it requires multipoint measurements the multichannel FLDV described here is ideal as it can be used to make simultaneous multipoint measurement on the shaft [13]. The multichannel FLDV in Section 5 is presented in this section for the application of rotational speed and torque measurements.

7.1 Rotational speed sensor

The geometry used for the determination of the angular rotation of adisc is shown Fig. 16, indicating the illumination point on the axis of the disc.

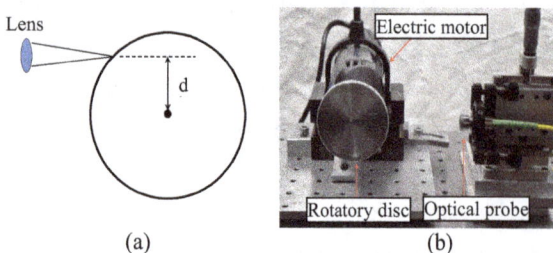

(a)   (b)

Fig. 16 Arrangement for measuring the axial rotation of a disc: (a) schematic diagram indicating the illumination point on the rim and its distance from the axis and (b) photographs of the disc and probe.

The Doppler frequency shift induced by the rotational motion is $2d\omega/\lambda$ [13], where ω is the angular frequency, d is the distance between the axis and center of the disc, and λ is the wavelength of the light. The formula is only correct if the back scattered beams are perpendicular to the axis of the disc. In this work, we only measured the angular velocity using one channel of the system. Normally, the torque was measured using 2 laser beams focused on the rim of the disc or shaft symmetrically either side of a diameter. A setup was prepared for rotational speed measurements using a rim of 6 cm in diameter with its axis connected to a high speed electric motor. The FLDV with the configuration shown in Fig. 7 was used in this experiment for the

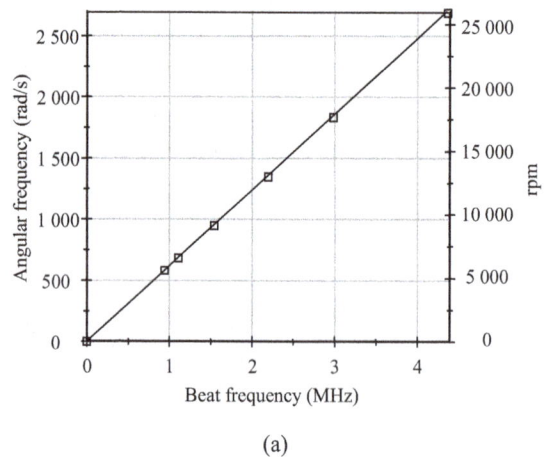

(a)

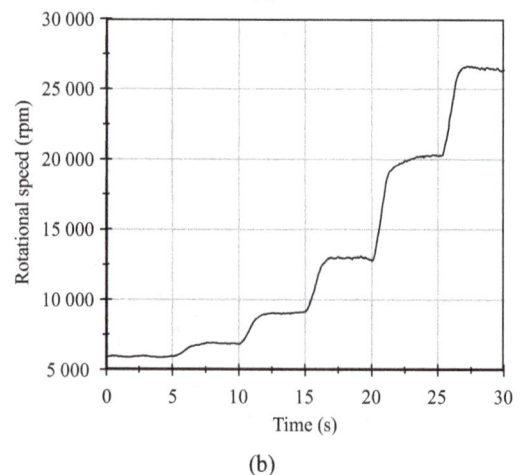

(b)

Fig. 17 Results from rotational speed measurement: (a) angular frequency of the disc as a function of the measured beat frequency and (b) step changes in rotational speed as its driving voltage is slowly stepped in time (measurement range from 6000 rpm to 26000 rpm).

rotational speed measurements. During the experiment, the speed of the electric motor was changed in steps, and the output of the FLDV was monitored. The results of Fig. 17(a) show the angular frequency of the disc as a function of the measured Doppler frequency shift, and Fig. 17(b) shows the step changes in rotational speed as it is slowly stepped in time.

7.2 Proposed optical torque sensor

An optical torque sensor requires that the angular velocity was measured at 2 locations at a specified distance apart. To eliminate non radial movement, 2 probes would be used at each location where the focused beams were set symmetrically about the axis of the shaft. Figure 18 shows how the top 2 probes would be used to make the compensated angular velocity measurement at Location 1. Similarly, the 2 bottom probes would be used to make measurements at the 2nd location. Subtraction of data yields the differential phase and hence the dynamic torque. The optical system shown in Fig. 7 requires no modifications for dynamic torque measurements on shafts.

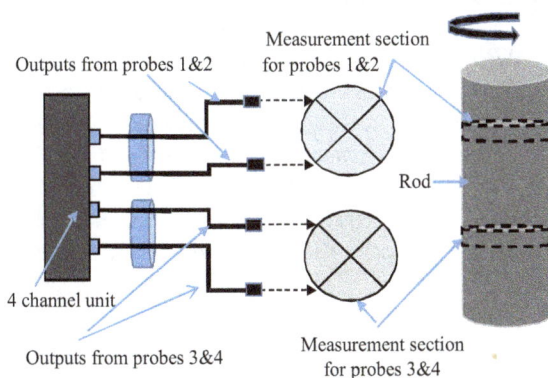

Fig. 18 Proposed torque sensor with the 4-channel power distribution unit.

8. Summary and conclusions

8.1 Multichannel fiber laser Doppler vibrometer

A 4-channel FLDV with a displacement resolution better than 1 nm at frequencies up to 60 kHz, limited by the shakers available, was

presented. The maximum frequency was set by the Bragg cell at ~40 MHz. A laser power of ~2 mW was sufficient to support the 4-channel FLDV; 8 channels – 16 channels could be supported with 4 mw – 8mw. Digital signal processing was used to recover and present the data. Absolute calibration of the FLDV was via the Bessel function ratio $J_0(m)=J_1(m)$, and other ratios could be used. An optical fiber switch was used to expand the number of channels. The design of the system based on the compact optical power transceiver unit makes it portable enabling application outside the benign laboratory environment. The system can also be used for angular velocity and torque measurements.

8.2 Common path vibrometer

An initial study of a common path vibrometer cavity indicated that it could be used to recover the dynamic motion of a target. When a target was subject to a constant amplitude sinusoidal signal, excellent optical fringes were observed allowing a fast Fourier transform (FFT) to recover the signal applied to the shaker; similarly for a multi-probe system. Excellent raw data were obtained for vibration studies in both local measurement and at large distances. A chirped laser was used to implement a PGC approach in the CPT. This enabled the recovery of the measurement using a digital quadrature demodulation.

8.3 Applications and impact tests

Applications for single channel LDV and FLDV for precision measurements were well reported in the literature. Multichannel FDLV can be used in situations where there is limited access and where many locations must be measured for example multi-valve diesel engines. The proposed topologies enhance the utility of FDLV for two major applications currently in progress. One is to study the effects of hypervelocity impacts in a light gas gun (University of Kent) that can project a single particle with velocities in excess of 10 km/s [14].

Targets are typically materials used for satellites or aerospace applications. The 4-channel FLDV could give 4 simultaneous high accuracy vibration signals generating data previously unobtainable. The second study, where common path vibrometers will be used, is related to the damage caused to high speed trains by stones sucked up from the track caused by passage of the trains.

Acknowledgment

Author J. E. Posada acknowledges a short term mobility grant from University Carlos III de Madrid to undertake research at the University of Kent, School of Physical Sciences, Canterbury, Kent.

References

[1] A. C. Lewin, A. D. Kersey, and D. A. Jackson, "Non-contact surface vibration analysis using a monomode fiber optic interferometer incorporating an open air path," *Journal of Physics E Scientific Instruments*, 1985, 18(7): 604–608.

[2] J. E. Posada, J. A. Garcia-Souto, and J. Rubio-Serrano, "Multichannel optical-fiber heterodyne interferometer for ultrasound detection of partial discharges in power transformers," *Measurement Science & Technology*, 2013, 24(9): 94015–94023.

[3] D. A. Jackson, J. E. Posada-Roman, and J. A. Garcia-Souto, "Calibration of laser Doppler vibrometer exploiting Bessel functions of the first kind," *Electronics Letters*, 2015, 51(14): 1100–1102.

[4] C. Yang, M. Guo, H. Liu, K. Yan, Y. Xu, H. Miao, *et al.*, "A multi-point laser Doppler vibrometer with fiber-based configuration.," *Review of Scientific Instruments*, 2013, 84(12): 121702-1–121702-6.

[5] Y. Fu, M. Guo, and P. B. Phua, "Multipoint laser Doppler vibrometry with single detector: principles, implementations, and signal analyses," *Applied Optics*, 2011, 50(10): 1280–1288.

[6] D. A. Jackson, "Monomode optical fiber interferometers for precision measurement," *Journal of Physics E Scientific Instruments*, 1985, 18(12): 981–1001.

[7] M. Schulz and P. Lehmann, "Measurement of distance changes using a fibre-coupled common-path interferometer with mechanical path length modulation," *Measurement Science & Technology*, 2013, 24(24): 48–49.

[8] T. O. H. Charrett, S. W. James, and R. P. Tatam, "Optical fibre laser velocimetry: a review," *Measurement Science & Technology*, 2012, 23(3): 32001–32032.

[9] K. M. Tan, M. Mazilu, T. H. Chow, W. M. Lee, K. Taguchi, B. K. Ng, *et al.*, "In-fiber common-path optical coherence tomography using a conical-tip fiber," *Optics Express*, 2009, 17(4): 2375–2384.

[10] M. J. Connelly, "Digital synthetic-heterodyne interferometric demodulation," *Journal of Optics A Pure & Applied Optics*, 2002, 4(6): S400–S405.

[11] D. A. Jackson, "High temperature Fabry-Perot probe interrogated with tunable fibre ring laser," *Electronics Letters*, 2008, 44(15): 898–899.

[12] J. Bush, A. Cekorich, and C. K. Kirkendall, "Multichannel interferometric demodulator," in *Proc. SPIE*, vol. 3180, pp. 19–29, 1997.

[13] T. Y. Liu, M. Berwick, and D. A. Jackson, "Novel fibre-optic torsional vibrometers," *Review of Scientific Instruments*, 1992, 63(4): 2164–2169.

[14] D. A. Jackson and M. J. Cole, "Fiber optic interrogation systems for hypervelocity and low velocity impact studies," *Photonic Sensors*, 2012, 2(1): 50–59.

27

A Phase Shift Demodulation Technique: Verification and Application in Fluorescence Phase Based Oxygen Sensors

Chuanwu JIA[1], Jun CHANG[1*], Fupeng WANG[1], Hao JIANG[1], Cunguang ZHU[2], and Pengpeng WANG[2]

[1]*School of Information Science and Engineering and Shandong Provincial Key Laboratory of Laser Technology and Application, Shandong University, Jinan, 250100, China*

[2]*School of Physics and Technology, University of Jinan, Jinan, 250022, China*

*Corresponding author: Jun CHANG E-mail: changjun@sdu.edu.cn

Abstract: A phase shift demodulation technique based on subtraction capable of measuring 0.03 phase degree limit between sinusoidal signals is presented in this paper. A self-gain module and a practical subtracter act the kernel parts of the phase shift demodulation system. Electric signals in different phases are used to verify the performance of the system. In addition, a new designed optical source, laser fiber differential source (LFDS), capable of generating mini phase is used to further verify the system reliability. R-square of 0.99997 in electric signals and R-square of 0.99877 in LFDS are achieved, and 0.03 degree measurement limit is realized in experiments. Furthermore, the phase shift demodulation system is applied to the fluorescence phase based oxygen sensors to realize the fundamental function. The experimental results reveal that a good repetition and better than 0.02% oxygen concentration measurement accuracy are realized. In addition, the phase shift demodulation system can be easily integrated to other applications.

Keywords: Phase shift detection system; self-gain, subtracter; LFDS; phase based oxygen sensor

1. Introduction

There is a considerable interest in the rapid and accurate measurement of phase shift for a variety of applications including biological, industrial, and environmental monitoring. Phase shift is a common parameter in the field of industrial measurement and controlling. Meanwhile, it is one of the basic tasks of the engineering signal analysis. In the practical work, we often need to measure the phase shift between two signals with the same frequency to solve variously practical problems. As the phase detection technique applied to scientific research, production practice, and other fields, the requirement of the phase detection technology turns to high precision and high intelligent. The main methods of phase shift measurement contain the variable delay-line method [1], phase-locked method [2], correlation method [3, 4], and the method of over-zero time [5]. In this paper, a system based on the subtraction measuring the phase shift is proposed. However, when most methods require the sophisticated equipment and careful calibration procedures, this phase shift demodulation system based on subtraction just relies on simple and cheap instrumentation and offers high sensitivity and fast response time.

Measurement of phase shift is different from the traditional detection of the voltage signal and current signal. First of all, phase difference is attached to the voltage and current signals. How to eliminate the changes in the voltage, current, and frequency is an important aspect of the phase detection. Secondly, measurement of the phase difference between two

signals not only need to keep the same frequency of two signals, but also need to exclude the impact of other factors such as the amplitude of the two signals. Obviously, making the amplitude of the two input signals same seems very important. Up to now, the common method making the amplitude of two sinusoidal signals same is the self-gain circuit [6]. In order to modulate the amplitude of the signals, a self-gain module is set up.

The self-gain module is shown in Fig. 1. As

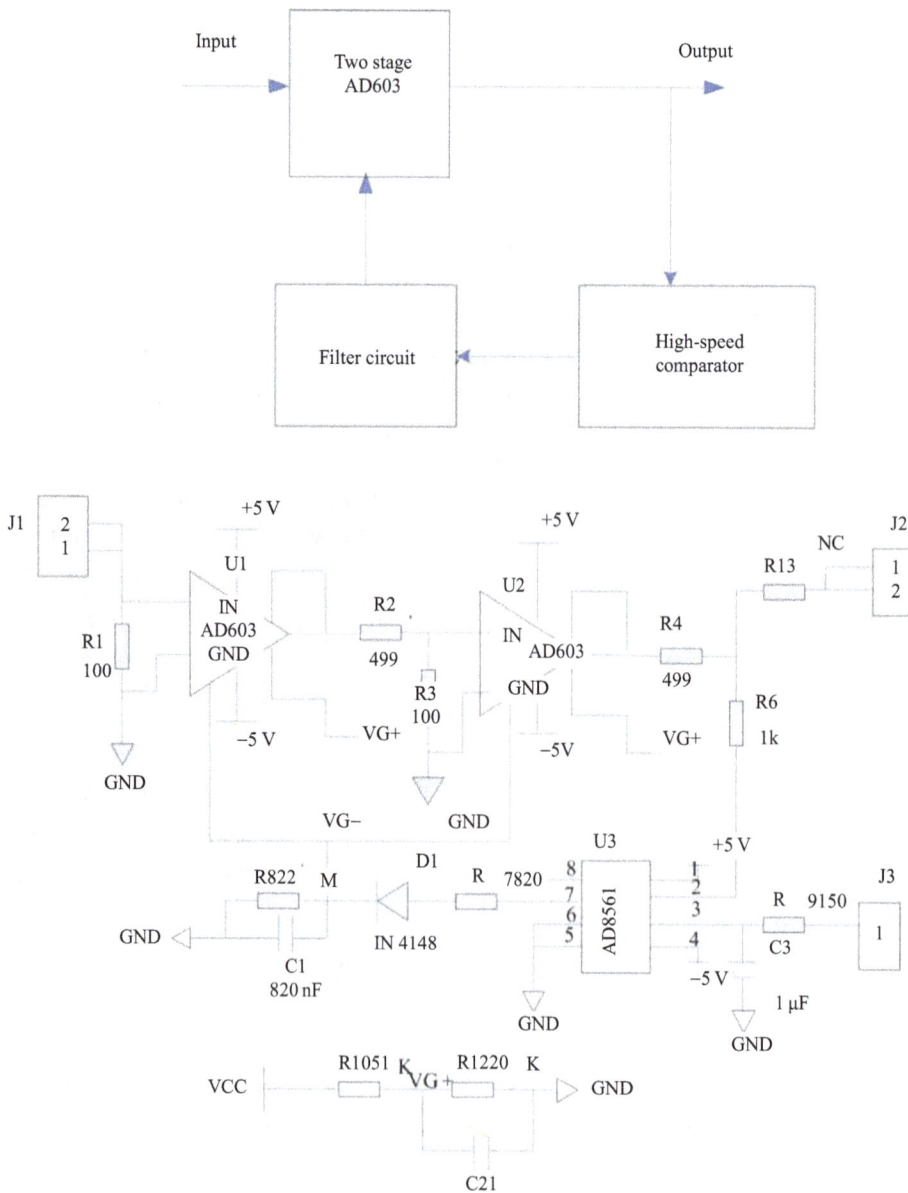

Fig. 1 Diagram of the self-gain module.

shown in Fig. 1(a), the self-gain module mainly contains three parts: the two-stage amplifier part, the high-speed comparator, and the filter circuit. Two commercial chips AD603 (Analog Devices, USA) are involved to build the two-stage amplifier part. In addition, a high-speed comparator constructed by AD8561 (Analog Devices, USA) is used to compare the output signals of the second AD603 with the setting voltage of the comparator. Diodes and RC circuits are used to build the filter circuit and remove the noise ratio of the output signal of the comparator. In Fig. 1(b), all the chips are powered by ±5 voltage. J1 and J2 are the interface of the input and output signals, respectively. Meanwhile, J3 is the interface of the setting voltage and controls the amplitude of the output signal. In this design, a two-stage AD603 amplification is used whose whole range of the gain will be −26 dB – 54 dB. The AD603 relies on the controlling voltage [(VG+)−(VG−)] to control the magnification. The range of the filter value VG− is 0 – 3.3 V. VG+ can be expressed as (1):

$$VG+ = V_{CC}\frac{R_{12}}{R_{10}+R_{12}} = 1.4V. \qquad (1)$$

So the range of the controlling voltage [(VG+)−(VG−))] is −1.9 V – 1.4 V, and it contains the linear range of AD603 (−500 mV – 500 mV). As a result, to make sure the value of the regulation voltage is in the range of the controlling voltage of AD603 (−500 mV – 500 mV) and make the AD603 in normal amplification, we set the VG+ to be 1.4 V. R9 and C3 is a first-order low-pass filter, the cutoff frequency (f_c) is

$$f_c = \frac{1}{2\pi R_9 C_3} = 719 Hz.$$

The negative feedback system provides the self-gain module with the ability of self-adjusting. Experimental results have shown that the fluctuation of the output voltage is less than 0.1 dB at the range of 1 kHz – 10 MHz.

2. Theory analysis

It is easy to understand that a sinusoidal signal minus another sinusoidal signal in the same frequency produces a new sinusoidal signal. The amplitude of the new sinusoidal signal is decided by the phase shift and amplitudes of the former sinusoidal signals as shown in Fig. 2.

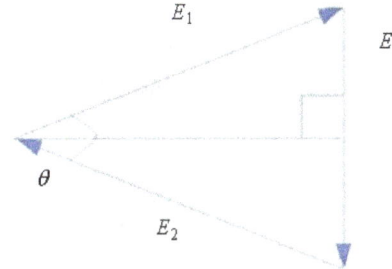

Fig. 2 Subtraction of two sinusoidal signals.

In this method, we assume that the amplitudes of the input sinusoidal signals are the same. It can be realized by a self-gain module as expressed in (2):

$$E_1 = E_2 \qquad (2)$$

where E_1 and E_2 are the amplitudes of the input sinusoidal signals. In the isosceles triangle as shown in Fig. 2, the output sinusoidal signal after subtraction is expressed by vector **E**. The phase shift between input sinusoidal signals θ can be expressed as (3):

$$\sin\frac{\theta}{2} = \frac{E}{2E_1} = \frac{E}{2E_2} \qquad (3)$$

where E is the amplitude of the output sinusoidal signal. Thus, the phase shift can be demodulated by

$$\theta = 2\arcsin\left(\frac{E}{2E_1}\right). \qquad (4)$$

As shown in (4), the phase shift of two sinusoidal signals can be obtained by measuring the amplitude of the output sinusoidal signal after subtraction. The phase shift θ is proportional to the amplitude of the output sinusoidal signal. Referencing the theoretical analysis, we can learn that when the phase shift is less than 0.1 degree, the amplitude of the output signal will be so small to be submerged in the noise. As a result, we should amplify the output signal and eliminate the noises to improve the measurement accuracy.

3. Phase shift demodulation system

The phase shift demodulation system is conducted as shown in Fig. 3. Signals 1 and 2 are the input sinusoidal signals under measurement for the phase shift. A self-gain module is used to tune the amplitudes of the input sinusoidal signals into the same after which E_1 and E_2 discussed in Section 2

are achieved. A commercial chip AD8221 (Analog Devices, USA) is involved to build the subtraction circuit. In addition, a band-pass filter circuit constructed by UAF42 (Texas Instruments, USA) is used to improve the signal to noise ratio (SNR) of the output sinusoidal signal. Then, a good sinusoidal signal with high SNR from the subtracter is sent to a lock-in amplifier for the accurate measurement of amplitude E. The locked amplitude E is collected by PC for some further process like (4), and then the phase shift θ between the input sinusoidal signals is demodulated.

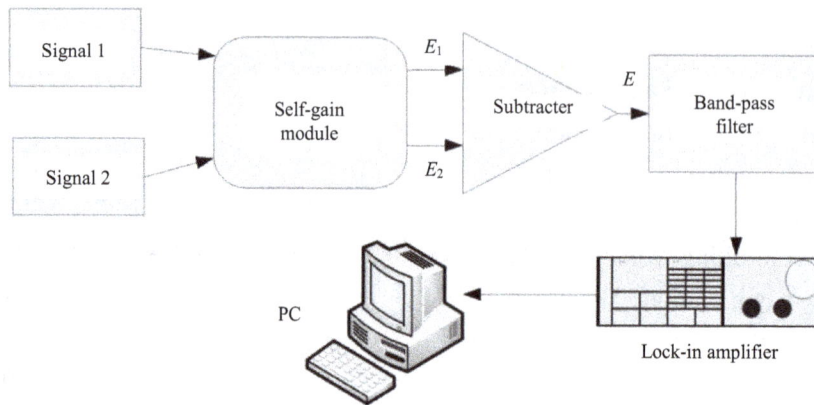

Fig. 3 Phase detection system.

4. Functional verification

4.1 The designed LFDS

In order to verify the performance of the phase shift demodulation system, it is easy for us to choose a signal generator to give the electric sinusoidal

signals. But most of the signal generators are always incapable of providing signals in mini phase differential. As shown in Fig. 4, an optical source called laser fiber differential source (LFDS) is designed to give the infinitesimal phase angle in theory, whose basic principle is the light propagation in fiber in different lengths.

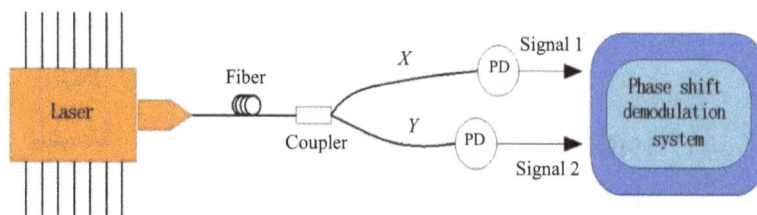

Fig. 4 Schematic of the LFDS.

In the LFDS, a distributed feedback laser diode driven by a 16 kHz sinusoidal signal is used as the light source in the experiment. A 1:1 coupler is used

to split the laser beam into two fibers in length of X meter and Y meter. The respective light is detected by the photo-detector, and the output signals can be expressed by Signals 1 and 2. Thus, a phase shift is generated between the output signals because of the optical path differences. The phase shift θ can be expressed as follows:

$$\begin{cases} \Delta l = X - Y \\ \varphi = \dfrac{\Delta l}{v} \\ \theta = 2\pi f \varphi = 2\pi f \dfrac{X-Y}{v} \end{cases} \quad (5)$$

where X and Y are the lengths of different fibers, v is the light speed propagating in the fiber, and f is the frequency of the laser driving signal which is set as 16 kHz in the experiment. Obviously, the phase shift θ is related to the optical path differences X–Y and laser modulation frequency f. In the experiment, it is simple but effective for us to get different phase shifts by changing the fiber length X or Y. In theory, one meter of the optical path difference corresponds to 0.03 phase degree approximately. To verify the reliability of the LFDS in the generation of the phase shift, a lock-in amplifier Model 7230 (Ametek advanced measurement technology, USA) is used to

calibrate the results of the LFDS as shown in Table 1.

Table 1 Calibration of the LFDS by Model 7230.

Δl (m)	θ (theory)	θ (Time 1)	θ (Time 2)	θ (Time 3)
1.04	0.03	0.03	0.03	0.03
2.08	0.06	0.06	0.06	0.07
3.12	0.09	0.09	0.10	0.09
4.16	0.12	0.12	0.12	0.12

The results in Table 1 reveal that the LFDS is reliable in mini phase shift production after multiple verifications.

4.2 Functional verification of the phase shift demodulation system

The mini phase shift can be generated by the LFDS as discussed in Section 4.1. Electric signals in a wide angle from 1 degree to 16 degree are produced by a signal generator. Our designed phase shift demodulation system is used to measure these signals in different phase angles from either the LFDS or the signal generator. The experimental results are plotted in Fig. 5.

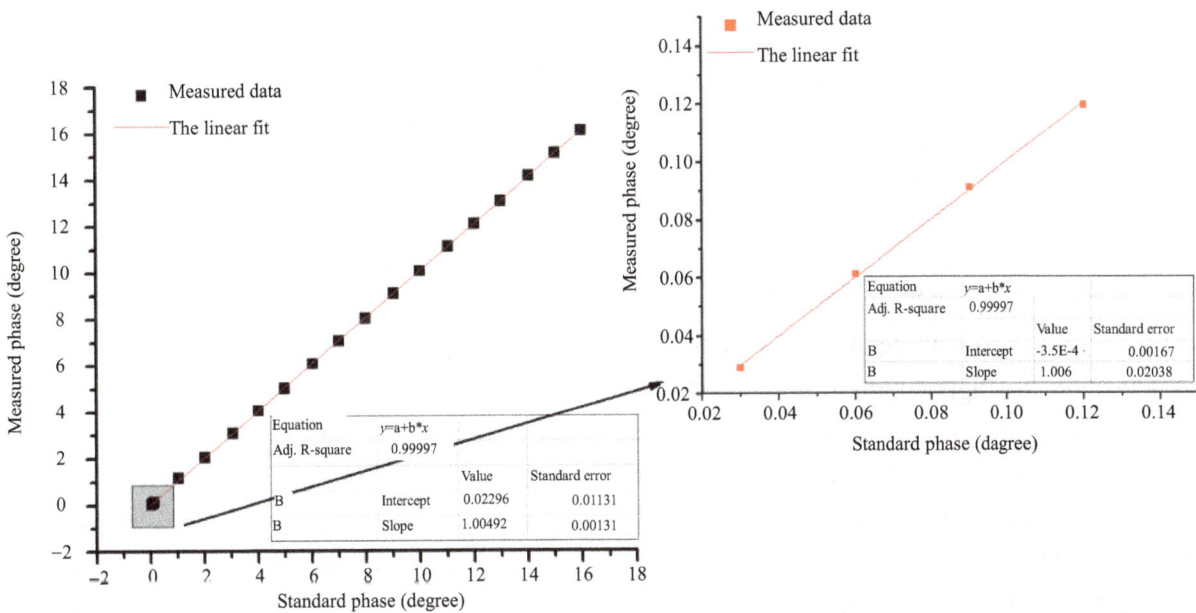

Fig. 5 Measured phases by the phase shift demodulation system.

A linear fit is applied to the measured data from   0.03 degree to 16 degree, and an R-square of 0.99997 is achieved as shown in Fig. 5. In addition, we plot the phase data 0.03 degree, 0.06 degree, 0.09 degree, and 0.12 degree from the LFDS separately in the right inner figure which applies a linear fit too. The R-square is 0.99877. So, we can draw a conclusion that our self-designed phase shift demodulation system is reliable and the new phase shift demodulation technique is feasible.

5. Application in fluorescence phase based oxygen sensor

The purpose of designing the phase shift demodulation system is for some practical applications. Simultaneously, it is also necessary to verify it in the specific application. Oxygen sensing is of major importance in many fields, and phase shift measurement is always the crucial step in the fluorescence phase based oxygen sensor. In the oxygen concentration detection, the fluorescence

intensity I and lifetime τ of the sensing dye are described by the Stern-Volmer equation [7−9]:

$$\frac{I_0}{I} = \frac{\tau_0}{\tau} = 1 + K_{SV}[O_2] \qquad (6)$$

where I_0 and τ_0 are the unquenched luminescence intensity and lifetime, respectively, K_{SV} is the Stern-Volmer constant, and O_2 is the concentration of oxygen. Meanwhile, the phase difference between the exciting light signal and the fluorescence signal can be described by (7) if we apply a sinusoidal modulation signal to excitation source [10]:

$$\tan \theta = 2\pi f \tau \qquad (7)$$

where θ is the phase shift between the exciting light and the fluorescence, f is the frequency of the sinusoidal modulation, and τ is the lifetime of the fluorescence signal. Obviously, the phase shift demodulation system can be used to measure the fluorescence phase θ in the oxygen sensor as shown in Fig. 6.

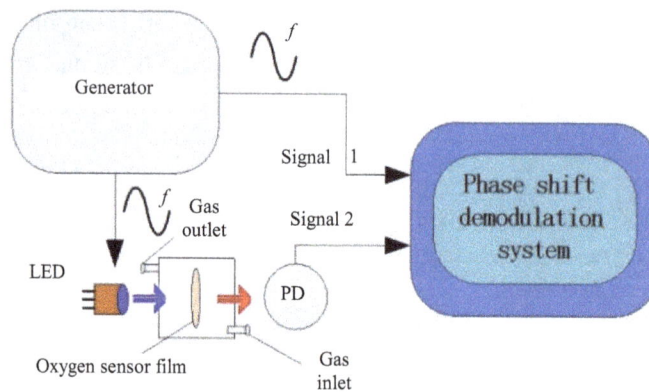

Fig. 6 Fluorescence-based optical oxygen sensor.

A sinusoidal signal with the frequency of 16 kHz is used to drive the light-emitting diode (LED) to excite the oxygen sensor film, and another sinusoidal signal in the same frequency is generated as Signal 1. The fluorescence signal is transformed by the photo-detector into Signal 2 whose phase shift θ is proportional to the concentration of oxygen. Sample gas in different oxygen concentrations is accessed for measurement. The fluorescence phases

in the concentration of 0.99%, 3%, 5%, and 8% are measured by our phase shift demodulation system, and the results are plotted in Fig. 7. It is clear that the measured data display the good SNR and repeatability. A phase detection accuracy of 0.03 degree is achieved in 0.99% oxygen, and the oxygen concentration detection accuracy better than 0.02% is realized by virtue of our phase shift demodulation system.

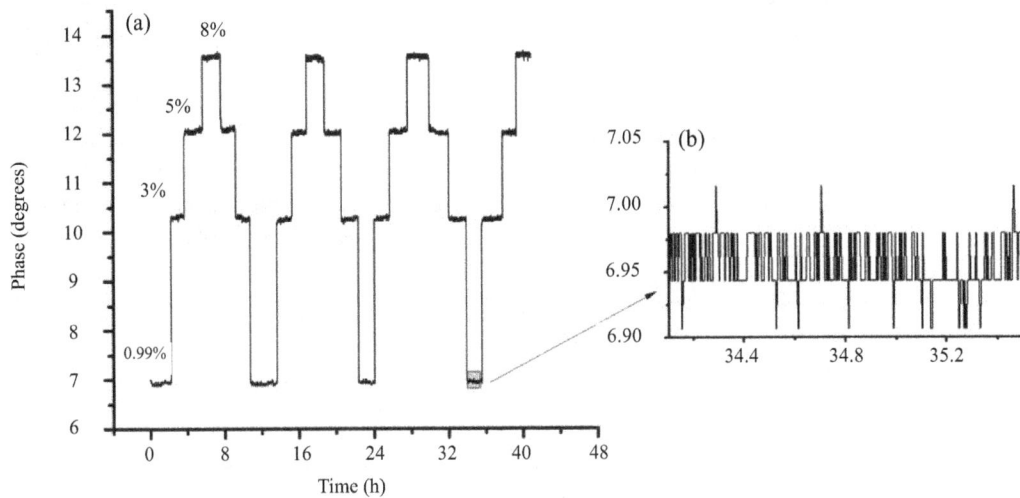

Fig. 7 Oxygen measurement in different sample gases with the concentration of 0.99%, 3%, 5%, and 8% oxygen by our phase shift demodulation system.

6. Conclusions

A new phase shift demodulation system based on the self-gain module and subtraction is presented in this paper. Electric signals in the wide angle from 1 degree to 16 degree produced by the signal generator and mini phase shift generated by the LFDS are specially chosen to verify the function of the system. The experimental results show that R-square of 0.99997 in the electric signals and R-square of 0.99877 in the LFDS are achieved, and 0.03 degree measurement limit is realized in the experiments. Furthermore, this phase shift demodulation system can be successfully applied in the fluorescence phase based oxygen sensors by virtue of the high stability and precision. Experimental results have demonstrated that a good repetition and better than 0.02% oxygen concentration measurement accuracy are realized. In conclusion, this phase shift demodulation system is reliable, practical, and stable. In addition, this phase detection system could be utilized in high-accuracy, high-precision portable measurement systems for sensing applications.

Acknowledgement

This work was supported by the National Natural Science Foundation of China (61475085), Science and Technology Development Project of Shandong Province (2014GGX101007), and the Fundamental Research Funds of Shandong University (2014YQ011).

References

[1] H. Gheidi and A. Banai, "Phase-noise measurement of microwave oscillators using phase – shifterless delay-line discriminator," *IEEE Transactions on Microwave Theory and Techniques*, 2010, 58(2): 468–477.

[2] W. Zhu, S. Zhu, and S. Ye, "Phase locked positioning method in heterodyne interferometer system," *Opto-Electronic Engineering*, 2014, 41(5): 35–39.

[3] G. Peng, M. Tang, H. Zou, C. Tian, and X. Chen, "Determination of TDC using digital correlation method for linear compressor," *Journal of the Institute of Brewing*, 2003, 694–697(5): 1608–1614.

[4] F. Chu, H. Cai, R. Q. Rong-Hui, Z. Fang, and J. Yang, "Study on optical fiber dissolved oxygen sensor based on fluorescence quenching," *Journal of Optoelectronics Laser*, 2009, 20(8): 1070–1072.

[5] J. Jia and M. Wang, "An optimisation method for the over-zero switching scheme," *Flow Measurement and Instrumentation*, 2012, 27(10): 47–52.

[6] X. Liu, X. Xu, and X. Tan, "The design of automatic gain control in self-excitation drive circuit of gyroscope," *Chinese Journal of Sensors and Actuators*, 2007, 26(11): 1662–1666.

[7] A. Campbell and D. G. Uttamchandani, "Optical dissolved oxygen lifetime sensor based on sol-gel immobilisation," *IEE Proceedings Science Measurement and Technology Technology*, 2004, 151(4): 291−297.

[8] X. Li, X. Lu, X. Luo, and G. Liu, "Development of the fluorescence quenching dissolved oxygen sensor," *Automation & Instrumentation*, 2013, 28(4): 17−20.

[9] L. Pirrami, J. Wicht, F. Debrot, A. Rosspeintner, E. Vauthey, J. N. Aebischer, *et al.*, "Oxygen sensor for indirect calorimetry bsaed on ruthenium fluorescence quenching," in *IEEE 2011 International Semiconductor Conference*, Sinaia, pp. 123−126, 2011.

[10] S. M. Borisov, G. Neuauter, C. Schroeder, I. Klimant, and O. S. Wolfbeis, "Modified dual lifetime referencing method for simultaneous optical determination and sensing of two analytes," *Applied Spectroscopy*, 2006, 60(10):1167−1173.

28

Liquid Level Sensor Based on CMFTIR Effect in Polymer Optical Fiber

Yulong HOU, Wenyi LIU[*], Huixin ZHANG, Shan SU, Jia LIU,
Yanjun ZHANG, Jun LIU, and Jijun XIONG

Key Laboratory of Instrumentation Science & Dynamic Measurement (North University of China), Ministry of Education, Taiyuan, 030051, China

[*]Corresponding author: Wenyi LIU   E-mail: liu_wenyi418@126.com

Abstract: The macro-bending induced optical fiber cladding modes frustrated total internal reflection effect is used to realize the liquid level probe with a simple structure of single macro-bend polymer optical fiber loop. The test results show that the extinction ratio reaches 1.06 dB. "First bath" phenomenon is not obvious (about 0.8%). The robustness of the sensor is better, and the ability of anti-pollution is stronger compared with the conventional sensors. The process of making this sensing probe is extremely easy, and the cost is very low.

Keywords: Water leakage measurement; polymer optical fiber; frustrated total internal reflection

1. Introduction

The techniques of liquid level measurement are widely required in the fields of oil and chemical engineering. At present, there are varieties of designs including the simple dipstick and kinds of liquid level sensors with complex structures such as capacitive type [1, 2], acoustic wave [3–5], and optical fiber based on different principles. But there are still prominent contradictions between the existing level measurement technologies and the market demands. Firstly, the demand for the level sensor is huge in the petrochemical field, but expensive and complex designs are difficult to be widely applied. Secondly, the complex measurement environments, such as flammable liquids, present a huge challenge on the safety and adaptability of existing level sensors. Due to non-electrical

measurements and immunity to electromagnetic interference, optical fiber liquid level sensors are suited to liquid level measurement in complex environments. Among them, pressure-sensitive sensors and refractive index-sensitive sensors are representative. With regard to pressure-sensitive types, there are fiber Bragg grating sensors [6–8], Fabry-Perot sensors [9–11], and so on. The production process of these sensors is difficult and expensive. And complex and expensive instruments, such as high-resolution spectrometer and tunable laser source, are also required for detecting the wavelength shift. So these sensors can hardly be applied and popularized in cost-sensitive areas. The refractive index-sensitive sensors mainly include long-period fiber grating (LPFG) sensors [12–14] and fiber frustrated total internal reflection (FTIR) sensors. The LPFG sensors are similar to the pressure-sensitive types in complexity and cost. The FTIR sensors just need to detect the light power changes, so the cost is quite low. And this advantage makes it possible that FTIR sensors play an important role in petroleum and chemical fields.

Currently, there are mainly two ways to produce FTIR sensors with fibers. Bottacini *et al.* [15] polished fiber end to form a 90° angle as the reflective surface on which the FTIR effect occurred. Using this method, 1.09-dB extinction ratio was achieved. But this method is difficult to process and make the fiber tip more slender, fragile, and susceptible to contamination. Golnabi *et al.* [16] using a prism as the reflective surface achieved an extinction ratio of 0.03 dB. Due to the sensor probe is large, the liquid level measurement accuracy of the sensor is affected. And the extinction ratio is relatively low, so erroneous judgment easily appears at noisy environments. Hou *et al.* [17] using plastic optical fibers with the twisted macro-bend coupling structure inspired cladding modes which had cladding modes frustrated total internal reflection (CMFTIR) effect and realized a liquid level probe with an extinction ratio of 4.18 dB. On one hand, because the coupling ratio is low and the sensing signal of sensor is weak, the signal will attenuate continuously in the long-distance propagation and hardly be detected in this situation. On the other hand, the gaps in the twisted structure result in more remaining liquids and evidently generate the "first bath" phenomenon (nearly 30% output difference).

In this paper, the CMFTIR effect is used to realize the liquid level probe with an extremely simple structure of single macro-bend polymer optical fiber loop (Fig. 1). By using this method, the process of making sensing probe is very easy, and the cost is very low. But the sensor may exhibit similar or even better performance than the existing sensors. The test results show that the extinction ratio reaches 1.06 dB. The robustness is better, and the ability of anti-pollution is stronger compared

with the conventional sensors. The sensing signal power is four orders higher than that of twisted macro-bend coupling structure (TMBCS), and the "first bath" phenomenon is not obvious (about 0.8%). It is foreseeable that the level probe has excellent competitive advantages in the future market.

2. Experiment and results

It can be seen in Fig. 1 that a commercial polymer optical fiber (POF) fiber (SK-40, Mitsubishi) is bended to form a small loop with the 2.5-mm macro-bend radius and is encapsulated with the silica gel in a rubber shield which is low-cost. Except for the bended head of fiber, all the naked parts are set into a black thermal casing to shield visible light interference. One end of the fiber is connected to a 660-nm light-emitting diode (LED) light source (Thorlabs, M660F1), while the other end connects the optical power meter (Thorlabs, PM100USB).

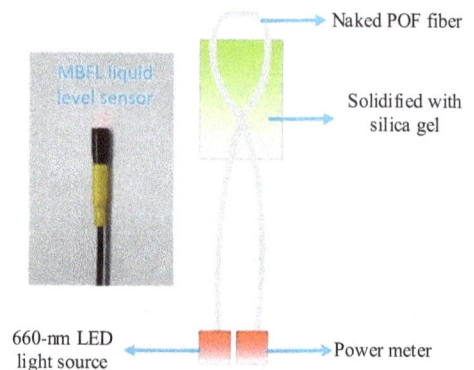

Fig. 1 Macro-bend optical fiber loop (MBFL) liquid level sensor.

As shown in Fig. 2, when the fiber is bent, lights originally propagating in the fiber core become leaky rays which lose their power through the mechanism of refraction and tunneling [18]. In the model of single-mode optical fiber, the thickness of cladding is usually seen as infinite to simplify calculation. Then only one reflection on the core-cladding interface needs to be considered. It means that the escaping lights at the core-cladding interface will lose forever [19]. This approximation is invalid when the bending radius is small enough especially in highly multimode SK-40 POF fibers, because more power is transferred to the cladding and forms cladding modes which cannot be neglected. The escaping lights at the core-cladding interface actually have the chance to reflect again at the cladding-environment interface and come back to the core. Similarly with the situation of the first reflection, at the cladding-environment interface, the lights dissatisfying the condition of total internal reflection become refraction rays, which will exhaust all their power in short-distance propagation. This part of lights is radiation mode. And the lights satisfying the condition of total internal reflection become tunneling rays, which form the cladding modes with slow leakage through the mechanism of tunneling. The difference between the first tunneling rays on core-cladding interface and the second tunneling rays on cladding-environment interface is that the evanescent fields of the second tunneling rays penetrate into the environment medium, and the loss of lights is influenced by the refractive index of the medium through the mechanism of the FTIR [20–23].

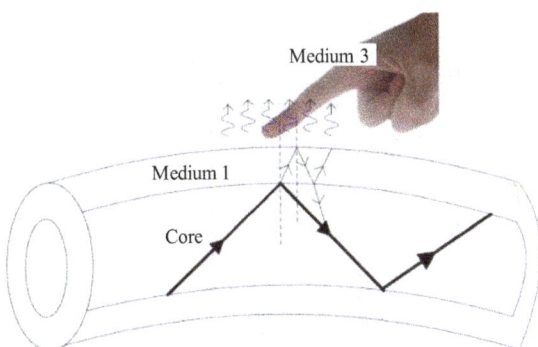

Fig. 2 Cladding mode frustrated total internal reflection in bending fiber.

Therefore, when the high refractive index medium (Medium 3) contacts with or is close to the fiber cladding (Medium 1), the FTIR effects cause partial cladding mode lights to transmit into the environment and lead to energy loss. In particular, we call this effect in fibers as CMFTIR. Specifically, when the MBFL level probe is immersed in water or touched by a finger (Fig. 2), the light loss is caused by the CMFTIR effect, and the power decline can be detected at the end of the fiber.

Under normal circumstances, the CMFTIR effect is difficult to be observed in fibers. Because the light energy in straight fibers is well confined within the core, the energy which can reach to the cladding-environment interface is almost negligible. Moreover, the quartz fiber cladding is also relatively thick and has external coating, so it prevents the energy field from contacting with the external environment. So we designedly choose POF fiber (SK-40, Mitsubishi) which has relatively thin cladding and no coating layer as the sensor material. The particular parameter of SK-40 fiber is shown in Table 1.

Table 1 Parameters of SK-40 fiber.

Core material	Core refractive index	Core diameter (μm)	Cladding diameter (μm)
Polymethyl-methacrylate-resin	1.49	980	1000

Another more important reason is that the POF fibers are flexible and can be bent to a small loop with no breakage. In the previous report [17], we used the TMBCS to acquire the dark-filed coupling signal and achieved the enhancement of the CMFTIR effect. In this paper, we find that as long as the macro-bending radius decreases to a certain extent, the CMFTIR effect can also be enhanced to a satisfactory degree.

To test the impact of bending radius of optical fiber on the enhancement of the CMFTIR effect, SK-40 optical fiber is used to produce fiber rings with different macro-bending radii. The LED light source is connected to one end of the fiber, and a power meter is connected to another end to measure the power change. The method of finger touch (Fig. 2) is used to simulate the effect of frustrated total internal reflection. The result is shown in Fig. 3.

As seen in Fig. 3, when the macro-bending radius is greater than 13 mm, there are almost no significant power changes produced by finger touch. So the CMFTIR effect is negligible under this situation. But with a gradual decrease in the bending radius, the CMFTIR effect becomes more obvious. When the bending radius reaches 2.5 mm, the power change caused by the CMFTIR effect is more than 50%. It is sufficient for the sensing applications.

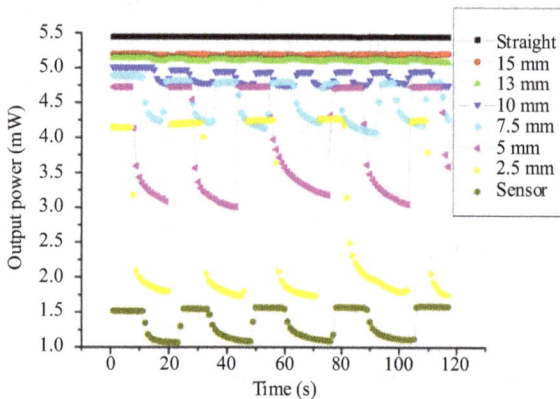

Fig. 3 Enhancement of the CMFTIR effect with different macro-bend radii.

In order to ensure the stability of the measurement results, the MBFL must be properly packaged. The MBFL is encapsulated in a structure shown in Fig. 1 and sealed with the silica gel. Flooding test of the packaged MBFL sensor is repeated 8 times, and the

output power changes of sensor probe are observed. The measurement results are shown in Fig. 4.

We can see from Fig. 4 that when the probe is in the air (completely dry), the output power is about 1.51 mW (Line 1 in Fig. 4). When the level probe is completely immersed in water, the energy loss is generated due to the CMFTIR effect, and the output power declines down to 1.175 mW nearby (Line 3 in Fig. 4).

Usually, the extinction ratio is used to estimate the performance of liquid level sensors [15], and the extinction ratio is defined as follows:

$$E_r = -10\lg(P_{liquid} / P_{air}) \qquad (1)$$

where P_{liquid} is the sensing signal output power when the sensor probe is immersed in liquid, and P_{air} is the output power in the air. It can be calculated that the extinction ratio of MBFL level probe reaches 1.06 dB. Due to the influence of the silica gel used for encapsulation, the output power of the sensor and extinction ratio decrease significantly compared with the values shown in Fig. 3 when the bending radius is 2.5 mm. So the silica gel is not an ideal choice, and there are actually spaces for further improvement, but the silica gel is cheap and already can fulfill the demands in many situations.

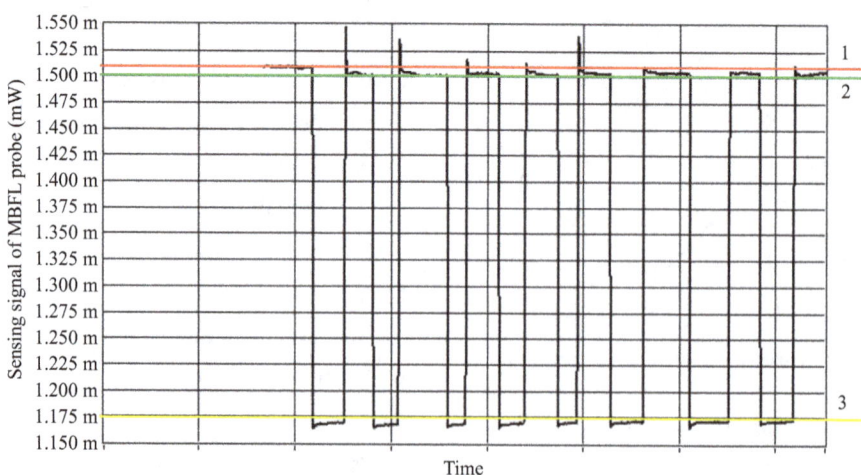

Fig. 4 Flooding test of the MBFL liquid level sensor probe.

From Fig. 4, we can also see that when the level probe is out of water and wet, the output power is about 1.50 mW (Line 2 in Fig. 4). There are some differences between the output powers of the wet

state (Line 2 in Fig. 4) and dry state (Line 1 in Fig. 4). This phenomenon which is called "first bath" is due to the remaining liquids on the sensor surface after the sensor coming out of the liquids. In the previous study [17], since gaps existed between the two twisted fibers of the TMBCS structure, water droplets easily remained in the cracks, resulting in more significant "first bath" phenomenon. The output power difference of TMBCS between the dry and wet states reached 30%. In Fig. 4, it can be seen that the difference caused by the "first bath" phenomenon is about 0.8%, far less than the difference of the TMBCS probe. Compared with the MBFL level sensor, the signal power of the TMBCS probe is relatively weak (hundreds of nW). So the signal is easily to be affected by environment lights, and the signal transmission distance limited by the attenuation in optical fiber is short. It is important that the MBFL probe has higher signal output power (several mW) when the signals need to be transmitted to a long distance.

3. Conclusions

Optical fiber liquid level sensors can be divided into two classes [24]: the sensors for punctual measurement and for continuous measurement. The continuous type can accomplish real-time measurement of the continuously changing liquid level [25]. In this situation, the different liquid levels will change the submerged lengths and the outputs of the sensors, and then the level values can be calculated. The sensors for the punctual measurement are usually arranged in different places with many basic sensing probes [26]. They can also be controlled by a stepping motor system to move up and down with only one probe [27]. The sensing probe designed in this paper is used for the punctual measurement, and many probes are needed to be fixed on different positions inside a liquid reservoir or just one probe is controlled by a stepping motor system for acquiring level values. So as a basic unit, the sensing probe is only used for estimating the existence of liquids.

In summary, by utilizing the MBFL structure, we have a smaller sensor size, smaller "first bath" error, and the not bad extinction ratio (1.06 dB), and we

get farther signal transmission distance and simpler production process. These make the MBFL level sensor probe have wider application scenarios and a more competitive market prospect.

Acknowledgment

This work was supported by the Major State Basic Research Development Program of China (Grant No. 2012CB723404) and the National Natural Science Foundation of China (No. 51275491 and No. 61275166).

References

[1] H. Canbolat, "A novel level measurement technique using three capacitive sensors for liquids," *IEEE Transactions on Instrumentation and Measure*, 2009, 58(10): 3762–3768.
[2] B. Kumar, G. Rajita, and N. Mandal, "A review on capacitive-type sensor for measurement of height of liquid level," *Measurement and Control*, 2014, 47(7): 219–223.
[3] V. E. Sakharov, S. A. Kuznetsov, B. D. Zaitsev, I. E. Kuznetsova, and S. G. Joshi, "Liquid level sensor using ultrasonic lamb waves," *Ultrasonics*, 2003, 41(4): 319–322.
[4] F. Lucklum and B. Jakoby, "Non-contact liquid level measurement with electromagnetic-acoustic resonator sensors," *Measurement Science and Technology*, 2009, 20(12): 314–317.
[5] D. Royer, L. Levin, and O. Legras, "Liquid level sensor using the absorption of guided acoustic waves," *IEEE Transactions on Ultrasonics Ferroelectrics and Frequency Control*, 1993, 40(4): 418–421.
[6] D. Li, Y. Gong, and Y. Wu, "Tilted fiber Bragg grating in graded-index multimode fiber and its sensing characteristics," *Photonic Sensors*, 2013, 3(2): 112–117.
[7] A. F. Obaton, G. Laffont, C. Wang, A. Allard, and P. Ferdinand, "Fiber Bragg gratings and phase sensitive-optical low coherence interferometry for refractometry and liquid level sensing," *Sensors and Actuators A Physical*, 2013, 189(2): 451–458.
[8] C. W. Lai, Y. L. Lo, W. F. Liu, and J. P. Yur, "Simultaneous measurement of liquid level and specific gravity based on fiber Bragg grating sensors," *Proc. SPIE*, 2009, 7503: 75033T.
[9] C. Yu, L. Liu, X. Chen, Q. Liu, and Y. Gong, "Fiber-optic Fabry-Perot hydrogen sensor coated with Pd-Y film," *Photonic Sensors*, 2015, 5(2): 142–145.

[10] T. Lü and S. Yang, "Extrinsic Fabry-Perot cavity optical fiber liquid-level sensor," *Applied Optics*, 2007, 46(18): 3682–3687.

[11] W. Wang and F. Li. "Large-range liquid level sensor based on an optical fiber extrinsic Fabry-Perot interferometer," *Optics and Lasers in Engineering*, 2014, 52(1): 201–205.

[12] J. N. Wang and C. Y. Luo, "Long-period fiber grating sensors for the measurement of liquid level and fluid-flow velocity," *Sensors*, 2012, 12(4): 4578–4593.

[13] A. A. Kazemi, C. Yang, and S. Chen, "Fiber optic cryogenic liquid level detection system for space applications," *Proc. SPIE*, 2009, 7314: 73140A.

[14] Y. Huang, B. Chen, G. Chen, H. Xiao, and S. U Khan, "Simultaneous detection of liquid level and refractive index with a long-period fiber grating based sensor device," *Measurement Science and Technology*, 2013, 24(9): 95303–95312.

[15] M. Bottacini, N. Burani, M. Foroni, F. Poli, and S. Selleri, "All-plastic optical-fiber level sensor," *Microwave and Optical Technology Letters*, 2005, 46(6): 520–522.

[16] H. Golnabi, "Design and operation of a fiber optic sensor for liquid level detection," *Optics and Lasers in Engineering*, 2004, 41(5): 801–812.

[17] Y. L. Hou, W. Y. Liu, S. Su, H. X. Zhang, J. W. Zhang, J. Liu, *et al.*, "Polymer optical fiber twisted macro-bend coupling system for liquid level detection," *Optics Express*, 2014, 22(19): 23231–22241.

[18] A. W. Snyder and J. D. Love, "Optical waveguide theory," *Chapman and Hall*, 1983, 12(3): 1–37.

[19] G. Durana, J. Zubia, J. Arrue, G. Aldabaldetreku, and J. Mateo, "Dependence of bending losses on cladding thickness in plastic optical fibers," *Applied Optics*, 2003, 42(6): 997–1002.

[20] F. P. Zanella, D. V. Magalhaes, and M. M. Oliveira, "Frustrated total internal reflection: a simple application and demonstration," *American Journal Physics*, 2003, 71(5): 494–496.

[21] K. Rahnavardy, V. Arya, A. Wang, and J. M. Weiss, "Investigation and application of the frustrated-total-Internal-reflection phenomenon in optical fibers," *Applied Optics*, 1997, 36(10): 2183–2187.

[22] A. A. Stahlhofen, "Photonic tunneling time in frustrated total internal reflection," *Physical Review A*, 2001, 98(62): 1108–1115.

[23] J. Y. Han, "Low-cost multi-touch sensing through frustrated total Internal reflection," in *ACM Symposium on User Interface Software and Technology*, Seattle, U. S. A., pp. 115–118, 2005.

[24] M. Lomer, A. Quintela, M. Lopez-Amo, J. Zubia, and J. M. Lopez-Higuera, "A quasi-distributed level sensor based on a bent side-polished plastic optical fibre cable," *Measurement Science and Technology*, 2007, 18(7): 2261–2267.

[25] C. Zhao, L. Ye, J. Ge, J.Zou, and X. Yu, "Novel light-leaking optical fiber liquid-level sensor for aircraft fuel gauging," *Optical Engineering*, 2013, 52(1): 177–182.

[26] K. E. Romo-Medrano and S. N. Khotiaintsev, "An optical-fiber refractometric liquid-level sensor for liquid nitrogen," *Measurement Science and Technology*, 2006, 17(5): 998–1004.

[27] B. Dong, Q. Zhao, J. Lu, T. Guo, L. Xue, S. Li, *et al.*, "A digital liquid level sensor system based on parallel fiber sensor heads," *Proc. SPIE*, 2007, 6595: 659541.

Permissions

The contributors of this book come from diverse backgrounds, making this book a truly international effort. This book will bring forth new frontiers with its revolutionizing research information and detailed analysis of the nascent developments around the world.

We would like to thank all the contributing authors for lending their expertise to make the book truly unique. They have played a crucial role in the development of this book. Without their invaluable contributions this book wouldn't have been possible. They have made vital efforts to compile up to date information on the varied aspects of this subject to make this book a valuable addition to the collection of many professionals and students.

This book was conceptualized with the vision of imparting up-to-date information and advanced data in this field. To ensure the same, a matchless editorial board was set up. Every individual on the board went through rigorous rounds of assessment to prove their worth. After which they invested a large part of their time researching and compiling the most relevant data for our readers.

The editorial board has been involved in producing this book since its inception. They have spent rigorous hours researching and exploring the diverse topics which have resulted in the successful publishing of this book. They have passed on their knowledge of decades through this book. To expedite this challenging task, the publisher supported the team at every step. A small team of assistant editors was also appointed to further simplify the editing procedure and attain best results for the readers.

Apart from the editorial board, the designing team has also invested a significant amount of their time in understanding the subject and creating the most relevant covers. They scrutinized every image to scout for the most suitable representation of the subject and create an appropriate cover for the book.

The publishing team has been an ardent support to the editorial, designing and production team. Their endless efforts to recruit the best for this project, has resulted in the accomplishment of this book. They are a veteran in the field of academics and their pool of knowledge is as vast as their experience in printing. Their expertise and guidance has proved useful at every step. Their uncompromising quality standards have made this book an exceptional effort. Their encouragement from time to time has been an inspiration for everyone.

The publisher and the editorial board hope that this book will prove to be a valuable piece of knowledge for researchers, students, practitioners and scholars across the globe.

List of Contributors

Rajneesh K. Verma
Department of Physics, Central University of Rajasthan (India), BandarSindri, Ajmer, 305817, India

Akhilesh K. Mishra
Department of Electrical Engineering, Technion - Israel Institute of Technology, Haifa 32000, Israel

Shangbin Tao, Deyuan Chen, Juebin Wang, Jing Qiao and Yali Duan
Nangjing University of Posts and Telecommunications, Nanjing, 210009, China

Qiuming Nan and Sheng Li
National Engineering Laboratory for Fiber Optic Sensing Technology, Wuhan University of Technology, Wuhan, 430070, China
Key Laboratory of Fiber Optic Sensing Technology and Information Processing, Ministry of Education, Wuhan, University of Technology, Wuhan, 430070, China

Bo Wang
School of Computer Science and Control Engineering, North University of China, Taiyuan, 030051, China

Guozhu Wu, Tao Guo and Qiulin Tan
Key Laboratory of Instrumentation Science and Dynamic Measurement, Ministry of Education, North University of China, Taiyuan, 030051, China
Science and Technology on Electronic Test and Measurement Laboratory, North University of China, Taiyuan, 030051, China

Qinpeng Liu, Xueguang Qiao, Zhen'an Jia and Haiwei Fu
Key Laboratory on Photoelectric Oil-Gas Logging and Detecting (Ministry of Education), Xi'an Shiyou University, Xi'an, 710065, China

Pinggang Jia
Postdoctoral Research Station of Optical Engineering, Chongqing University, Chongqing, 400030, China
Key Laboratory of Instrumentation Science and Dynamic Measurement of the Ministry of Education of China, North University of China, Taiyuan, 030051, China

Guocheng Fang
Key Laboratory of Instrumentation Science and Dynamic Measurement of the Ministry of Education of China, North University of China, Taiyuan, 030051, China

Daihua Wang
Key Laboratory of Optoelectronic Technology and Systems of the Ministry of Education of China, Chongqing University, Chongqing, 400030, China

Jinyu Wang, Binxin Hu, Faxiang Zhang and Guangdong Song
Key Laboratory of Optical Fiber Sensing Technology of Shandong Province, Laser Institute of Shandong Academy of Science, Jinan, 250014, China

Tongyu Liu
Key Laboratory of Optical Fiber Sensing Technology of Shandong Province, Laser Institute of Shandong Academy of Science, Jinan, 250014, China
Shandong Micro-Sensor Photonics Ltd, Jinan, 250014, China

Long Jiang
Shandong Micro-Sensor Photonics Ltd, Jinan, 250014, China

Zengrong Sun
Shandong Shenglong Safe Technology Co. Ltd, Jinan, 250032, China

Junfeng Qi and Longping Zhang
Laiwu Mining Co. Ltd of Laiwu Steel Group, Laiwu, 271100, China

Gang Wang, Shuqiang Chen and Huajun Yang
School of Physical Electronics, University of Electronic Science and Technology of China, Chengdu, 610054, China

Catarina S. Monteiro, Marta S. Ferreira, Susana O. Silva and Orlando Frazão
Department of Physics and Astronomy, Sciences Faculty, Porto University, Rua do Campo Alegre 687, 4169-007 Porto, Portugal
Inesc-Tec, Rua do Campo Alegre 687, 4169-007 Porto, Portugal

Jens Kobelke, Kay Schuster and Jörg Bierlich
IPHT Jena – Leibniz Institute of Photonic Technology,
Albert-Einstein-Str. 9, 07745 Jena, Germany

Meiqi Ren, Ping Lu, Liang Chen and Xiaoyi Bao
Department of Physics, University of Ottawa,
Ottawa, ON, K1S5G5, Canada

Chunliu Zhao, Yanru Wang, Dongning Wang and Zhewen Ding
Institute of Optoelectronic Technology, China
Jiliang University, Hangzhou, 310018, China

Yubin Shi, Jianmin Zhang and Zhen Zhang
State Key Laboratory of Laser Interaction with
Matter, Northwest Institute of Nuclear Technology,
Xi'an, 710024, China

Md. Ibadul Islam, Shuvo Sen, Sawrab Chowdhury, Md. Shadidul Islam and Mohammad Badrul Alam Miah
Department of Information and Communication
Technology (ICT), Mawlana Bhashani Science and
Technology University, Santosh, Tangail 1902,
Bangladesh

Sayed Asaduzzaman
Department of Information and Communication
Technology (ICT), Mawlana Bhashani Science and
Technology University, Santosh, Tangail 1902,
Bangladesh
Group of Bio-photomatiχ, Tangail, 1902, Bangladesh
Department of Software Engineering (SWE),
Daffodil International University, Sukrabad, Dhaka,
1207, Bangladesh

Kawsar Ahmed and Bikash Kumar Paul
Department of Information and Communication
Technology (ICT), Mawlana Bhashani Science and
Technology University, Santosh, Tangail 1902,
Bangladesh
Group of Bio-photomatiχ, Tangail, 1902, Bangladesh

Zhengyi Zhang and Chuntong Liu
Department Two, Rocket Force University of
Engineering, Xi'an, 710025, China

Guodong Wang and Xiaolian Liu
School of Physics and Electronic Information
Engineering, Henan Polytechnic University,
No.2001, Shiji road, Jiaozuo, 454003, China
Junling SHEN, Lu NI and Saili Wang

School of Electrical Engineering and Automation,
Henan Polytechnic University, No.2001, Shiji road,
Jiaozuo, 454003, China

Xuezhi Jia and Zongxuan Li
National and Local United Engineering Research
Center of Small Satellite Technology, Changchun
Institute of Optics, Fine Mechanics and
Physics,Chinese Academy of Sciences, Changchun,
130033, China

Changcheng Deng
National and Local United Engineering Research
Center of Small Satellite Technology, Changchun
Institute of Optics, Fine Mechanics and
Physics,Chinese Academy of Sciences, Changchun,
130033, China
University of Chinese Academy of Sciences, Beijing,
100039, China

Deqiang Mu
National and Local United Engineering Research
Center of Small Satellite Technology, Changchun
Institute of Optics, Fine Mechanics and
Physics,Chinese Academy of Sciences, Changchun,
130033, China
Changchun University of Technology, Changchun,
130012, China

Jie Tian, Shuhui Liu and Peigang Deng
Laboratory of Optical Information Technology,
Wuhan Institution of Technology, Wuhan, 430205,
China

Wenbing Yu
School of Electronic Information, Shanghai DianJi
University, Shanghai, 201306, China

Sing Yee Chua, Xin Wang and Ningqun Guo
School of Engineering, Monash University Malaysia,
Jalan Lagoon Selantan, Bandar Sunway, 47500
Selangor, Malaysia

Ching Seong Tan
Faculty of Engineering, Multimedia University,
Jalan Multimedia, 63000 Cyberjaya, Selangor,
Malaysia

Yaozhang Sai and Dianli Hou
School of Information and Electrical Engineering,
Ludong University, Yantai, 264025, China

Xiuxia Zhao
Tianrun Crankshaft Co., Ltd, Wendeng, 264400,
China

Mingshun Jiang
School of Control Science and Engineering,
Shandong University, Jinan, 250061, China

**Zhengyi Zhang, Chuntong Liu, Hongcai Li,
Zhenxin He and Xiaofeng Zhao**
Department Two, Rocket Force University of
Engineering, Xi'an, 710025, China

**Mahmoud Nikoufard, Masoud Kazemi Alamouti
and Alireza Adel**
Department of Electronics, Faculty of Electrical
and Computer Engineering, University of Kashan,
Kashan, Iran

Binxin Hu and Jinyu Wang
Key Laboratory of Optical Fiber Sensing Technology
of Shandong Province, Laser Institute of Shandong
Academy of Sciences, Jinan, 250014, China

Guangxian Jin and Tongyu Liu
Shandong Micro-sensor Photonics Co. Ltd, Jinan,
250014, China

Dong Zhong and Zhongming Li
School of Electronic and Information, Hubei
University of Science and Technology, Xianning,
437100, China

**Meng Jiang, Zhongze Zhao, Kun Li, Zeming
Wang, Yage Zhan, Hongying Zhou and Fu Yang**
College of Science Donghua University of China,
Shanghai, 201620, China

**Tao Yi, Chuanke Wang, Weiming Yang, Shenye
Liu, Shaoen Jiang and Yongkun Ding**
Laser Fusion Research Center, Chinese Academy
of Engineering Physics, Mianyang, 621900, China

Jinwen Yang
Laser Fusion Research Center, Chinese Academy
of Engineering Physics, Mianyang, 621900, China
School of Energy Science and Engineering,
University of Electronic Science and Technology
of China, Chengdu, 611731, China

Ming Yang and Tingshuai Li
School of Energy Science and Engineering,
University of Electronic Science and Technology
of China, Chengdu, 611731, China

Shaoqiu Xiao
School of Physical Electronics, University of
Electronic Science and Technology of China,
Chengdu, 611731, China

Julio E. Posada-Roman and Jose A. Garcia-Souto
Department of Electronics Technology, GOTL
University Carlos III de Madrid, Butarque 15,
28911, Leganés, Madrid, Spain

David A. Jackson
Applied Optics Group, School of Physical Sciences,
University of Canterbury, Kent, CT2 7NH, UK

**Chuanwu Jia, Jun Chang, Fupeng Wang and Hao
Jiang**
School of Information Science and Engineering
and Shandong Provincial Key Laboratory of Laser
Technology and Application, Shandong University,
Jinan, 250100, China

Cunguang Zhu and Pengpeng Wang
School of Physics and Technology, University of
Jinan, Jinan, 250022, China

**Yulong Hou, Wenyi Liu, Huixin Zhang, Shan Su,
Jia Liu, Yanjun Zhang, Jun Liu and Jijun Xiong**
Key Laboratory of Instrumentation Science and
Dynamic Measurement (North University of China),
Ministry of Education, Taiyuan, 030051, China

Index